The Theory of Distributions

Series Editor
Nikolaos Limnios

The Theory of Distributions

Introduction

El Mustapha Ait Ben Hassi

ISTE

WILEY

First published 2023 in Great Britain and the United States by ISTE Ltd and John Wiley & Sons, Inc.

ISTE Ltd
27-37 St George's Road
London SW19 4EU
UK

www.iste.co.uk

John Wiley & Sons, Inc.
111 River Street
Hoboken, NJ 07030
USA

www.wiley.com

Library of Congress Control Number: 2023938463

British Library Cataloguing-in-Publication Data
A CIP record for this book is available from the British Library
ISBN 978-1-78630-937-2

Contents

Preface

This book is a collection of course notes written in the spirit of presenting distributions in as clear a sense as possible, without overburdening them with even more interesting notions, and not to deviate from the main objective, which is to provide students with relatively simple course supporting material that contains the most essential and important results on distributions. The reader interested in more details on distributions will find a bibliographic list given at the end of the book.

P.1. Summary

This book is organized into nine chapters and an appendix, and almost all of them comprise a theoretical foundation in the form of definitions, propositions and theorems illustrated by examples, and a series of exercises with their answers.

In Chapter 1, the essentials on topological vector spaces are presented, which will allow building topology on the space of test functions.

Chapter 2 is dedicated to the presentation and study of $\mathcal{C}_c^\infty$: the space of test functions ends with some density results that will be useful later on.

Chapter 3 introduces the notion of distributions through the presentation of equivalent definitions and various examples.

Chapter 4 addresses some operations on distributions, namely multiplication by a $\mathcal{C}^\infty$ class function, differentiation of distributions, and classical transformations on distributions.

Chapter 5 is dedicated to the notion of distribution support, its properties allowing a classification of distributions according to their support.

In Chapter 6, the fundamental principle of convolution regularization is presented. More precisely, distribution convolution is defined and its properties are exposed.

In order to define temperate distributions in Chapter 7, Schwartz spaces $\mathcal{S}(\mathbb{R}^d)$ are first defined and some properties are given that will be needed in the second part of this chapter, where the space of temperate distributions $\mathcal{S}'(\mathbb{R}^d)$ is presented as well as its fundamental properties.

Given the importance of the Fourier transform, Chapter 8 is dedicated to this notion in different functional spaces, namely the space $L^1(\mathbb{R}^d)$, the Schwartz space $\mathcal{S}(\mathbb{R}^d)$ and finally the dual space $\mathcal{S}'(\mathbb{R}^d)$ of temperate distributions.

Chapter 9 presents an opportunity to apply all this theory exposed for the solution of some differential equations.

An appendix concludes the book with a series of synthesis exercises that can be used for self-evaluation.

P.2. Targeted audience

This book is intended for master's students in mathematics as well as for students of the physical sciences. It is also intended for students preparing for post-graduate degrees for the teaching of mathematics. It equips them with the most elementary definitions and notions that will help them familiarize with distributional calculus such as the notions of derivation, limit of a sequence or series of distributions as well as other operations on distributions. Through a range of well-chosen examples, they will be able to grasp the difference between what is classical for ordinary functions and what is related to distributions when considering what is usually referred to as "generalized functions", which are themselves distributions.

Finally, I would like to thank all the people who helped me, in one way or another, in the preparation of this book. I would especially like to thank my students and their contribution to preparing these notes. I will be grateful to those of my readers who will send me their comments about this first edition.

April 2023

Introduction

The emergence of the theory of distributions can be traced back to 1945–1950 and is due to Laurent Schwartz. This theory provides a general setting with a pleasant formalism for studying functional spaces and partial differential equations. Since then, distributions are the natural framework that many analysts use basing themselves on Schwartz's notations and ideas. A good familiarity with the language of distributions has become almost essential to any analyst. Schwartz himself was well aware that the main merit of his approach was not in the introduction of new tools, but in a clear and accessible synthesis of multiple recipes that had already been employed in various contexts.

There are several reasons for introducing the concept of distribution. Some of them are purely physical (even experimental) reasons, while others are more mathematical ones; these latter, on the one hand, consist of giving meaning to some objects manipulated in physics, as for example the mysterious Dirac function introduced by Dirac (1929) which equals 0 anywhere, except at 0 where it equals $+\infty$, and whose integral is equal to 1 (contrarily to all of the rules of Lebesgue's theory of integration). Not only did Dirac use this function for formal computation purposes, but he would also derive it at will, simply observing that the successive derivatives were *increasingly singular*. On the other hand, they consist of exposing the operations, especially differentiation operations, carried out within the setting of partial differential equations.

In order to motivate the generalization of the point aspect of functions and make it possible to move from the notion of function to the notion of *functional*, a classical physical example is given here related to the measurement of the temperature of a straight wire "at a given point". Understandably and for obvious reasons, such a measurement is never perfectly feasible. Any thermometer, regardless of the physical principle used for the measurement, presents a *spatial extension* that cannot possibly be reduced to that of a point: what should be achieved to be able to measure the temperature at *a point* x_0. It can nonetheless be admitted that, in the case of a realistic temperature measurement, the thermometer takes into account all the temperatures in a "neighborhood", of the point x_0 according to a sensitivity function ϕ_0 so that for a distribution function $T(x)$ of the temperature along the bar (a function about which it is a priori not known what it is *really* equal to at a specific point of the bar), it can be said that the measured temperature T will be in fact $T_0 = \displaystyle\int T(x)\phi_0(x)\,\mathrm{d}x$. If the measurement would be taken at another point x_1, it would yield $T_1 = \displaystyle\int T(x)\phi_1(x)\,\mathrm{d}x$, etc.

It can be seen that the measured temperature T, assuming that the functions $T(x)$, ϕ_0, ϕ_1, etc., are sufficiently regular, appears as a linear expression in the sensitivity function ϕ. This expression can generously be expressed in the form:

$$\langle T, \phi \rangle = \int T(x)\phi(x)\,\mathrm{d}x$$

When the product $x \mapsto T(x)\phi(x)$ is Lebesgue integrable, this notation is perfectly justified. However, in many practical cases, the physical quantity being considered (which here would be $T(x)$) proves to be too *singular* for the written integral to have any meaning with a realistic choice of the function ϕ. The set of functions ϕ is called the set of *test* functions or *trial* functions. The measure of a physical quantity T is then represented by the "bracket": $\langle T, \phi \rangle$ independently or not of an integral form for this expression. The set of *measurable* quantities T by the test functions is then generically called a *distribution*.

What should be the required minimum for the objects thus considered?

1) The set of test functions constitutes a *vector space of functions*. That is, any linear combination with complex coefficients of test functions is still a test function.

2) The distribution set is the set of continuous linear forms on the vector space of the test functions.

3) The result of the measure of T by ϕ is then the real or complex number $\langle T, \phi \rangle$.

One of the other main interests of the theory of distributions is to allow the construction of a differential calculus that extends the ordinary differential calculus and for which any distribution is indefinitely differentiable. This theory has become an essential tool, especially in the study of partial differential equations. It has also allowed for a mathematical modernization for many physical phenomena.

1

Topological Vector Spaces

The purpose of this chapter is to present some of the tools needed to make sense of the notion of the distribution. In order to define a distribution as a continuous linear form on a space of test functions, it is necessary that a topology on this space be defined. To this end, a semi-norm is first defined, and then a topology on this space is deduced, namely the one constructed from a family of semi-norms.

1.1. Semi-norms

DEFINITION 1.1.– *Let E be a vector space over a field $\mathbb{K} = \mathbb{R}, \mathbb{C}$.*

We call semi-norm on E any mapping $P : E \longrightarrow \mathbb{R}^+$ verifying:

– $\forall \lambda \in \mathbb{K}, \forall x \in E, P(\lambda x) = |\lambda| P(x)$;

– $\forall x, y \in E, P(x+y) \leq P(x) + P(y)$.

EXAMPLE 1.1.– Any norm on E is a semi-norm.

EXAMPLE 1.2.– We consider $\mathcal{K}_{\mathbb{R}}$ the set of compact sets of $\mathbb{R}$. Let $K \in \mathcal{K}_{\mathbb{R}}$.

– For $f \in \mathcal{C}(\mathbb{R}, \mathbb{R})$, we consider $P_K(f) = \sup\limits_{x \in K} |f(x)| = \max\limits_{x \in K} |f(x)|$. The mapping P_K defines a semi-norm on the space $\mathcal{C}(\mathbb{R}, \mathbb{R})$.

– Let $n \in \mathbb{N}^*$. For $f \in \mathcal{C}^n(\mathbb{R}, \mathbb{R})$, we consider

$$P_{n,K}(f) = \sup_{x \in K, m \leq n} |f^{(m)}(x)| = \max_{x \in K, m \leq n} |f^{(m)}(x)|.$$

The mapping $P_{n,K}$ defines a semi-norm on the space $\mathcal{C}^n(\mathbb{R}, \mathbb{R})$.

– The family $(P_{n,K})_{n\in\mathbb{N}^*}$ is a family of semi-norms on the space $\mathcal{C}^\infty(\mathbb{R},\mathbb{R})$.

PROPOSITION 1.1.– Let P be a semi-norm. We have

$$\forall x,y \in E, |P(x)-P(y)| \leq P(x-y).$$

Proof. Let $x, y \in E$. From the triangle inequality, we have

$$P(x) = P(x-y+y) \leq P(x-y) + P(y).$$

Therefore, $P(x) - P(y) \leq P(x-y)$. Similarly, we have

$$P(y) = P(x-y+y) \leq P(y-x) + P(x).$$

Here, $P(y) - P(x) \leq P(y-x)$. Since $P(y-x) = P(x-y)$, then

$$P(y) - P(x) \leq P(x-y).$$

Consequently, $|P(x) - P(y)| \leq P(x-y)$. □

DEFINITION 1.2.– *A family of semi-norms $(P_\alpha)_{\alpha\in I}$ is said to be separating if for all $x \in E$:*

$$x \neq 0 \Longrightarrow \exists \alpha \in I, P_\alpha(x) \neq 0.$$

EXAMPLE 1.3.– The families of semi-norms $(P_K)_{K\in\mathcal{K}_\mathbb{R}}$, $(P_{n,K})_{K\in\mathcal{K}_\mathbb{R}}$ and $(P_{n,K})_{p\in\mathbb{N}^*,K\in\mathcal{K}_\mathbb{R}}$ previously defined, respectively, on the spaces $\mathcal{C}(\mathbb{R},\mathbb{R})$, $\mathcal{C}^p(\mathbb{R},\mathbb{R})$ and $\mathcal{C}^\infty(\mathbb{R},\mathbb{R})$ are separating.

DEFINITION 1.3.– *A family of semi-norms $(P_\alpha)_{\alpha\in I}$ is said to be filtering if*

$$\forall \alpha,\beta \in I, \sup(P_\alpha, P_\beta) \in \{P_\gamma, \gamma \in I\}.$$

EXAMPLE 1.4.– The family $(P_{n,K})_{n\in\mathbb{N}^*,K\in\mathcal{K}_\mathbb{R}}$ of semi-norms previously defined on the space $\mathcal{C}^\infty(\mathbb{R},\mathbb{R})$ is filtering.

1.2. Topological vector space: definition and properties

In this section, a topology is defined from a family of semi-norms.

1.2.1. *Topology and semi-norms*

Let $P := (P_\alpha)_{\alpha \in I}$ be a family of semi-norms on E.

DEFINITION 1.4.– *We call a "P–ball" of center x_0 and radius $r > 0$ the set*

$$V_n(x_0, r) := \{x \in E, P_i(x - x_0) < r, \text{ for } i \in J_n \subseteq I, J_n \text{ finite}\}.$$

We also write that

$$V_n(x_0, r) = \{x \in E, \max_{i \in J_n} P_i(x - x_0) < r, J_n \subseteq I, J_n \text{ finite}\}.$$

DEFINITION 1.5.– *We call "P–topology", associated with the family of semi-norms $(P_\alpha)_{\alpha \in I}$, the topology formed by any finite intersection or union of P–balls.*

This is the topology generated by the family $(P_\alpha)_{\alpha \in I}$ (i.e. the smallest one making the P_α continuous).

DEFINITION 1.6.– *A subset O of E is said to be open if it is empty or if, for each $x_0 \in O$, there exists a finite non-empty subset J_n of I and a real $r > 0$ such that $V_n(x_0, r) \subset O$.*

Sets such as $V_n(x_0, r)$ are open and their role is analogous to that of open balls in metric spaces. The subsets of E such as

$$\overline{V}_n(x_0, r) := \{x \in E, \max_{i \in J_n} P_i(x - x_0) \leq r, J_n \subseteq I, J_n \text{ finite}\}$$

are closed and their role is analogous to that of closed balls in metric spaces.

REMARK 1.1.– *P–balls form a basis for the P–topology.*

THEOREM 1.1.– *If the family of semi-norms $P = (P_\alpha)_{\alpha \in I}$ is separating, then the associated P–topology is separated.*

In other words, $\forall x, y \in E$, if $x \neq y$, then there exists U an open set containing x and V an open set containing y such that $U \cap V = \varnothing$.

Proof. Let $x, y \in E$ such that $x \neq y$. We have $x \neq y$, therefore $x - y \neq 0$.

Then, $\exists \alpha \in I, r := P_\alpha(x - y) > 0$.

We consider $U = \{z \in E, P_\alpha(x - z) < \frac{r}{2}\}$ and $V = \{z \in E, P_\alpha(y - z) < \frac{r}{2}\}$.

U and V are two open sets of the $P-$topology.

Let us assume that there exists $z \in U \cap V$. Then $P_\alpha(x-z) < \frac{r}{2}$ and $P_\alpha(y-z) < \frac{r}{2}$.

By triangular inequality, we get $P_\alpha(x - y) < r$, which is absurd. Consequently, $U \cap V = \varnothing$. □

1.2.2. *Topological vector space*

DEFINITION 1.7.– *A $\mathbb{K}-$vector space E endowed with topology $\mathcal{T}$ is a topological vector space, abbreviated as T.v.s if:*

– (A_1)*: $H_1 : (x, y) \longmapsto x + y$ is a continuous mapping from $E \times E$ to E;*

– (A_2)*: $H_2 : (\lambda, x) \longmapsto \lambda x$ is a continuous mapping from $K \times E$ to E.*

It is obvious that in both axioms, we consider the topology produced in vector spaces produced. The vector space structure of E is said to be compatible with the topology $\mathcal{T}$ if the axioms (A_1) and (A_2) are verified.

EXAMPLE 1.5.– The normed vector space $(E, \|.\|)$ is a topological vector space.

THEOREM 1.2.– *A vector space E endowed with the topology generated by a separating family of semi-norms is a topological vector space.*

Proof. Let us establish the continuity of the addition at a point (a, b).

We set an index $i_0 \in I$ and a real $\varepsilon > 0$. For $(x, y) \in E \times E$, we have

$$P_{i_0}[(x + y) - (a + b)] \leq P_{i_0}(x - a) + P_{i_0}(y - b).$$

It follows that $P_{i_0}[(x + y) - (a + b)] \leq \varepsilon$ as soon as $P_{i_0}(x - a) \leq \frac{\varepsilon}{2}$ and $P_{i_0}(y - b) \leq \frac{\varepsilon}{2}$. Hence, the continuity of the addition at point (a, b).

For the external multiplication $(\lambda, x) \longmapsto \lambda x$, let us establish continuity at a point (α, a). We set an index $i_0 \in I$ and a real $\varepsilon > 0$. For $(\lambda, x) \in \mathbb{K} \times E$, we have

$$P_{i_0}[(\lambda x) - (\alpha a)] \leq |\lambda - \alpha| P_{i_0}(x) + |\alpha| P_{i_0}(x - a).$$

We set a real $r > 0$ such that $|\alpha| r \leq \frac{\varepsilon}{2}$. It is observed that for x verifying $P_{i_0}(x-a) \leq r$, we have $P_{i_0}(x) \leq P_{i_0}(a) + r$ and thus,

$$|\lambda - \alpha| P_{i_0}(x) \leq |\lambda - \alpha| (P_{i_0}(a) + r).$$

Let us set a real η such that $\eta\big(P_{i_0}(a) + r\big) \leq \frac{\varepsilon}{2}$. For $(\lambda, x) \in \mathbb{K} \times E$ verifying $|\lambda - \alpha| < \eta$ and $P_{i_0}(x-a) < r$, we have $P_{i_0}\left[(\lambda x) - (\alpha a)\right] \leq \varepsilon$, hence continuity of the multiplication at the point (α, a). □

It is deduced that

– translations and homotheties, of ratios $k \neq 0$, are homeomorphisms of E;

– the set $\mathcal{V}(a)$ of the neighborhoods of a point $a \in E$ is the image by the translation τ_a of the set of neighborhoods $\mathcal{V}(0)$ of 0. The topology of a vector space is known as soon as the neighborhoods of 0 are known: $\mathcal{V}(a) = \tau_a(\mathcal{V}(0))$. In other words, we have the following equivalence:

$$V \in \mathcal{V}(0) \iff a + v \in \mathcal{V}(a). \qquad [1.1]$$

1.2.3. *Locally convex topological vector space*

Reminder: let A be a subset of E.

A is convex if, for every $x, y \in A$ and for every $\lambda \in [0,1]$, we have

$$\lambda x + (1-\lambda)\, y \in A.$$

DEFINITION 1.8.– *Let E be a topological vector space.*

E is locally convex if there exists a fundamental system (basis) of convex neighborhoods of the origin.

THEOREM 1.3.– *Let E be a vector space.*

If the topology of E is defined by a family of semi-norms $\left(P_i\right)_{i\in I}$, *then E is a locally convex topological space.*

Proof. If the topology of E is defined by a family of semi-norms $(P_i)_{i\in I}$, then the sets

$$V_n(\varepsilon) := \{x \in E, P_i(x) < \varepsilon, \text{ for } i \in J_n \subseteq I, J_n \text{ finite}\},$$

where $0 < \varepsilon < 1$, form a fundamental system of convex neighborhoods of the origin. □

EXAMPLE 1.6.– Let Ω be an open set of $\mathbb{R}^d$ and $1 \leq p < +\infty$. We denote by $L^p_{loc}(\Omega)$ the space of measurable functions $p-$locally integrable on Ω, that is, for any compact subset K of Ω, we have $\int_K |f(x)|^p < +\infty$. A semi-norm is defined by

$$P_K(f) = \left[\int_K |f(x)|^p\right]^{\frac{1}{p}} < +\infty.$$

The family of semi-norms $(P_K)_{K\in\mathcal{K}_\Omega}$ determines a topology which makes $L^p_{loc}(\Omega)$ a locally convex vector space.

1.2.4. *Topological metrizable vector space*

DEFINITION 1.9.– *A metric d on a vector space E is said to be translation-invariant if, for each $x, y, a \in E$, we have $d(x, y) = d(x + a, y + a)$.*

Two translation-invariant metrics d and d' on the same vector space E, which determine the same topology, are uniformly equivalent (i.e. the mappings $I_E : (E, d) \longrightarrow (E, d')$ and $I_E : (E, d') \longrightarrow (E, d)$ are uniformly continuous). When d and d' are two translation-invariant and topologically equivalent metrics on E, then (E, d) is complete if and only if (E, d') is complete.

DEFINITION 1.10.– *The topology $\mathcal{T}$ of a topological vector space E is said to be metrizable if there exists a translation-invariant metric on E which determines the topology $\mathcal{T}$.*

THEOREM 1.4.– *The topology on a vector space determined by a separating* ***sequence*** *of semi-norms (i.e. I is at most countable) is metrizable.*

Proof. Let $(P_k)_k$ be a separating sequence of semi-norms on a vector space E.

For $x, y \in E$, we define $d(x,y) = \sum_{k=1}^{+\infty} 2^{-k} \frac{P_k(x-y)}{1+P_k(x-y)}$. It is easy to verify that d is a translation-invariant distance on E and that this distance d determines the same topology as that defined by the sequence $(P_k)_k$. □

EXAMPLE 1.7.– When $(P_k)_{1\le k\le m}$ is a separating finite sequence of semi-norms on E, then

$$\max_{1\le k\le m} P_k \text{ and } \left(\sum_{k=1}^{m} P_k^r\right)^{\frac{1}{r}}, 1 \le r < +\infty$$

are equivalent semi-norms that determine the same topology as the sequence $(P_k)_{1\le k\le m}$.

DEFINITION 1.11.– *We mention that E is a Frechet space if E is a locally convex topological vector space whose associated topology is metrizable and for the associated metric, E is* ***complete****.*

EXAMPLE 1.8.– The spaces $\mathcal{C}(\mathbb{R},\mathbb{R})$, $\mathcal{C}^n(\mathbb{R},\mathbb{R})$ and $\mathcal{C}^\infty(\mathbb{R},\mathbb{R})$ are Frechet spaces.

1.2.5. *Convergence in a topological vector space*

DEFINITION 1.12.– *We state that a sequence $(x_n)_n$ of E converges to an element $x \in E$ if, for each neighborhood V of 0, there exists an integer m_0 such that, for every integer $m > m_0$, we have $x_m - x \in V$.*

Since the topology is separated, a convergent sequence possesses a single limit. In terms of semi-norms, we have the following equivalent definition.

DEFINITION 1.13.– *A sequence $(x_n)_n$ of E converges to an element $x \in E$ if, for each index $i \in I$ and each $\varepsilon > 0$, there exists an integer m_0 such that for each integer $m > m_0$, we have $P_i(x_m - x) \le \varepsilon$.*

We can introduce the notion of Cauchy sequence in topological vector spaces.

DEFINITION 1.14.– *We say that a sequence $(x_n)_n$ of E is a Cauchy sequence if, for each neighborhood V of 0, there exists an integer m_0 such that, for each integer $m, m' > m_0$, we have $x_m - x_{m'} \in V$.*

This is expressed in terms of semi-norms as the following definition.

DEFINITION 1.15.– *A sequence* $(x_n)_n$ *of* E *is a Cauchy sequence if, for each index* $i \in I$ *and each* $\varepsilon > 0$*, there exists an integer* m_0 *such that for each integer* $m, m' > m_0$*, we have* $P_i(x_m - x_{m'}) \leq \varepsilon$.

REMARK 1.2.– *Analogously, the topological notions met in undergraduate studies can be extended to the setting of vector spaces whose topology is determined by a separating family of semi-norms.*

1.2.6. *Applications and linear forms*

E and F denote two vector spaces respectively endowed with topologies determined by the families of semi-norms $(P_i)_{i\in I}$ and $(q_l)_{l\in L}$.

DEFINITION 1.16.– *A mapping* f *from* E *into* F *is continuous at point* $a \in E$ *if, for each* $l \in L$ *and each real* $\varepsilon > 0$*, there exists a finite non-empty subset* $J_n \subset I$ *and a real* $r > 0$ *such that*

$$\max_{i\in J_n} P_i(x-a) < r \Longrightarrow q_l\big(f(x) - f(a)\big) < \varepsilon.$$

If f is a linear mapping, this definition can be reformulated as follows.

THEOREM 1.5.– *Let* f *be a linear mapping from* E *into* F*. The following statements are equivalent:*

1) f *is continuous on* E*;*

2) f *is continuous at* 0*;*

3) for all $l \in L$*, there exists* $J_n \subset I$*, finite, and there exists* $c > 0$*, such that for each* $x \in E$*, we have*

$$q_l(f(x)) \leq c \max_{i\in J_n} P_i(x).$$

The characterization [1.1] of the neighborhood of a point a by the neighborhood of 0 makes it possible to easily see that f being continuous on E is equivalent to f being continuous at 0.

When $F = \mathbb{K}$ and f is a linear form on E, we have the following result.

THEOREM 1.6.– *Let f be a linear form on E. The following statements are equivalent:*

1) f is continuous on E;

2) f is continuous at 0;

3) there exists finite $J_n \subset I$, and there exists $c > 0$, such that for each $x \in E$, we have $|f(x)| \leq c \max_{i \in J_n} P_i(x)$.

Moreover, if the family of semi-norms $(P_i)_{i \in I}$ is filtering, we have the following result.

THEOREM 1.7 (Fundamental theorem).– *Let E be a locally convex topological vector space whose topology is defined by a separating and filtering family of semi-norms $(P_i)_{i \in I}$. Let f be a linear form on E. The following statements are equivalent:*

1) f is continuous on E;

2) f is continuous at 0;

3) there exists $\alpha \in I$, $\exists\ c > 0$, such that for each $x \in E$, we have

$$|f(x)| \leq cP_\alpha(x).$$

EXERCISE 1.1.– Show the fundamental theorem 1.7.

1.3. Inductive limit topology

This section addresses the notion of inductive limit topology from a point of view that will be used as another tool for defining the notion of distributions.

Let $(E_i, \mathcal{T}_i)_{i \in \mathbb{N}^*}$ be an increasing sequence of locally convex topological spaces such that the identity mapping *id* : $E_i \hookrightarrow E_{i+1}$ is continuous for each $i \in \mathbb{N}^*$. We set $E = \bigcup_{i=1}^{+\infty} E_i$. Let us define on E the least fine-grained locally convex topology $\mathcal{T}$ making the identities $id : E_i \hookrightarrow E_{i+1}$ continuous for $i = 1, 2, \cdots$. It is called the inductive limit topology of E defined by the subspaces E_i.

The space E endowed with this topology $\mathcal{T}$ is said to be **inductive limit** of spaces $(E_i)_{i \in \mathbb{N}^*}$.

For a convex set V to be a neighborhood of 0 for the inductive limit topology, it is necessary and sufficient that each intersection $V \cap E_i$ for all $i = 1, 2, \cdots$ is a neighborhood of 0 in E_i. Therefore, a fundamental system of neighborhoods of the origin in E is obtained by taking all convex hulls of the form $V = \text{Conv}(\bigcup_{i=1}^{+\infty} V_i)$ where each V_i belongs to the fundamental system of convex neighborhoods of each E_i, for all $i \in \mathbb{N}^*$.

PROPOSITION 1.2.– Let E be the inductive limit of $(E_i)_{i\in\mathbb{N}^*}$ and F a locally convex topological space. A linear map $u : E \longrightarrow F$ is continuous if and only if the restriction $u_i = u|_{E_i}$ of u is continuous from E_i to F for all $i \in \mathbb{N}^*$.

Proof. If u is continuous, then every restriction u_i is continuous since, by definition, the identity $E_i \hookrightarrow E$ is continuous.

Conversely, we assume that each u_i by: $E_i \longrightarrow F$ is continuous. Let U be a convex neighborhood of 0 in F. If there exists a convex neighborhood of 0, denoted by V_i, in E_i such that $u_i(V_i) \subset U$, then, $V = \text{Conv}(\bigcup_{i=1}^{+\infty} V_i)$ is a neighborhood of 0 in E and we have $u(V) \subset U$. Therefore, u is continuous from E into F. □

THEOREM 1.8.– *Let $(E, \mathcal{T})$ be the union of an increasing sequence $(E_i, \mathcal{T}_i)_{i\in\mathbb{N}^*}$ of locally convex topological spaces such that:*

– for all i, the identity $E_i \hookrightarrow E_{i+1}$ is continuous;

– the topology induced by E_{i+1} on E_i coincides with the topology of E_i, for all $i \in \mathbb{N}^*$;

– E_i is a closed subspace of E_{i+1}, for all $i \in \mathbb{N}^*$.

Therefore,

1) the inductive limit topology of E induces on each E_i its original topology;

2) a subset A is bounded in the inductive topology of E if and only if there exists an index j such that A is contained and bounded in E_j.

A completeness result for the inductive limit of complete spaces here follows.

PROPOSITION 1.3.– Let $(E, \mathcal{T})$ be the inductive limit of an increasing sequence $(E_i, \mathcal{T}_i)_{i\in\mathbb{N}^*}$ of locally convex separated topological vector spaces. If each E_i is complete, then E is also complete.

Regarding the continuity of linear forms on an inductive limit space, we have the following result.

PROPOSITION 1.4.– Let $(E, \mathcal{T})$ be the inductive limit of an increasing sequence $(E_i, \mathcal{T}_i)_{i \in \mathbb{N}^*}$ of locally convex topological vector spaces, and let T be a linear form on E. If each E_i is metrizable, then T is continuous if and only if it is sequentially continuous.

2

Spaces of Test Functions

Before discussing the notion of distributions, we give a few functional spaces on which are defined topological structures that will help making sense of the notions of limit and continuity within the context of distributions. In this chapter, Ω denotes an open set of $\mathbb{R}^d$.

2.1. Multi-index notations

A multi-index α is a $d-$uplet of natural numbers $\alpha = (\alpha_1, \alpha_2, \cdots, \alpha_d) \in \mathbb{N}^d$:

$$\forall x \in \mathbb{R}^d, x^\alpha = x_1^{\alpha_1} x_2^{\alpha_2} \cdots x_d^{\alpha_d}.$$

The length of a multi-index α is $|\alpha| = \alpha_1 + \alpha_2 + \cdots + \alpha_d$.

Let α and β denote two multi-indexes. We have

$$\alpha \leq \beta \iff \forall i \in \{1, \cdots, d\}, \alpha_i \leq \beta_i$$

$$\binom{\alpha}{\beta} = \frac{\alpha!}{\beta!(\alpha-\beta)!} \quad \text{with } \alpha! = \alpha_1!\alpha_2! \cdots \alpha_d!$$

$$\partial^\alpha \varphi = \frac{\partial^{\alpha_1+\alpha_2+\cdots+\alpha_d}}{\partial x_1^{\alpha_1} \partial x_2^{\alpha_2} \cdots \partial x_d^{\alpha_d}} \varphi = \frac{\partial^{\alpha_1}}{\partial x_1^{\alpha_1}} \circ \frac{\partial^{\alpha_2}}{\partial x_2^{\alpha_2}} \circ \cdots \circ \frac{\partial^{\alpha_d}}{\partial x_d^{\alpha_d}} \varphi.$$

We also use the notation $D^\alpha \varphi$ which designates $\partial^\alpha \varphi$, that is $\partial^\alpha \varphi = D^\alpha \varphi$.

2.1.1. *Leibniz formula*

Binomial formula:

$$(x+y)^\alpha = \sum_{0\leq\beta\leq\alpha} \binom{\alpha}{\beta} x^{\alpha-\beta} y^\beta$$

Leibniz's formula is an important one. Let $k \geq 1$ and $\varphi, \psi \in \mathcal{C}^k(\Omega)$.

Multinomial formula: for $x \in \mathbb{R}^d$ and $k \in \mathbb{N}$, we have

$$(x_1 + x_2 + \cdots + x_n)^k = \sum_{|\alpha|=k} \frac{k!}{\alpha!} x^\alpha.$$

For every multi-index α of length less than or equal to k, we have

$$\partial^\alpha(\varphi\psi) = \sum_{0\leq\beta\leq\alpha} \binom{\alpha}{\beta} \partial^\beta \varphi \partial^{\alpha-\beta} \psi.$$

2.2. $\mathcal{C}^\infty$ function with compact support

This section begins by defining the support of a function.

2.2.1. *Support of a continuous function*

DEFINITION 2.1.– *Let $f \in \mathcal{C}^0(\Omega)$ be a continuous function on Ω.*

We call support of the function f the set supp $(f) := \overline{\{x \in \Omega, f(x) \neq 0\}}$.

Otherwise, $(\text{supp}\,(f))^c$ is the largest open set where f is zero.

REMARK 2.1.– *When working with Borel functions (not necessarily continuous), it is no longer reasonable to say that the support is the closure of the set of points where the function is non-zero. Actually, if we consider, for example, the function $\mathbb{1}_\mathbb{Q}$ indicative of $\mathbb{Q}$, it is zero almost everywhere, but as it has been defined before, its support will be $\mathbb{R}$, whereas its support should reasonably be empty. This leads to redefining the support of a function, which is not necessarily continuous, as: the smallest closed set* *such that $f = 0$ a.e., on its complement. The support is then the complement of the largest cancellation open set almost everywhere.*

REMARK 2.2.– *If f and g coincide almost everywhere on Ω, then supp $(f) =$ supp(g). We can therefore refer to the support of an element of $L^p(\Omega)$ without having to specify which representative to choose in the equivalence class.*

EXAMPLE 2.1.–

1) supp $\mathbb{1}_{[0,1]} =$ supp $\mathbb{1}_{[0,1[} =$ supp $\mathbb{1}_{]0,1[} = [0,1]$;

2) supp $\mathbb{1}_{\mathbb{Q}\cap[0,1]} = \varnothing$;

3) supp $\mathbb{1}_{\mathbb{Q}} = \varnothing$;

4) supp $\mathbb{1}_{\mathbb{Q}^c} = \mathbb{R}$;

5) supp $\sin = \mathbb{R}$;

6) supp $(x \longmapsto x) = \mathbb{R}$.

PROPOSITION 2.1.– Let f and g denote two continuous functions on Ω. We have

– supp $(f) = \varnothing \iff f = 0$ on Ω;

– supp $(fg) \subset$ supp $(f) \cap$ supp (g);

– if $f \in \mathcal{C}^k(\Omega)$, then for all $\alpha \in \mathbb{N}^d$, $|\alpha| \leq k$, supp $(\partial^\alpha f) \subset$ supp (f).

2.2.2. *Spaces of test functions*

We choose a space of test functions that possesses all the necessary physical and mathematical characteristics.

DEFINITION 2.2.– *We call space of test functions on Ω, denoted by $\mathcal{D}(\Omega)$ or $\mathcal{C}_c^\infty(\Omega)$ the set of functions $\varphi \in \mathcal{C}^\infty(\Omega)$ such that there exists a compact $K \subset \Omega$ verifying $K =$ supp (φ):*

$$\text{that is } \mathcal{D}(\Omega) = \{\varphi \in \mathcal{C}^\infty, \exists K \subset \Omega \text{ compact}, K = \text{supp}\,(\varphi)\}.$$

Let $K \subset \Omega$ be a fixed compact subset. We denote by $\mathcal{D}_K(\Omega)$ or $\mathcal{C}_K^\infty(\Omega)$ the set

$$\mathcal{D}_K(\Omega) = \{\varphi \in \mathcal{C}^\infty(\Omega), \text{supp}\,(\varphi) \subset K\}.$$

Question: Is $\mathcal{D}(\Omega) \neq \varnothing$?

EXAMPLE 2.2.– We consider the function ρ defined by

$$\begin{cases} \rho(x) = \exp(-\frac{1}{1-\|x\|^2}) & \text{if } \|x\| < 1 \\ \rho(x) = 0 \quad \text{if } \|x\| \geq 1. \end{cases}$$

$\|\cdot\|$ designates the Euclidean norm of $\mathbb{R}^d$.

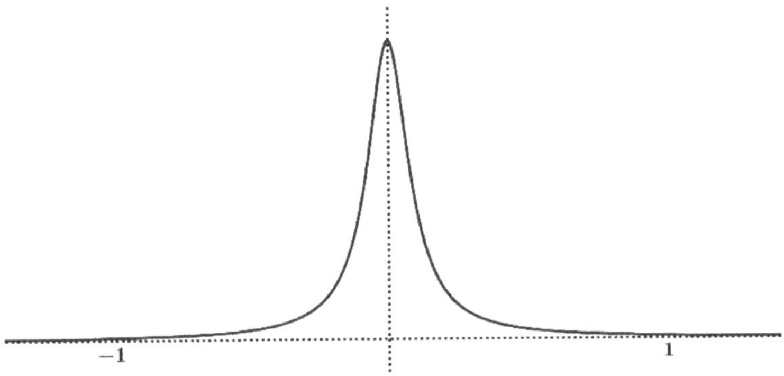

Figure 2.1. *Function* $\rho, d = 1$

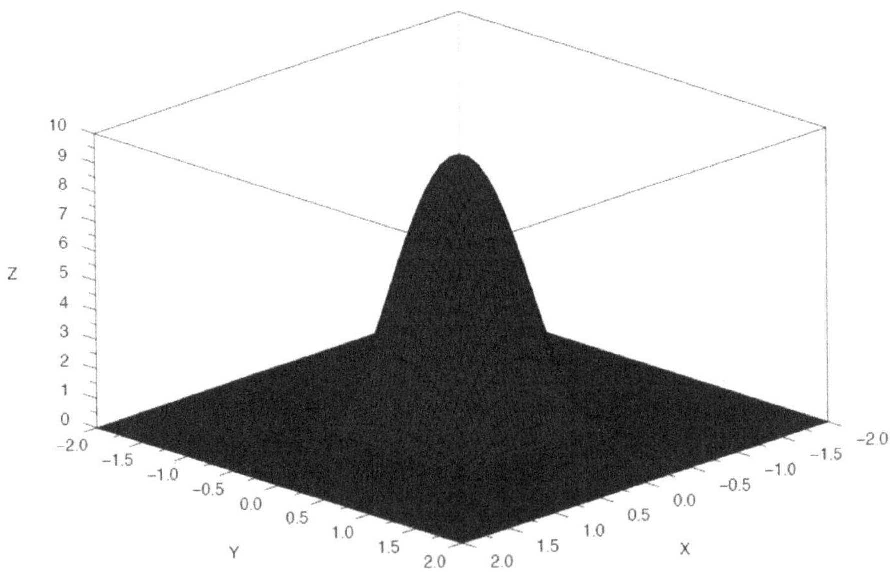

Figure 2.2. *Function* $\rho, d = 2$

We have $\rho \in \mathcal{C}^\infty(\mathbb{R}^d)$ and $\text{supp}\,(\rho) = \overline{B}(0,1)$. Then $\rho \in \mathcal{D}(\mathbb{R}^d)$.

Indeed, the last point is clear and the class of ρ is $\mathcal{C}^\infty$ for $\|x\| \neq 1$; since ρ depends on $\|x\|$ only and that the class of $x \to \|x\|$ is $\mathcal{C}^\infty$ for $x \neq 0$, we merely have to prove that ρ is of class $\mathcal{C}^\infty$ in the case $d = 1$, and for $x = 1$ (by parity).

Since $\rho^{(n)}(t)$ has the form $\dfrac{P_n(t)}{(t^2-1)^{2n}} e^{\frac{1}{t^2-1}}$ for $n \in \mathbb{N}$, $|t| < 1$, and 0 for $|t| > 1$, it can be seen by induction on n, that the class of ρ is $\mathcal{C}^n$ and $\rho^{(n)}(1) = 0$.

REMARK 2.3.– *The function ρ is called the "Pic" function.*

For $n \in \mathbb{N}^*$, we then set $\rho_n(x) = \dfrac{n^d}{I}\rho(nx)$, with $I = \displaystyle\int_{\mathbb{R}^d} \rho(x)\mathrm{d}x$.

PROPOSITION 2.2.– The sequence of functions $(\rho_n)_{n>0}$ verifies the following conditions:

a) $\rho_n \in \mathcal{D}(\mathbb{R}^d)$ and $\rho_n \geq 0$;

b) $\text{supp}\,(\rho_n) \subset \overline{B}\left(0, \frac{1}{n}\right)$;

c) $\displaystyle\int_{\mathbb{R}^d} \rho_n(x)\mathrm{d}x = 1$.

DEFINITION 2.3.– *Any sequence of functions satisfying (a), (b) and (c) of proposition 2.2 is called a regularizing sequence.*

EXAMPLE 2.3.– In this example, a method is given for constructing a $\mathcal{C}^\infty$-class function with compact support.

Let a, b, c and d be real numbers such that $a < b < c < d$.

– We consider the function ψ defined by

$$\begin{cases} \psi(x) = e^{\frac{-1}{t}}, & t > 0 \\ \psi(x) = 0 & \text{otherwise.} \end{cases}$$

It can be seen that $\psi \in \mathcal{C}^\infty(\mathbb{R})$.

– We consider the function $\alpha : t \longmapsto \psi(t)\psi(1-t) > 0$.

– We consider the function $\beta : t \longmapsto \dfrac{\displaystyle\int_0^t \alpha(s)ds}{\displaystyle\int_0^1 \alpha(s)ds} \leq 1$.

Then, the class of function $\varphi : x \longmapsto \beta\left(\frac{x-a}{b-a}\right)\beta\left(\frac{d-x}{d-c}\right)$ is $\mathcal{C}^\infty$ on $\mathbb{R}$ and equal to 1 on $[b, c]$ and with compact support contained in $[a, d]$. In addition, $0 \leq \varphi \leq 1$.

This function is called *plateau function.*

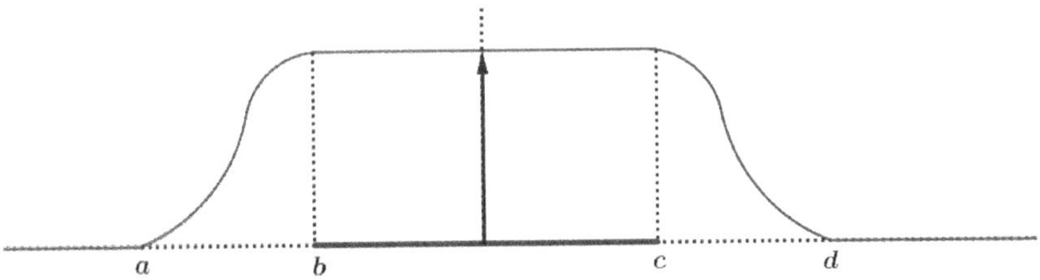

Figure 2.3. *Plateau function. For a color version of this figure, see www.iste.co.uk/aitbenhassi/distribution.zip*

EXAMPLE 2.4.– If α is an indefinitely differentiable function on $\mathbb{R}$ and if φ is a function of $\mathcal{D}(\mathbb{R})$, then their product $\alpha\varphi$ is a function of $\mathcal{D}(\mathbb{R})$.

As a result, we have the following proposition.

PROPOSITION 2.3.– Let α be a function defined on $\mathbb{R}$. We have the following equivalence:

$$\forall\varphi \in \mathcal{D}(\mathbb{R}), \alpha\varphi \in \mathcal{D}(\mathbb{R}) \iff \alpha \in \mathcal{C}^\infty(\mathbb{R}).$$

EXAMPLE 2.5.– Let φ be a function of $\mathcal{D}(\mathbb{R})$. Then, the translated function and the dilated function $d_\lambda\varphi$ are also in $\mathcal{D}(\mathbb{R})$ with $\tau_a\varphi$ defined by

$$(\tau_a\varphi)(x) = \varphi(x-a),$$

where a is real and $d_\lambda\varphi$ is defined by

$$(d_\lambda\varphi)(x) = \varphi\left(\frac{x}{\lambda}\right),$$

where λ is a real number other than 0.

EXAMPLE 2.6.– The test functions of $\mathcal{D}(\mathbb{R})$ and all their derivatives are integrable and bounded.

REMARK 2.4.–

1) We have $\mathcal{D}(\Omega) = \bigcup_{K \in \mathcal{K}_\Omega} \mathcal{D}_K(\Omega)$.

2) The topology of $\mathcal{D}_K(\Omega)$ *is defined by the family of semi-norms:*

$$P_{n,K}(\varphi) = \max_{x \in K, |\alpha| \leq n} |\partial^\alpha \varphi(x)|, \quad |\alpha| = \alpha_1 + \alpha_2 + \cdots + \alpha_d.$$

LEMMA 2.1.– *The space* $\mathcal{D}_K(\Omega)$ *is a Fréchet space.*

LEMMA 2.2 (Decomposition of Ω).– *There exists a sequence* $(K_n)_{n\geq 1}$ *of compacts of* Ω *such that:*

1) $\forall n \geq 1, K_n \subset \overset{\circ}{K}_{n+1}$*;*

2) $\Omega = \bigcup_{n\geq 1} K_n = \bigcup_{n\geq 2} \overset{\circ}{K}_n$*;*

3) for any compact K *of* Ω*, there exists* $n_0 \geq 1$ *such that* $K \subset K_{n_0}$.

Proof. It suffices to take

$$K_n = \{x \in \mathbb{R}^d, \|x\| \leq n\} \cap \{x \in \Omega, d(x, \Omega^c) \geq \frac{1}{n}\}.$$

□

A family of semi-norms is then defined on $\mathcal{D}(\Omega)$ as follows:

$$P_n(\varphi) = \max_{x \in K_n, |\alpha| \leq n} |\partial^\alpha \varphi(x)|.$$

This sequence of semi-norms makes it possible to give $\mathcal{D}(\Omega)$ a locally convex and metrizable topological vector space structure but which **is not complete**.

Counterexample: Let f be a function defined by $f(x) = e^{-x^2}$. We consider the sequence of functions $(\psi_n)_{n\geq 1}$ defined by

$$\begin{cases} \psi_n(x) = 1, & |x| \leq n \\ \psi_n(x) = 0 & |x| > 2n. \end{cases}$$

The sequence of functions $(\psi_n f)_{n\geq 1}$ is a Cauchy sequence in $\mathcal{D}(\mathbb{R})$; it simply converges to f but does not converge in $\mathcal{D}(\mathbb{R})$ because f does not belong to $\mathcal{D}(\mathbb{R})$ due to the fact that supp $(f) = \mathbb{R}$, which is not compact.

2.2.3. *Convergence in* $\mathcal{D}(\Omega)$

DEFINITION 2.4.– *It is said that a sequence* $(\varphi_n)_{n\geq 0}$ *of elements of* $\mathcal{D}(\Omega)$ *converges to* $\varphi \in \mathcal{D}(\Omega)$ *if:*

i) there exists a compact subset K *of* Ω *such that* $\forall n \in \mathbb{N}$, *supp* $(\varphi_n) \subset K$*;*

ii) (φ_n) *and* $(\partial^\alpha \varphi_n)$ *uniformly converge on* K*, respectively, to* φ *and* $\partial^\alpha \varphi$ *for any multi-index* $\alpha \in \mathbb{N}^d$.

EXAMPLE 2.7.– Let $\varphi \in \mathcal{D}(\mathbb{R})$ be fixed. We consider the sequence $(\varphi_n)_{n\geq 0}$ defined by

$$\varphi_n(x) = \varphi(x + \frac{1}{n+1}) - \varphi(x), \quad \forall x \in \mathbb{R}.$$

Let us show that $(\varphi_n) \longmapsto 0$ in $\mathcal{D}(\mathbb{R})$.

We have $\varphi \in \mathcal{D}(\mathbb{R})$, therefore there exists $M > 0$ such that supp $(\varphi) \Subset [-M, M]$.

Then $\forall n \in \mathbb{N}$, supp $(\varphi_n) \Subset [-M-1, M+1]$. Indeed, for all $x \in \mathbb{R}$, if $x \notin [-M-1, M+1]$, then $\varphi(x) = 0$ and $\varphi(x + \frac{1}{n+1}) = 0$. Consequently, we have $\varphi_n(x) = 0$.

According to the finite increment theorem, we have

$$\|\varphi_n^{(k)}(x)\| = \|\varphi^{(k)}(x + \frac{1}{n+1}) - \varphi^{(k)}(x)\| \leq \frac{1}{n+1} \sup_{x\in\mathbb{R}} \|\varphi^{(k+1)}(x)\|.$$

Since supp $\varphi^{(k)} \subset$ supp φ, then, we have

$$\begin{aligned}\|\varphi_n^{(k)}(x)\| &= \|\varphi^{(k)}(x + \frac{1}{n+1}) - \varphi^{(k)}(x)\| \\ &\leq \frac{1}{n+1} \sup_{x\in\mathbb{R}} \|\varphi^{(k+1)}(x)\| \\ &\leq \frac{1}{n+1} \sup_{x\in[-M,M]} \|\varphi^{(k+1)}(x)\|.\end{aligned}$$

Consequently, $(\varphi_n^{(k)})$ uniformly converges to 0 on $K = [-M-1, M+1]$.

Hence, $(\varphi_n) \longmapsto 0$ in $\mathcal{D}(\mathbb{R})$.

2.2.4. *Convolution product*

2.2.4.1. *Definitions*

Consider the measured space $\big(\mathbb{R}^d, \mathcal{M}_L(\mathbb{R}^d), \lambda_d\big)$. We want to define the convolution product of two functions f and g by the formula

$$\forall x \in \mathbb{R}^d, f * g(x) = \int_{\mathbb{R}^d} f(x-t)g(t)\mathrm{d}t.$$

In the case where the functions f and g are positive, their measurability is enough for this formula to make sense. Without assuming this positivity, the convolution product of f and g can still be defined provided that, in addition to their measurability, a regularity of order L^p is assumed.

PROPOSITION 2.4.– Let $p \in [1, +\infty]$. Let $f \in L^p(\mathbb{R}^d)$ and $g \in L^1(\mathbb{R}^d)$.

For almost any $x \in \mathbb{R}^d$, the function $y \longmapsto f(x-y)g(y)$ is integrable. Then the convolution product $f * g$ is defined almost everywhere, moreover $f * g \in L^p(\mathbb{R}^d)$ and $\|f * g\|_p \leq \|f\|_p\|g\|_1$. In addition, we have $f * g = g * f$.

Proof. The proof is given for the case where $p = 1$.

We start it by applying the Fubini–Tonelli theorem and the change in variables theorem that the function $(x, y) \longmapsto \big|f(x-y)g(y)\big|$ is summable. In fact,

$$\begin{aligned}\int_{\mathbb{R}^d}\int_{\mathbb{R}^d} \big|f(x-y)g(y)\big|dxdy &= \int_{\mathbb{R}^d} \big|g(y)\big|\big(\int_{\mathbb{R}^d} \big|f(x-y)\big|dx\big)dy \\ &= \int_{\mathbb{R}^d} \big|g(y)\big|\big(\int_{\mathbb{R}^d} \big|f(x)\big|dx\big)dy = \big\|f\big\|_{L^1}\big\|g\big\|_{L^1}.\end{aligned}$$

The function $(x, y) \mapsto f(x-y)g(y)$ is therefore summable on $\mathbb{R}^{2d}$, and Fubini's theorem ensures that the function $y \longmapsto f(x-y)g(y)$ is summable for almost all x and that the function defined almost everywhere by $f * g(x) = \displaystyle\int_{\mathbb{R}^d} f(x-y)g(y)dy$ is itself summable on $\mathbb{R}^d$. Finally, we have

$$\big\|f * g\big\|_{L^1} = \int_{\mathbb{R}^d} \big|f * g\big|(x)dx \leq \int_{\mathbb{R}^d} \big|g(y)\big|\big(\int_{\mathbb{R}^d} \big|f(x-y)\big|dx\big)dy = \big\|f\big\|_{L^1}\big\|g\big\|_{L^1}.$$

□

REMARK 2.5.–

1) A sufficient condition for defining the convolution product $f * g$ is that the function $y \mapsto f(x-y)g(y)$ be integrable on $\mathbb{R}^d$.

2) To define the convolution product $f * g$ where $f \in L^p(\mathbb{R}^d)$ and $g \in L^q(\mathbb{R}^d)$, we simply have to assume that $\frac{1}{p} + \frac{1}{q} > 1$.

A result generalizing proposition 2.4 is given by the following Hausdorf–Young inequality theorem.

THEOREM 2.1 (Hausdorf–Young inequality).– *Let $p, q, r \in [1, +\infty]$ such that*

$$\frac{1}{p} + \frac{1}{q} = 1 + \frac{1}{r}, \text{ with the convention } \frac{1}{\infty} = 0,$$

and let $f \in L^p(\mathbb{R}^d)$ and $g \in L^q(\mathbb{R}^d)$. Then f and g are convolvable on $\mathbb{R}^d$ and $f * g \in L^r(\mathbb{R}^d)$. Moreover, $f * g$ verifies the inequality:

$$\|f * g\|_{L^r(\mathbb{R}^d)} \leq \|f\|_{L^p(\mathbb{R}^d)} \|g\|_{L^q(\mathbb{R}^d)}.$$

REMARK 2.6.– *If $f \in L^p(\mathbb{R}^d)$ and $g \in L^q(\mathbb{R}^d)$ with $\frac{1}{p} + \frac{1}{q} = 1$, then $f * g \in L^\infty(\mathbb{R}^d)$.*

2.2.4.2. *Some properties of convolution*

The goal here is to give the main properties of convolution that are often used.

2.2.4.2.1. Translation of a convolution product

Let $a \in \mathbb{R}^d$ and f a function. We denote by $\tau_a f$ the $a-$vector translation of f, that is, $\tau_a f(x) = f(x - a)$.

PROPOSITION 2.5.– Let f and g denote two functions such that $(f * g)(x)$ is defined for almost all $x \in \mathbb{R}^d$ and let $a \in \mathbb{R}^d$. Therefore,

$$\tau_a(f * g) = (\tau_a f) * g = f * (\tau_a g).$$

2.2.4.2.2. Differentiation of a convolution product

PROPOSITION 2.6.– Let $f \in L^\infty(\mathbb{R}^d)$, $g \in L^1(\mathbb{R}^d)$ and $k \in \mathbb{N} \cup \{\infty\}$. It is assumed that f is of class C^k and that its partial derivatives of any order are bounded. Then $f * g$ is of class C^k and for any $\alpha \in \mathbb{N}^d$ such that $|\alpha| \leq k$, we have

$$\partial^\alpha (f * g) = (\partial^\alpha f) * g.$$

Proof. For almost every $y \in \mathbb{R}^d$, the function $x \longmapsto f(x-y)g(y)$ is in $\mathcal{C}^k(\mathbb{R}^d)$. Moreover, for all $x \in \mathbb{R}^d$,

$$|\partial^\alpha(f(x-y)g(y))| = |(\partial^\alpha(f(x-y))g(y)| \leq \|\partial^\alpha f\|_\infty |g(y)|.$$

Since $g \in L^1(\mathbb{R}^d)$, the theorem for the differentiation under the integral sign can be applied to obtain the result. □

COROLLARY 2.1.– *If $f \in \mathcal{C}^p(\mathbb{R}^d)$ and $g \in \mathcal{C}^q(\mathbb{R}^d)$, then $f * g \in \mathcal{C}^{p+q}(\mathbb{R}^d)$ and in addition:*

$$\forall \alpha, \beta \in \mathbb{N}^*, |\alpha| \leq p, |\beta| \leq q : \partial^{\alpha+\beta}(f * g) = (\partial^\alpha f) * (\partial^\beta g).$$

In particular, if $f \in \mathcal{C}^\infty(\mathbb{R}^d)$ and $g \in \mathcal{C}^0(\mathbb{R}^d)$, then $f * g \in \mathcal{C}^\infty(\mathbb{R}^d)$.

It should be noted that the condition $g \in L^1_{loc}(\mathbb{R}^d)$ is sufficient, as partial primitives are then continuous.

2.2.4.2.3. Convolution product support

PROPOSITION 2.7.– Let f and g be two functions such that $(f * g)(x)$ is defined for almost any x. Therefore,

$$\text{supp}(f * g) \subset \overline{\text{supp}(f) + \text{supp}(g)}.$$

Proof. Let $x \notin \text{supp}(f) + \text{supp}(g)$. Then $(x - \text{supp}(f)) \cap \text{supp}(g) = \emptyset$ and therefore,

$$\begin{aligned} f * g(x) &= \int_{\mathbb{R}^d} f(x-y)g(y)dy \\ &= \int_{(x-\text{supp}(f))\cap\text{supp}(g)} f(x-y)g(y)dy \\ &= 0 \end{aligned}$$

Therefore, $f * g(x) = 0$ at any point not belonging to $\text{supp}(f) + \text{supp}(g)$ and therefore also at any point not belonging to its closure $\overline{\text{supp}(f) + \text{supp}(g)}$, which gives the result. □

It should be recalled that, $A + B = \{x + y \ / \ x \in A, y \in B\}$, the sum of two closed sets is closed if one of them is compact and it is compact if both are compact.

EXAMPLE 2.8.– Let us consider $\chi = \mathbb{1}_{[0,1]}$. We have

$$\text{supp}\,(\chi) = [0,1]\,, \quad \text{supp}\,(\chi * \chi) = [0,2]\,, \text{supp}\,(\chi * \chi * \chi) = [0,3]$$

$$\text{supp}\,(\underbrace{\chi * \ldots * \chi}_{ntimes}) = [0,n]\,.$$

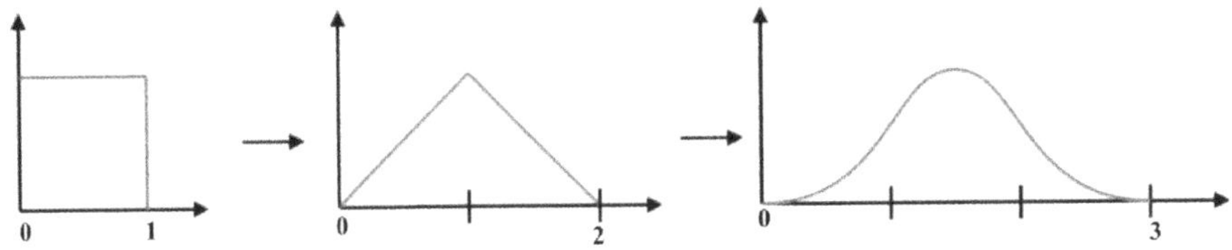

Figure 2.4. $\mathbb{1}_{[0,1]} \to \mathbb{1}_{[0,1]} * \mathbb{1}_{[0,1]} \to \mathbb{1}_{[0,1]} * \mathbb{1}_{[0,1]} * \mathbb{1}_{[0,1]}$

2.2.4.2.4. Convolution and $\mathcal{D}(\mathbb{R}^d)$

PROPOSITION 2.8.– Let $f \in L^p_c(\mathbb{R}^d) = \{u \in L^p(\mathbb{R}^d), u = 0$ outside a compact of $\mathbb{R}^d\}$. For all $\varphi \in \mathcal{D}(\mathbb{R}^d)$, we have $\varphi * f \in \mathcal{D}(\mathbb{R}^d)$.

Proof. Let a and b be two real numbers such that $\text{supp}\,\varphi \subset B(0,a)$ and $f(y) = 0$ for $|y| \geq b$. Let $x \in \mathbb{R}^d$ such that $|x| > a + b$. As in the integral which defines $\varphi * f$, it can be assumed that $x - y \in \text{supp}\,\varphi$, from which $|x - y| \leq a$. Then, $|y| \geq |x| - |x - y| > a + b - a = b$. So, $f(y) = 0$. Therefore, $(\varphi * f)(x) = 0$ for all x such that $|x| > a + b$. Consequently, $\text{supp}\,(\varphi * f) \subset B\,(0, a + b)\,.$ □

The previous results will be used to show that it is possible to *regularize* a non-regular function by "convolving" it by a regular function. We start by considering a function Pic, ρ whose support is included in $B(0,1)$ and whose integral over $\mathbb{R}^d$ is 1.

For $\varepsilon > 0$, we set $\rho_\varepsilon(x) = \frac{1}{\varepsilon^d}\rho\left(\frac{x}{\varepsilon}\right)$. The sequence $\left(\rho_\varepsilon\right)$ is called an approximation of the identity. More precisely, the convolution operators by ρ_ε converge to the identity operator. By convolving with the sequence (ρ_ε), the following approximation result is obtained:

PROPOSITION 2.9 (Bony 2001).–

1) If $u \in C_c^k(\mathbb{R}^d)$ and $k \in \mathbb{N}$, then for all $\alpha \in \mathbb{N}^d$ such that $|\alpha| \leq k$, the sequence $\big(\partial^\alpha(\rho_\varepsilon * u)\big)$ uniformly converges to $\partial^\alpha u$ on $\mathbb{R}^d$ when ε tends to 0.

2) If $u \in L_c^p(\mathbb{R}^d)$, then the sequence $\big(\rho_\varepsilon * u\big)$ converges to u in $L_c^p(\mathbb{R}^d)$ for all $1 \leq p < +\infty$ when ε tends to 0.

2.2.5. *A few density results*

Some density results now follow that will be useful later on.

We endow $\mathcal{C}_c(\Omega)$, the space of continuous functions with compact support in Ω, with the following topology: it is said that the sequence (u_n) tends to u if for any $\varepsilon > 0$ there exists n_0 such that for $n \geq n_0$, we have $\text{supp}(u_n) \subset \text{supp}(u) + B(0, \varepsilon)$ and $\sup\limits_{x \in \Omega} \big|u_n(x) - u(x)\big| < \varepsilon$.

PROPOSITION 2.10.– Let Ω be an open set of $\mathbb{R}^d$. The space $\mathcal{D}(\Omega)$ is dense in $\mathcal{C}_c(\Omega)$.

Proof. Let $f \in \mathcal{C}_c(\Omega)$, by proposition 2.8, we indeed have $\rho_\varepsilon * f \in \mathcal{D}(\mathbb{R}^d)$. Moreover, since $f \in \mathcal{C}_c(\Omega)$, then it is uniformly continuous and $\rho_\varepsilon * f$ tends to f uniformly. In addition, by proposition 2.7, $\text{supp}(\rho_\varepsilon * f) \subset \text{supp}(f) + B(0, \varepsilon)$. Therefore, $\rho_\varepsilon * f$ indeed tends to f in $\mathcal{C}_c(\Omega)$. The density of $\mathcal{D}(\Omega)$ in $\mathcal{C}_c(\Omega)$ is deduced thereof. □

THEOREM 2.2.– *Let Ω be an open set of $\mathbb{R}^d$.*

The space $\mathcal{D}(\Omega)$ is dense in $L^p(\Omega)$ for any $1 \leq p < \infty$.

Proof. We consider $\Omega_r = \big\{x, B(x,r) \subset \Omega\big\} \cap B(0, \frac{1}{r})$ the inner bounded neighborhood of Ω. Since $\Omega = \bigcup\limits_{r>0} \Omega_r$, by Lebesgue's theorem, it follows that:

$$\forall \varepsilon > 0, \exists r > 0, \int_\Omega |f(x) - f(x)\mathbb{1}_{\Omega_r}|^p dx < \varepsilon^p.$$

Then we choose $r' < r$ such that $\big\|(f\mathbb{1}_{\Omega_r}) * \rho_{r'} - f\mathbb{1}_{\Omega_r}\big\|_{L^p(\Omega)} < \varepsilon$. This thus leads to $\big\|f - (f\mathbb{1}_{\Omega_r}) * \rho_{r'}\big\|_{L^p(\Omega)} < 2\varepsilon$. Since $r' < r$, propositions 2.7 and 2.8 imply that $\big(f\mathbb{1}_{\Omega_r}\big) * \rho_{r'} \in \mathcal{D}(\Omega)$. □

PROPOSITION 2.11.– The space $\mathcal{D}(\Omega)$ is dense in $\mathcal{C}^k(\Omega)$ for all $k \in \mathbb{N}$.

As an application of these two density results, lemma 2.3 is given, which is known as the Dubois–Reymond lemma, the proof of which is the subject of a corrected exercise.

LEMMA 2.3 (Dubois-Reymond).– *Let* $f \in L^1_{loc}(\Omega)$. *It is assumed that, for all* $\varphi \in \mathcal{D}(\Omega)$, $\int_\Omega f(x)\varphi(x)dx = 0$. *Then,* $f = 0$ *almost everywhere.*

Proof. See exercise 2.5. □

2.3. Exercises with solutions

EXERCISE 2.1.– Consider the sequence of functions $f_n(x) = x(1-\frac{x}{n})^n \ln x \ \chi_{[0,n]}$, for $n \geq 1$ and the function $f(x) = xe^{-x} \ln x$.

1) Show that f is integrable on $[0, +\infty[$.

2) Show that $\lim_{n \mapsto +\infty} \int_0^n x(1-\frac{x}{n})^n \ln x \ dx = \int_0^\infty xe^{-x} \ln x \ dx.$

SOLUTION 2.1.–

1) We have that f is continuous on $]0, 1]$ and extendable by continuity to the right at 0. Therefore, f is integrable on $[0, 1]$. On the other hand, f is continuous on $[1, +\infty[$ and $f(x) = O\left(e^{-\frac{x}{2}}\right)$. Since the function $x \mapsto e^{-\frac{x}{2}}$ is integrable on $[1, +\infty[$, then f also is. Consequently, f is integrable on $[0, +\infty[$.

2) Let $n \in \mathbb{N}^*$. For all $x \in [0, n]$, we have $(1-\frac{x}{n})^n \leq e^{-x}$. Thereafter, $\forall x \in [0, n], |f_n(x)| \leq |f(x)|$. On the other hand, $\forall x \notin [0, n], 0 = |f_n(x)| \leq |f(x)|$. Hence:

$$\forall x \in \mathbb{R}, |f_n(x)| \leq \left|f(x)\chi_{[0,+\infty[}\right|.$$

- The functions f_n are integrable because they are upper bounded in absolute value by an integrable function.

- We have $\forall x \in \mathbb{R}, |f_n(x)| \leq \left|f(x)\chi_{[0,+\infty[}\right|$.

- The function $f\chi_{[0,+\infty[}$ is integrable on $\mathbb{R}$ because f is integrable on $[0, +\infty[$.

- For all $x \in \mathbb{R}$, we have $\lim_{n \mapsto +\infty} (1-\frac{x}{n})^n = e^{-x}$ and $\lim_{n \mapsto +\infty} \chi_{[0,n]} = \chi_{[0,+\infty[}$, thus $\lim_{n \mapsto +\infty} f_n(x) = f(x)\chi_{[0,+\infty[}$.

The assumptions of Lebesgue's dominated convergence theorem are all verified. Consequently, it follows that

$$\lim_{n\mapsto+\infty}\int_{\mathbb{R}} f_n(x)\mathrm{d}x = \int_{\mathbb{R}} f(x)\chi_{[0,+\infty[}\mathrm{d}x.$$

Hence, $\lim\limits_{n\mapsto+\infty}\int_0^n x(1-\frac{x}{n})^n \ln x\mathrm{d}x = \int_0^\infty xe^{-x}\ln x\mathrm{d}x.$

EXERCISE 2.2.– Is there a function g integrable on $\mathbb{R}$ such that

$$\forall n\in\mathbb{N}, \forall x\in\mathbb{R}, ne^{-n|x|}\leq g(x).$$

SOLUTION 2.2.– We assume by contradiction that there exists a function $g\in L^1(\mathbb{R})$ such that

$$\forall n\in\mathbb{N},\ \forall x\in\mathbb{R},\ ne^{-n|x|}\leq g(x).$$

Since the functions $x\mapsto ne^{-n|x|}$ are measurable and upper bounded by an integrable function, the dominated convergence theorem can be applied:

$$\lim_{n\to\infty}\int_{\mathbb{R}} ne^{-n|x|}\,\mathrm{d}x = \int_{\mathbb{R}}\lim_{n\to\infty} ne^{-n|x|}\mathrm{d}x = \int_{\mathbb{R}} 0\,\mathrm{d}x = 0.$$

Yet, $\forall n\in\mathbb{N},\quad \int_{\mathbb{R}} ne^{-n|x|}\mathrm{d}x = \int_{-\infty}^0 ne^{nx}\mathrm{d}x + \int_0^{+\infty} ne^{-nx}\mathrm{d}x = 1+1=2.$

It is a contradiction!

EXERCISE 2.3.– Let $f\in L^1(\mathbb{R})$. Consider the mapping:

$$\begin{aligned} T_f : \mathbb{R} &\longrightarrow \mathbb{R}\\ x &\mapsto \int_0^1 f(x-y)\,\mathrm{d}y. \end{aligned}$$

1) Show that if f is continuous and has compact support, T_f is continuous.

2) Let $(f_n)_{n\in\mathbb{N}}$ be a sequence of continuous functions with compact support that converges to f in $L^1(\mathbb{R})$. Show that $(T_{f_n})_{n\in\mathbb{N}}$ uniformly converges to T_f on $\mathbb{R}$. Thereof, deduce that T_f is continuous on $\mathbb{R}$.

3) Deduce that the convolution product on $L^1(\mathbb{R})$ has no unit element.

SOLUTION 2.3.–

1) Let $f \in C_c(\mathbb{R})$ such that $\operatorname{supp} f \subset [a,b]$. Then $f \in L^1(\mathbb{R})$ and for all fixed $x \in \mathbb{R}$, $y \mapsto f(x-y)$ also is in $L^1(\mathbb{R})$.

So, T_f is well defined. Moreover, as f is continuous, so is $x \mapsto f(x-y)$, we have

$$\forall x \in \mathbb{R}, \forall y \in [0,1], \mid f(x-y) \mid \leq M \mathbb{1}_{[a-1,b+1]}(y).$$

Since $y \mapsto M\mathbb{1}_{[a-1,b+1]}(y) \in L^1(\mathbb{R})$ and this increase is uniform in x, then by the continuity theorem under the integral sign, T_f is continuous.

2) Since $C_c(\mathbb{R})$ is dense in $L^1(\mathbb{R})$, for any $f \in L^1(\mathbb{R})$, there exists a sequence of functions $(f_n)_{n\in\mathbb{N}}$ of continuous functions with compact support that converges to f in $L^1(\mathbb{R})$. Therefore,

$$\forall x \in \mathbb{R}, |T_{f_n}(x) - T_f(x)| \leq \int_0^1 |f_n(x-y) - f(x-y)| \, \mathrm{d}y \leq \| f_n - f \|.$$

Moreover, $\sup_{x\in\mathbb{R}} |T_{f_n}(x) - T_f(x)| \leq \| f_n - f \| \longrightarrow 0$ if n tends to ∞. Therefore, (T_{f_n}) converges uniformly to T_f on $\mathbb{R}$. Since every T_{f_n} is continuous, then T_f is continuous.

3) We have $T_f = f * \mathbb{1}_{[0,1]}$. We assume by contradiction that there exists $u \in L^1(\mathbb{R})$ such that $f * u = f$, then $T_u = u * \mathbb{1}_{[0,1]} = \mathbb{1}_{[0,1]}$. However, according to Question 2, T_u is continuous on $\mathbb{R}$, whereas $\mathbb{1}_{[0,1]}$ is not. It is a contradiction!

EXERCISE 2.4.– Let $h \in \mathbb{R}^d$. The translation operator is defined by $h \in \mathbb{R}^d$, denoted by τ_h, acting on a function $f : \mathbb{R}^d \to \mathbb{R}$ by $\forall x \in \mathbb{R}^d$, $(\tau_h f)(x) = f(x-h)$. The purpose of this exercise is to show the following result: if $f \in L^p(\mathbb{R}^d)$ with $1 \leq p < +\infty$, then $\lim_{h\to 0} \| \tau_h f - f \|_p = 0$ (i.e. $\tau_h f$ tends to f in $L^p(\mathbb{R}^d)$ when h tends to 0). Let $1 \leq p < +\infty$.

1) Show that if f is continuous and compactly supported included in the ball $B(0,M)$ centered at 0 and of radius M and if $\mid h \mid \leq 1$, then

$$\forall x \in \mathbb{R}^n, \mid f(x-h) - f(x) \mid^p \leq \mathbb{1}_{B(0,M+1)}(x) 2^p \| f \|_\infty^p .$$

2) Deduce that, for f continuous with compact support, we have $\lim_{h\to 0} \| \tau_h f - f \|_p = 0$.

3) Prove the above stated result for any function in $L^p(\mathbb{R}^d)$, $1 \leq p < +\infty$.

4) What happens if $p = \infty$?

SOLUTION 2.4.–

1) If f is a continuous function with compact support in the ball $B(0,M)$ and if $|\, h \,|\leq 1$, then $\forall x \in \mathbb{R}^d$:

$$
\begin{aligned}
|\, f(x-h) - f(x) \,|^p &\leq \left(|\, f(x-h) \,| + |\, f(x) \,|\right)^p \\
&\leq \left(2 \,\|\, f \,\|_\infty \, \mathbb{1}_{B(0,M+1)(x)}\right)^p \\
&\leq \mathbb{1}_{B(0,M+1)}(x) 2^p \,\|\, f \,\|_\infty^p \,.
\end{aligned}
$$

2) Let f be a continuous function with compact support. Since f is continuous, then $\lim_{h\to 0} |\, f(x-h) - f(x) \,| = 0$.

Then, since $x \mapsto \mathbb{1}_{B(0,M+1)}(x)2^p \,\|\, f \,\|_\infty^p$ is in $L^1(\mathbb{R}^d)$ and dominates the functions $f_h : x \mapsto |\, f(x-h) - f(x) \,|^p$, Lebesgue's dominated convergence theorem can be applied:

$$
\begin{aligned}
\lim_{h\to 0} \|\, \tau_h f - f \,\|_p^p &= \lim_{h\to 0} \int_{\mathbb{R}^d} |\, f(x-h) - f(x) \,|^p \, dx \\
&= \int_{\mathbb{R}^d} \lim_{h\to 0} |\, f(x-h) - f(x) \,|^p \, dx \\
&= \int_{\mathbb{R}^d} 0 \, dx \\
&= 0.
\end{aligned}
$$

3) Let $p \in [1, +\infty[$ and $f \in L^p(\mathbb{R}^d)$. Since $C_c(\mathbb{R}^d)$ is dense in $L^p(\mathbb{R}^d)$ for the norm $\|\, . \,\|_p$, for all $\varepsilon > 0$ there exists f_ε continuous with compact support such that $\|\, f - f_\varepsilon \,\|_p \leq \varepsilon$. Then,

$$
\begin{aligned}
\|\, \tau_h f - f \,\|_p &= \|\, \tau_h(f - f_\varepsilon) - (f - f_\varepsilon) + \tau_h f_\varepsilon - f_\varepsilon \,\|_p \\
&\leq \|\, \tau_h(f - f_\varepsilon) \,\|_p + \|\, f - f_\varepsilon \,\|_p + \|\, \tau_h f_\varepsilon - f_\varepsilon \,\|_p \\
&= 2 \,\|\, f - f_\varepsilon \,\|_p + \|\, \tau_h f_\varepsilon - f_\varepsilon \,\|_p \\
&\leq 2\varepsilon + \|\, \tau_h f_\varepsilon - f_\varepsilon \,\|_p \,.
\end{aligned}
$$

Since $f_\varepsilon \in C_c(\mathbb{R}^d)$, from Question 2, there exists $\delta > 0$ such that

$$
|\, h \,| < \delta \Longrightarrow \|\, \tau_h f_\varepsilon - f_\varepsilon \,\|_p \leq \varepsilon.
$$

Then $|\, h \,| < \delta \Rightarrow \|\, \tau_h f - f \,\|_p \leq 3\varepsilon$. Therefore, $\lim_{h\to 0} \|\, \tau_h f - f \,\|_p = 0$.

4) The theorem does not hold for $p = +\infty$.

As a matter of fact, for $f = \mathbb{1}_{B(0,1)}$, $h \neq 0$, we have $\| \tau_h f - f \|_\infty = 1$.

However, for $h = 0$, $\| \tau_h f - f \|_\infty = 0$. As such, $\| \tau_h f - f \|_\infty$ does not tends to 0 when h tends to 0.

EXERCISE 2.5 (Dubois-Reymond).– Let I be an open set of $\mathbb{R}$ and $f \in L^1_{\text{loc}}(I)$. It is assumed that $\forall \varphi \in \mathcal{D}(I)$, $\displaystyle\int_I f(x)\varphi(x)\,\mathrm{d}x = 0$. Let $[a, b] \subset I$ be an interval.

1) Using a density argument, show that

$$\forall g \in \mathcal{C}^0(I), \text{ supp } g \subset [a, b], \int_{[a,b]} f(x)g(x)\,\mathrm{d}x = 0$$

2) Let $\varepsilon > 0$. Let $h_\varepsilon \in \mathcal{C}^0(]a, b[)$ be such that $\displaystyle\int_{[a,b]} | f(x) - h_\varepsilon | \,\mathrm{d}x \leq \varepsilon$. Show that $\forall g \in \mathcal{C}^0(]a, b[)$, $\left|\displaystyle\int_{[a,b]} h_\varepsilon(x)g(x)\,\mathrm{d}x\right| \leq \varepsilon \| g \|_\infty$.

3) Properly chosing $g \in \mathcal{C}^0(]a, b[)$, show that $\displaystyle\int_{[a,b]} | h_\varepsilon(x) | \,\mathrm{d}x \leq \varepsilon$.

4) Derive therefrom that $f = 0$ almost everywhere on I.

SOLUTION 2.5.–

1) The space $\mathcal{D}(I) = C_c^\infty(I)$ is dense in $C_c(I)$ for the infinite norm. More precisely, let $g \in C_c(I)$ with $\text{supp}\, g \subset [a, b] \subset I$.

Then, there exists a sequence $\varphi_n \in \mathcal{D}(I)$ with $\text{supp}\, \varphi_n \subset [a', b'] \subset I$; it can be specified that $[a, b] \subset [a', b']$, such that φ_n uniformly converges to g on $[a, b]$. Then, $\left|\displaystyle\int_I f(x)(g(x) - \varphi_n(x))\mathrm{d}x\right| \leq \|g - \varphi_n\|_\infty \displaystyle\int_{a'}^{b'} |f(x)|\,\mathrm{d}x$ is locally integrable on I, and finally $\left|\displaystyle\int_I f(x)(g(x) - \varphi_n(x))\mathrm{d}x\right| \leq M \|g - \varphi_n\|_\infty$, which tends to 0 when n tends to $+\infty$. By hypothesis, for all $n \in \mathbb{N}$, we have $\displaystyle\int_I f(x)\varphi_n(x)\mathrm{d}x = 0$, therefore $\displaystyle\int_{[a,b]} f(x)g(x)\mathrm{d}x = \int_I f(x)g(x)\mathrm{d}x = 0$.

2) Since $f \in L^1([a, b])$, for all $\varepsilon > 0$ there exists $h_\varepsilon \in C_c(]a, b[)$ such that

$$\int_a^b |f(x) - h_\varepsilon(x)|\,\mathrm{d}x \leq \varepsilon.$$

Let $g \in \mathcal{C}_c(]a,b[)$. Then,

$$\int_a^b h_\varepsilon(x)g(x)\mathrm{d}x = \int_a^b (h_\varepsilon(x) - f(x))g(x)\mathrm{d}x + \int_a^b f(x)g(x)\mathrm{d}x.$$

Therefore,

$$\left|\int_a^b h_\varepsilon(x)g(x)\mathrm{d}x\right| \leq \int_a^b |h_\varepsilon(x) - f(x)|\,|g(x)|\,\mathrm{d}x \leq \varepsilon \parallel g \parallel_\infty .$$

3) We set $g_n(x) = \dfrac{h_\varepsilon}{|h_\varepsilon(x)| + \frac{1}{n}}$. For all $n \in \mathbb{N}$, we have $g_n \in \mathcal{C}_c(]a,b[)$ and $\|g_n\|_\infty \leq 1$.

Let $\psi_n(x) = h_\varepsilon(x)g_n(x) = \dfrac{h_\varepsilon^2}{\mid h_\varepsilon(x) \mid + \frac{1}{n}}$.

We have $\lim\limits_{n\to\infty} \psi_n(x) = \mid h_\varepsilon(x) \mid$ and $|\psi_n(x)| \leq |h_\varepsilon(x)|$. According to Lebesgue's dominated convergence theorem, we have $\lim\limits_{n\to+\infty} \displaystyle\int_{[a,b]} \psi_n(x)\mathrm{d}x = \int_{[a,b]} |h_\varepsilon(x)|\,\mathrm{d}x$. Hence the result.

4) From the definition and Question 3, it can be derived that

$$\begin{aligned}\int_{[a,b]} |f(x)|\,\mathrm{d}x &= \int_{[a,b]} |f(x) - h_\varepsilon(x) + h_\varepsilon(x)|\,\mathrm{d}x \\ &\leq \int_{[a,b]} |f(x) - h_\varepsilon(x)|\,\mathrm{d}x + \int_{[a,b]} |h_\varepsilon(x)|\,\mathrm{d}x \\ &\leq 2\varepsilon.\end{aligned}$$

Therefore, this integral is zero. This holds for all $[a,b] \subset I$, so f is zero almost everywhere on I.

EXERCISE 2.6.– Let $\varphi \in \mathcal{D}(\mathbb{R})$ non-identically zero. For all $n \in \mathbb{N}^*$, and for all $x \in \mathbb{R}$, we set $\varphi_n(x) = \frac{1}{n}\varphi(nx)$. Study the convergence of the sequence $(\varphi_n)_{n\geq 1}$ in $\mathcal{D}(\mathbb{R})$.

SOLUTION 2.6.– It is easy to see that the sequence $(\varphi_n)_{n\geq 1}$ simply converges to 0. If we prove that the sequence $(\varphi_n)_{n\geq 1}$ converges in $\mathcal{D}(\mathbb{R})$, it can only be toward the zero function. It can also be observed that if φ has its support in $[-M, M]$, then any

φ_n has its support in $[-M/n, M/n] \subset [-M, M]$. Therefore, all the functions φ_n have their support contained in the same compact $[-M, M]$. It will be interesting to see whether for all $k \in \mathbb{N}$, the sequence $(\varphi_n^{(k)})_{n\geq 1}$ converges uniformly to 0 on the compact $[-M, M]$. By taking the first derivative, it can be seen that $\varphi_n'(x) = \varphi'(nx)$, consequently $\|\varphi_n'\|_\infty = \|\varphi'\|_\infty$. It does not necessarily tend to 0 when n tends to $+\infty$.

EXERCISE 2.7 (Borel's theorem).– Let $(a_n)_{n\in\mathbb{N}}$ be a sequence of complex numbers. The goal of this exercise is to show that there exist functions $f : \mathbb{R} \to \mathbb{C}$ indefinitely differentiable and such that

$$\forall n \in \mathbb{N}, f^{(n)}(0) = a_n.$$

1) Using a function $\phi \in \mathcal{D}(\mathbb{R})$ equal to 1 in a neighborhood of 0, solve the problem when the integer series $\sum_{n\in\mathbb{N}} \frac{a_n}{n!} x^n$ has a non-zero convergence radius.

2) Let $\phi \in \mathcal{D}(\mathbb{R})$ with support included in $[-1, 1]$ and equal to 1 on $\left[-\frac{1}{2}, \frac{1}{2}\right]$. We define a sequence α_n as $\alpha_n = 1$ if $| a_n |\leq 1$ and $\alpha_n =| a_n |$ otherwise. We set, for each $n \in \mathbb{N}$,

$$f_n(x) = \frac{a_n}{n!} x^n \phi(\alpha_n x), f(x) = \sum_{n\in\mathbb{N}} f_n(x).$$

a) Verify that the series defining f converges normally on $\mathbb{R}$ (bounds can be set for each $nf_n(x)$ by separating the cases $| x\alpha_n |\geq 1$ and $| x\alpha_n |< 1$).

b) Show that f is $\mathcal{C}^\infty$, and calculate $f^{(k)}(0)$.

3) Is it always possible to obtain a function verifying the same property if we require that f be integer?

SOLUTION 2.7.–

1) Let $R > 0$ be the radius of convergence of the power series $\sum_{n\in\mathbb{N}} \frac{a_n}{n!} x^n$, and let θ be a plateau function element of $\mathcal{D}(\mathbb{R})$, which is equal to 1 in the neighborhood of 0 and 0 outside $\left]-\frac{R}{2}, \frac{R}{2}\right[$. We set

$$f(x) = \theta(x) \sum_{n\in\mathbb{N}} \frac{a_n}{n!} x^n \quad \textit{if } | x |< R, f(x) = 0 \textit{ otherwise}.$$

The function f is clearly $\mathcal{C}^\infty$ in $]-R, R[$, and is zero in $\mathbb{R} \setminus \left]-\frac{R}{2}, \frac{R}{2}\right[$. It is thus of class $\mathcal{C}^\infty$ on $\mathbb{R}$, and since it coincides with the power series in the neighborhood of 0, it correctly verifies that $f^{(n)}(0) = a_n$.

2) a) It should be noted that if $|x|\alpha_n \geq 1$, then $f_n(x) = 0$. On the other hand, if $|x|\alpha_n \leq 1$, we have

$$| f_n(x) | \leq \frac{| a_n |}{n!} \frac{\| \phi \|_\infty}{\alpha_n^n},$$

which is an inequality that is in fact verified for all x in $\mathbb{R}$. Now $\frac{|a_n|}{\alpha_n^n} \leq 1$ and thus $| f_n(x) | \leq \frac{1}{n!}$, which is the general term of a convergent series. Consequently, the series $\sum_{n \in \mathbb{N}} f_n(x)$ converges normally on $\mathbb{R}$.

b) It is simply necessary to show that for each k, the series $\sum f_n^{(k)}$ converges normally on $\mathbb{R}$. We set $M_k = \sup_{x \in \mathbb{R}} | \phi^{(k)} |$ (this exists because the function has compact support). Using the Leibniz formula leads to

$$f_n^{(k)}(x) = \sum_{l=0}^{k} \binom{k}{l} \frac{x^{n-l}}{(n-l)!} a_n \alpha_n^{k-l} \phi^{(k-l)}(\alpha_n x).$$

Again, distinguishing between $| x | \alpha_n \leq 1$ and $| x | \alpha_n > 1$, yields for $n > k$:

$$| f_n^{(k)}(x) | \leq \sum_{l=0}^{k} \frac{M_{k-l} \binom{k}{l}}{(n-1)!}.$$

The left-hand side of this inequality is the general term of a convergent series. This proves that f is of class $\mathcal{C}^{infty}$, and that we have

$$f^{(k)}(x) = \sum_{n \geq 0} f_n^{(k)}(x).$$

In particular, since in the neighborhood of 0, $f_n(x) = \frac{a_n}{n!} x^n$, we have $f^{(k)}(0) = a_k$, which correctly gives the desired function.

3) No, because if the function f is integer, the series $\sum a_n x^n$ must converge for all x.

EXERCISE 2.8.–

1) Let $f \in \mathcal{D}(\mathbb{R}^n), a, b \in \mathbb{R}^n$ and $m \geq 0$. Prove the Taylor formula with integral remainder

$$f(b) = \sum_{|\alpha|\leq m} \frac{(b-a)^\alpha}{\alpha!}\partial^\alpha f(a) +$$
$$(m+1) \sum_{|\alpha|=m+1} \frac{(b-a)^\alpha}{\alpha!} \int_0^1 (1-t)^m \partial^\alpha f(a+t(b-a))dt.$$

2) Let $f \in \mathcal{D}(\mathbb{R}^n)$ which cancels out at the origin. Show that there exists functions $g_1, \cdots, g_n \in \mathcal{D}(\mathbb{R}^n)$ such that

$$\forall x \in \mathbb{R}^n, f(x) = x_1 g_1(x) + \cdots + x_n g_n(x).$$

3) Generalize to the case where f and all its (partial) derivatives up to order $m-1$ cancel out at 0.

SOLUTION 2.8.–

1) Let g be the function defined on $\mathbb{R}$ by $g(t) = f(a+t(b-a))$. The Taylor formula with integral remainder for g (in one variable) applied between 0 and 1 is written as

$$g(1) = \sum_{k=0}^{m} \frac{g^k(0)}{k!} + \int_0^1 \frac{(1-t)^m}{m!} g^{(m+1)}(t)dt.$$

We then show by induction on k that

$$\frac{g^k(t)}{k!} = \sum_{|\alpha|=k} \frac{(b-a)^\alpha}{\alpha!}\partial^\alpha f(a+t(b-a)).$$

This is clear if $k=1$. Let us assume that the formula is proved at rank k, and prove it at rank $k+1$. We indeed have

$$\frac{d}{dt}\partial^\alpha f(a+t(b-a)) = \sum_{j=1}^{n}(b_j - a_j)\partial^{\alpha+e_j} f(a+t(b-a)),$$

where e_j designates $(0, \cdots, 0, 1, 0, \cdots, 0)$.

We have

$$\begin{aligned}\frac{g^{(k+1)}(t)}{(k+1)!} &= \frac{1}{k+1}\frac{d}{dt}\frac{g^{(k)}(t)}{(k)!} \\ &= \sum_{|\alpha|=k}\frac{(b-a)^\alpha}{\alpha!}\frac{d}{dt}\partial^\alpha f(a+t(b-a)) \\ &= \sum_{|\alpha|=k}\frac{(b-a)^\alpha}{\alpha!}\sum_{j=1}^{n}(b_j-a_j)\partial^{\alpha+e_j} f(a+t(b-a)) \\ &= \sum_{|\alpha|=k}\sum_{j=1}^{n}\frac{(b-a)^\alpha}{\alpha!}(b_j-a_j)\partial^{\alpha+e_j} f(a+t(b-a)).\end{aligned}$$

Let β such that $|\beta| = k+1$.

It should be noted that each term of the sum

$$\sum_{|\alpha|=k}\sum_{j=1}^{n}\frac{(b-a)^\alpha}{\alpha!}(b_j-a_j)\partial^{\alpha+e_j} f(a+t(b-a))$$

corresponds to expressing β in the form $\alpha + e_k$. By grouping the terms equal to $(b - a)^\beta \partial^\beta f(a + t(b - a))$ whose number is $\frac{1}{k+1}\left(\frac{1}{(\beta_1-1)!\cdots\beta_n!}+\cdots+\frac{1}{\beta_1!\cdots(\beta_n-1)!}\right)$ the following result is obtained: $\frac{g^{k+1}(t)}{(k+1)!} = \sum_{|\beta|=k+1}\frac{(b-a)^\beta}{\beta!}\partial^\beta f(a+t(b-1))$.

This completes the inductive reasoning.

The requested formula results immediately therefrom.

2) Applying the previous formula at order $0!$ yields

$$f(x) = \sum_{i=1}^{n} x_i \int_0^1 (1-t)\frac{\partial f}{\partial x_i}(tx)dt.$$

Let us thus set $h_i(x) = \int_0^1 (1-t)\frac{\partial f}{\partial x_i}(tx)dt.$

Each function h_i is clearly of class C^∞. But caution should be taken, and there is no requirement that h_i be compactly supported. Truncating with a plateau function will help us achieve what we are looking for. Therefore, let $[-A, A]^n$ containing the support of f, and let θ in $\mathcal{D}(\mathbb{R}^n)$ equal to 1 on $[-A, A]^n$. Finally, we set $g_i = \theta h_i \in \mathcal{D}(\mathbb{R}^n)$. For $x \in [-A, A]^n$, the relation $f(x) = x_1 g_1(x) + \cdots + x_n g_n(x)$ is clearly satisfied. If x is not in $[-A, A]^n$, both members are equal to zero and the relation sought for is satisfied as well.

3) In this case, f is written as

$$f(x) = \sum_{|\alpha|=m} x^\alpha g_\alpha(x),$$

where the function g_α is in $\mathcal{D}(\mathbb{R}^n)$.

3

Distributions on an Open Set of $\mathbb{R}^d$

This chapter introduces the notion of distribution as a continuous linear form on $\mathcal{D}(\Omega)$: the space of test functions is discussed in Chapter 2. After giving the order of a distribution, a series of examples illustrating this notion is presented. In this chapter, Ω denotes an open set of $\mathbb{R}^d$.

3.1. Definitions

Two equivalent definitions of the notion of distribution are given, one functional and theoretical in which continuity is sequentially expressed and the other effective in which continuity is directly expressed by estimates.

3.1.1. *Functional definition*

DEFINITION 3.1.– *We call distribution on the open set Ω, any linear form $T : \mathcal{D}(\Omega) \longrightarrow \mathbb{K}(= \mathbb{R}$ or $\mathbb{C})$ continuous in 0 (for any sequence $(\varphi_n)_{n\in\mathbb{N}}$ of elements of $\mathcal{D}(\Omega)$ that converges to 0, the sequence $\langle T, \varphi_n\rangle \longmapsto 0$ in $\mathbb{K}$).*

We will denote by $\mathcal{D}'(\Omega)$ the set of distributions on Ω.

The symbol $\langle T, \varphi_n\rangle$ refers here to a duality bracket, and it simply means the action of T on φ_n : $T(\varphi_n)$. The set $\mathcal{D}'(\Omega)$ is none other than the topological dual of the space $\mathcal{D}(\Omega)$. Convergence in $\mathcal{D}(\Omega)$ being a very restrictive condition, the continuity condition with respect to this topology is very strong and this will thus imply many properties for distributions.

Another definition that can be deduced from this functional definition is as follows.

PROPOSITION 3.1.– *A linear form T on $\mathcal{D}(\Omega)$ is a distribution on Ω if and only if for any compact subset $K \in \mathcal{K}_\Omega$, the linear form $T|_K$ is continuous.*

The proof is mainly based on the fact that $\mathcal{D}(\Omega) = \bigcup_{K \in \mathcal{K}_\Omega} D_K(\Omega)$.

These abstract definitions of distributions can be used for theoretical questions, but in order to show in practice that a linear form on $\mathcal{D}(\Omega)$ is a distribution, the following characterization is preferable, which also gives an idea about the order of the distribution.

3.1.2. *Definition of order*

PROPOSITION 3.2.– A linear form T on $\mathcal{D}(\Omega)$ is a distribution on Ω if and only if, for any compact subset K of Ω, there exists $m \in \mathbb{N}$ and $C_K > 0$ such that, for any test function $\varphi \in \mathcal{D}(\Omega)$ such that $\operatorname{supp} \varphi \subset K$,

$$| \langle T, \varphi \rangle | \leq C_K \max_{x \in K, |\alpha| \leq m} |\partial^\alpha \varphi(x)|.$$

NOTATION.– We keep denoting $P_m(\varphi) = \max_{x \in K, |\alpha| \leq m} |\partial^\alpha \varphi(x)|$, which implicitly depends on the compact subset K initially.

Proof. We assume that $T \in \mathcal{D}'(\Omega)$ and will follow a reasoning by contradiction. For this, we assume that there exists a compact $K \subset \Omega$ on which $\forall m \in \mathbb{N}, \forall c > 0$, $\exists\, \varphi_m \in \mathcal{D}(\Omega)$ such that $\operatorname{supp}(\varphi) \subset K$ and $| \langle T, \varphi_m \rangle | > cP_m(\varphi)$. We consider that $\tilde{\varphi}_m = \frac{\varphi_m}{\langle T, \varphi_m \rangle}$. Since T is linear, then $\langle T, \tilde{\varphi}_m \rangle = \frac{\langle T, \varphi_m \rangle}{\langle T, \varphi_m \rangle} = 1$. Moreover, $\operatorname{supp} \tilde{\varphi}_m \subset K$ because $\operatorname{supp} \tilde{\varphi}_m = \operatorname{supp} \varphi_m$. Consequently, $\tilde{\varphi}_m \in \mathcal{D}(\Omega)$.

On the other hand, we have $P_m(\tilde{\varphi}_m) = \frac{P_m(\varphi)}{|\langle T, \varphi_m \rangle|} < \frac{1}{m}$. Thus, $P_m(\tilde{\varphi}) \underset{m \longrightarrow +\infty}{\longmapsto} 0$.

Let $k \in \mathbb{N}$ be fixed, and let us show that $P_k(\tilde{\varphi}_m) \underset{m \longrightarrow +\infty}{\longmapsto} 0$.

We have $\forall m \geq k, P_k(\tilde{\varphi}_m) \leq P_m(\tilde{\varphi}_m) \underset{m \longrightarrow +\infty}{\longmapsto} 0$. This exactly means that the sequence $(\tilde{\varphi}_m)$ tends to 0 in $\mathcal{D}(\Omega)$. Now, $\langle T, \tilde{\varphi}_m \rangle = 1$ does not tend to 0, which contradicts the fact that $T \in \mathcal{D}'(\Omega)$.

Let us show the converse. Let (φ_n) be a sequence that converges to 0 in $\mathcal{D}(\Omega)$. Let K be a compact subset that contains all $\operatorname{supp}\varphi_n$. By the definition of convergence in $\mathcal{D}(\Omega)$ we have, for any $m \in \mathbb{N}$, $P_m(\varphi_n) \underset{n\longrightarrow+\infty}{\longmapsto} 0$. Then

$$|\langle T, \varphi_n\rangle| \leq C_{K,m} P_m(\varphi_n) \underset{n\longrightarrow+\infty}{\longmapsto} 0.$$

Therefore, $T \in \mathcal{D}'(\Omega)$. □

This characterization of distributions will be constantly used in the future. It directly leads to the notion of order of a distribution.

3.1.3. *Order of a distribution*

In the definition of a distribution, the integer m depends a priori on the choice of the compact K. If an integer m that is suitable for all compacts K of Ω can be found, it will be said that the distribution is of finite order.

DEFINITION 3.2.– *A linear form T on $\mathcal{D}(\Omega)$ is a distribution of finite order m at most on Ω when there exists $m \in \mathbb{N}$ such that, for any compact K on Ω, there exists $C_K > 0$ such that, for any test function $\varphi \in \mathcal{D}(\Omega)$, supp $\varphi \subset K$, we have*

$$|\langle T, \varphi\rangle| \leq C_K \max_{x\in K, |\alpha|\leq m} |\partial^\alpha \varphi(x)|.$$

The smallest possible integer m is called the order of the distribution T.

3.2. Examples of distributions

In this section, some basic examples of distributions are presented.

3.2.1. *Regular distributions*

One of the first things to verify is that the theory of distributions properly generalizes the theory of classical functions, typically integrable functions. We will therefore show how the functions $L^1_{loc}(\Omega)$ are injected in $\mathcal{D}'(\Omega)$.

PROPOSITION 3.3.– Let $f \in L^1_{loc}(\Omega)$. A distribution on $\mathcal{D}(\Omega)$, denoted by T_f, can be associated with it, such that $\forall \varphi \in \mathcal{D}(\Omega)$, $\langle T_f, \varphi\rangle = \int_\Omega f(x)\varphi(x)\mathrm{d}x$. This distribution is of order 0.

Proof. 1) **Linearity**: The linearity of the mapping $T_f : \varphi \longmapsto \int_\Omega f(x)\varphi(x)\mathrm{d}x$ is trivial.

2) **Continuity**: Since f is in $L^1_{loc}(\Omega)$, its restriction to any compact set is integrable on this compact.

Let K be a compact of subset Ω and let $\varphi \in \mathcal{D}(\Omega)$, such that $\operatorname{supp}\varphi \subset K$. The function f is integrable on K. Since f is bounded, because it is continuous on the compact K, it can be inferred that

$$\left|\int_\Omega f(x)\varphi(x)\mathrm{d}x\right| \leq \max_{x\in K}|\varphi(x)| \int_K |f(x)|\,\mathrm{d}x.$$

The linear form $T_f : \varphi \longmapsto \int_\Omega f(x)\varphi(x)\mathrm{d}x$ is thus a distribution, which is at most of order 0. □

REMARK 3.1.– *The mapping* $f \in L^1_{loc}(\Omega) \longmapsto T_f \in \mathcal{D}'(\Omega)$ *is linear and injective. It can be written that* $L^1_{loc}(\Omega) \hookrightarrow \mathcal{D}'(\Omega)$.

EXAMPLE 3.1.– Let f and g be two locally integrable functions. If f and g are equal almost everywhere (i.e. $f(x) = g(x)$ a.e.), then their associated distributions are equal (i.e. $T_f = T_g$).

Reciprocally, using the Dubois–Reymond lemma, we have the following result.

THEOREM 3.1.– *Let f be a locally integrable function. We have*

$$T_f = 0 \Longleftrightarrow f = 0 \text{ a.e.}$$

DEFINITION 3.3.– *A distribution T is regular if it is associated with a locally integrable function f.*

NOTATION.– When there is no ambiguity in the context, it is very often simpler to write T_f simply as f.

On the other hand, the notation $f(x)$, which is tolerated for a function f, is less so for the distribution T_f, because it gives the false impression that it is possible to know the value of a distribution at a precise point x.

Some authors also write it as $T_f = [f]$.

EXAMPLE 3.2 (Constant distribution).– The constant (regular) distribution C is defined by

$$\forall \varphi \in \mathcal{D}(\mathbb{R}), \quad \langle C, \varphi \rangle = \int_{\mathbb{R}} C\varphi(x)\,\mathrm{d}x = C \int_{\mathbb{R}} \varphi(x)\,\mathrm{d}x.$$

In particular, $C = 0$ defines the (regular) null distribution.

EXAMPLE 3.3 (Heaviside).– Similarly, the Heaviside function

$$H(x) = \begin{cases} 1, & \text{if } x \geq 0 \\ 0, & \text{otherwise} \end{cases}$$

defines the Heaviside distribution H as

$$\forall \varphi \in \mathcal{D}(\mathbb{R}), \quad \langle H, \varphi \rangle = \int_{\mathbb{R}} H(x)\varphi(x)\,\mathrm{d}x = \int_{\mathbb{R}+} \varphi(x)\,\mathrm{d}x,$$

and the function $2H(x) - 1 = \mathrm{sign}(x)$ defines a distribution and is denoted by sign_x.

3.2.2. *Non-regular distributions*

DEFINITION 3.4.– *Any distribution that is not regular is said to be* ***singular*** *or non-regular.*

3.2.2.1. *Dirac distribution*

EXAMPLE 3.4 (Dirac distribution).– Let $a \in \Omega$. The linear form $\delta_a : \mathcal{D}(\Omega) \longrightarrow \mathbb{C}$ defined by $\forall \varphi \in \mathcal{D}(\Omega), \quad \langle \delta_a, \varphi \rangle = \varphi(a)$ is a distribution on Ω of order 0. This distribution is called Dirac mass at point a.

Proof. Let K be a compact of Ω. Let $\varphi \in \mathcal{D}(\Omega)$, such that $\mathrm{supp}\,\varphi \subset K$. Then,

$$|\langle \delta_a, \varphi \rangle| = |\varphi(a)| \leq \max_{x \in K} |\varphi(x)|.$$

The linear form δ_a is thus indeed a distribution, which is of order 0. □

The distribution δ_a is not associated with any function of $L^1_{loc}(\Omega)$.

As a matter of fact, let us assume that there exists $f \in L^1(\Omega)$ such that $\delta_a = T_f$.

That is, $\langle \delta_a, \varphi \rangle = \displaystyle\int_\Omega f(x)\varphi(x)\mathrm{d}x = \varphi(a), \forall \varphi \in \mathcal{D}(\Omega)$.

We set $\tilde{\Omega} = \Omega - \{a\}$. Then, $\langle \delta_a, \varphi \rangle = \displaystyle\int_\Omega f(x)\varphi(x)\mathrm{d}x = 0, \forall \varphi \in \mathcal{D}(\tilde{\Omega})$.

Therefore, according to the Dubois–Reymond lemma, $f = 0$ almost everywhere in $\tilde{\Omega}$ and thus almost everywhere in Ω. Therefore, $\displaystyle\int_\Omega f(x)\varphi(x)\mathrm{d}x = 0$ for all $\varphi \in \mathcal{D}(\Omega)$.

Consequently, $\forall \varphi \in \mathcal{D}(\Omega), \varphi(a) = 0$, which does not make sense.

NOTE OF CAUTION.– Books about physics and electronics in their great majority employ the notation $\delta(x)$ for the distribution δ and $\delta(x - x_0)$ for the distribution δ_{x_0}. This notation is a constant source of error in the calculations, since it implies that these distributions have a given value at a given point. To avoid errors, it is therefore strongly recommended that the computations are performed in the sense of the distributions, even if it requires presenting the final formulas as physicists and experts in electronics traditionally do.

3.2.2.2. *The principal value of* $\frac{1}{x}$

EXAMPLE 3.5 (The principal value of $\dfrac{1}{x}$).– Consider the function $x \longmapsto \dfrac{1}{x}$ from $\mathbb{R}$ in $\mathbb{R}$. This function is not locally integrable on $\mathbb{R}$; but it is on $\mathbb{R}^*$. We will see how it can be extended to $\mathbb{R}$, the distribution that this function defines on $\mathbb{R}^*$.

PROPOSITION 3.4.– The linear form denoted $\mathrm{Vp}\dfrac{1}{x}$ defined by:

$$\begin{aligned}\left\langle \mathrm{Vp}\frac{1}{x}, \varphi \right\rangle &= \lim_{\varepsilon \to 0+} \left[\int_{x<-\varepsilon} \frac{\varphi(x)}{x}\, \mathrm{d}x + \int_{x>\varepsilon} \frac{\varphi(x)}{x}\, \mathrm{d}x \right] \\ &= \lim_{\varepsilon \to 0+} \int_{|x|>\varepsilon} \frac{\varphi(x)}{x}\, \mathrm{d}x, \quad \forall \varphi \in \mathcal{D}(\mathbb{R}),\end{aligned}$$

is a first-order distribution. This distribution is called the principal value of $\dfrac{1}{x}$.

Proof. **First method**: Let $\varphi \in \mathcal{D}(\mathbb{R})$ be such that $\operatorname{supp}(\varphi) \Subset [-A, A]$. There exists $\psi \in \mathcal{C}^\infty(\mathbb{R})$ such that $\varphi(x) = \varphi(0) + x\psi(x)$ with $\psi(0) = \varphi'(0)$. We have

$$\left\langle \mathrm{Vp}\left(\frac{1}{x}\right), \varphi \right\rangle = \lim_{\epsilon \to 0^+} \left\{ \varphi(0) \left[\int_{-A}^{-\epsilon} \frac{\mathrm{d}x}{x} + \int_{\epsilon}^{A} \frac{\mathrm{d}x}{x} \right] + \int_{-A}^{-\epsilon} \psi(x)\mathrm{d}x + \int_{\epsilon}^{A} \psi(x)\mathrm{d}x \right\}$$

$$= \int_{-A}^{A} \psi(x)\mathrm{d}x.$$

Hence

$$\left|\left\langle \mathrm{Vp}\left(\frac{1}{x}\right), \varphi \right\rangle\right| \leq 2A \max_{x\in[-A,A]} |\psi| \leq 2A \max_{x\in[-A,A]} |\varphi'|,$$

because for $x \neq 0$, we have $\psi(x) = \dfrac{\varphi(x) - \varphi(0)}{x} = \varphi'(\theta x), 0 < \theta < 1$ and $\psi(0) = \varphi'(0)$. Consequently, $\mathrm{Vp}\dfrac{1}{x}$ is at most a first-order distribution on $\mathbb{R}$.

Second method: It can be observed that by using a change in variable, we obtain

$$\left\langle \mathrm{Vp}\left(\frac{1}{x}\right), \varphi \right\rangle = \lim_{\varepsilon \to 0^+} \int_{x>\varepsilon} \frac{\varphi(x) - \varphi(-x)}{x} \mathrm{d}x.$$

We have

$$\int_{|x|>\varepsilon} \frac{\varphi(x)}{x} \mathrm{d}x = \int_{\varepsilon}^{A} \frac{\varphi(x) - \varphi(-x)}{x} \mathrm{d}x$$

$$= \left[(\varphi(x) - \varphi(-x)) \ln x \right]_{\varepsilon}^{A} - \int_{\varepsilon}^{A} (\varphi'(x) + \varphi'(-x)) \ln x \mathrm{d}x.$$

But $\varphi(x) = \varphi(0) + x\psi(x)$, on the one hand, $\varphi(\varepsilon) = \varphi(0) + \varepsilon\psi(\varepsilon)$ and $\varphi(-\varepsilon) = \varphi(0) - \varepsilon\psi(-\varepsilon)$, on the other hand.

Hence $(\varphi(\varepsilon) - \varphi(-\varepsilon)) \ln(\varepsilon) = \varepsilon \ln(\varepsilon)\psi(\varepsilon) + \varepsilon \ln(\varepsilon)\psi(-\varepsilon)$.

Since $\varepsilon \ln(\varepsilon) \underset{\varepsilon \longmapsto 0}{\longmapsto} 0$ and $\psi(\varepsilon) \underset{\varepsilon \longmapsto 0}{\longmapsto} \psi(0)$, we have $\varepsilon \ln(\varepsilon)\psi(\varepsilon) \underset{\varepsilon \longmapsto 0}{\longmapsto} 0$ and $\varepsilon \ln(\varepsilon)\psi(-\varepsilon) \underset{\varepsilon \longmapsto 0}{\longmapsto} 0$. Therefore,

$$\left\langle \mathrm{Vp}\left(\frac{1}{x}\right), \varphi \right\rangle = -\int_{0}^{A} \varphi'(x) \ln |x| \mathrm{d}x - \int_{0}^{A} \varphi'(-x) \ln |x| \mathrm{d}x$$

$$= -\int_{-A}^{+A} \varphi'(x) \ln |x| \mathrm{d}x.$$

Let K be compact subset of $\mathbb{R}$, and $\varphi \in \mathcal{D}(\mathbb{R})$ such that supp $(\varphi) \subset K$. We have

$$\left\langle \text{Vp}\left(\frac{1}{x}\right), \varphi \right\rangle = -\int_K \varphi'(x) \ln|x| \mathrm{d}x.$$

Hence,

$$\left|\left\langle \text{Vp}\left(\frac{1}{x}\right), \varphi \right\rangle\right| \leq \max_{x \in K} |\varphi'(x)| \int_K \ln|x| \mathrm{d}x.$$

Then, $\text{Vp}\frac{1}{x}$ is a first-order distribution, at most.

Show that $\text{Vp}\frac{1}{x}$ is not a zeroth-order distribution. □

REMARK 3.2.–

1) Since Vp$\frac{1}{x}$ *is a first-order distribution, it can then never be regular.*

2) It is essential, in the definition, to specify that we integrate in the complementary of symmetrical neighborhoods of the origin. To prove this point, we simply consider h_ε *which takes* 0 *in the interval* $[-\varepsilon, \varepsilon^2]$ *and takes* $\frac{1}{x}$ *outside. It can be shown that the sequence* (h_ε) *does not have a* $\mathcal{D}'(\mathbb{R})$ *limit.*

3) In this example, the imparity of the function $\frac{1}{x}$ *was used to define the distribution Vp*$\frac{1}{x}$ *as the limit, when* $\varepsilon \longrightarrow 0$*, of the distribution associated with the function* $\frac{\mathbb{1}_{\{x:|x|>\varepsilon\}}(x)}{x}$*. For a non-impair function, such a procedure does not work and this makes that a corrective (divergent) term has to be introduced to compensate for this defect by using the finite part method.*

EXAMPLE 3.6 (The finite part of $\frac{H(x)}{x}$).– Consider the linear form denoted by $\text{Pf}(\frac{H(x)}{x})$ defined on $\mathcal{D}(\mathbb{R})$ by

$$\forall \varphi \in \mathcal{D}(\mathbb{R}), \left\langle \text{Pf}(\tfrac{H(x)}{x}), \varphi \right\rangle = \lim_{\varepsilon \longrightarrow 0} \left(\int_\varepsilon^{+\infty} \frac{\varphi(x)}{x} \mathrm{d}x + \varphi(0) \log(\varepsilon) \right).$$

$\text{Pf}(\frac{H(x)}{x})$ is a first-order distribution.

EXAMPLE 3.7 (The finite part of $\frac{1}{x^2}$).– The function $x \longmapsto \frac{1}{x^2}$ is not locally integrable on $\mathbb{R}$, so a regular distribution cannot be associated therewith. To eliminate the divergent part of the integral $\int_0^{+\infty} \frac{\varphi(x)}{x^2}\mathrm{d}x$ for $\varphi \in \mathcal{D}(\mathbb{R})$, the finite part of this integral is defined by

$$\left\langle \mathrm{Pf}(\frac{1}{x^2}), \varphi \right\rangle = \lim_{\varepsilon \longrightarrow 0} \left(\int_{|x|\geq\varepsilon} \frac{\varphi(x)}{x^2}\mathrm{d}x - 2\frac{\varphi(0)}{\varepsilon} \right) = \int_0^{+\infty} \frac{\varphi(x)+\varphi(-x)-2\varphi(0)}{x^2}\mathrm{d}x.$$

$\mathrm{Pf}(\frac{1}{x^2})$ indeed defines a distribution.

EXAMPLE 3.8 (The finite part of x^α).– We can attempt to continue to integrate non-integrable functions, for example x^α for $-2 < \alpha < -1$ and $x > 0$. It is verified that

$$\forall \varphi \in \mathcal{D}(\mathbb{R}), \quad \int_\varepsilon^{+\infty} x^\alpha \varphi(x)\mathrm{d}x = \int_\varepsilon^{+\infty} x^\alpha \varphi(0)\mathrm{d}x + \int_\varepsilon^{+\infty} x^{\alpha+1}\varphi'(0)\mathrm{d}x + \cdots .$$

The first term is $-\frac{\varepsilon^{\alpha+1}}{\alpha+1}\varphi(0)$ and tends to $+\infty$ when $\varepsilon \longmapsto 0$.

This is the infinite part of x^α. More precisely, we have the following equality:

$$\forall \varphi \in \mathcal{D}(\mathbb{R}), \quad \int_\varepsilon^{+\infty} x^\alpha \varphi(x)\mathrm{d}x = -\frac{\varepsilon^{\alpha+1}}{\alpha+1}\varphi(\varepsilon) - \int_\varepsilon^{+\infty} \frac{x^{\alpha+1}}{\alpha+1}\varphi'(x)\mathrm{d}x.$$

On the other hand, the function $\frac{x^{\alpha+1}}{\alpha+1}$ is integrable because $\alpha + 1 > -1$; it therefore defines a distribution. The finite part thus becomes visible.

DEFINITION 3.5.– *The finite part of x^α, denoted by Pf(x^α), is the distribution defined by:*

$$\forall \varphi \in \mathcal{D}(\mathbb{R}), \quad \langle Pf(x^\alpha), \varphi \rangle = \lim_{\varepsilon \longmapsto 0} \left(\int_\varepsilon^{+\infty} x^\alpha \varphi(x)dx + \frac{\varepsilon^{\alpha+1}}{\alpha+1}\varphi(\varepsilon) \right)$$
$$= -\int_0^{+\infty} \frac{x^{\alpha+1}}{\alpha+1}\varphi'(x)dx.$$

Likewise, $\mathrm{Pf}(x^\alpha)$ can be defined for $\alpha \in]-n-1, -n[$, based on the following expression:

$$\forall \varphi \in \mathcal{D}(\mathbb{R}), \quad \langle \mathrm{Pf}(x^\alpha), \varphi \rangle = (-1)^n \int_0^{+\infty} \frac{x^{\alpha+n}}{(\alpha+1)(\alpha+1)\cdots(\alpha+n)}\varphi^{(n)}(x)\mathrm{d}x.$$

In this case, $\mathrm{Pf}(x^\alpha)$ defines a nth-order distribution.

3.2.3. *Other examples*

This section starts by establishing the link between the theory of distributions and the theory of measures. Then another example of distributions is given, namely, the positive distribution.

3.2.4. *Radon measure*

EXAMPLE 3.9 (Radon measure).– Let μ be a Radon measure on Ω.

The linear form $\varphi \in \mathcal{D}(\Omega) \longmapsto \int_\Omega \varphi d\mu$ is a distibution of order 0 on Ω.

THEOREM 3.2 (Admitted).– *Let $T \in \mathcal{D}'(\Omega)$ be a distribution of order 0. Then, there exists a Radon measure μ on Ω such that*

$$\forall \varphi \in \mathcal{D}(\Omega), \quad \langle T, \varphi \rangle = \int_\Omega \varphi d\mu.$$

The proof of this theorem is based on the fact that, from the Riesz representation theorem, positive Radon measures on Ω identify with positive linear forms on $\mathcal{C}_c^0(\Omega)$ as $\varphi : \mu \longrightarrow \varphi(\mu)$ with $\varphi(\mu) : f \in \mathcal{C}_c^0(\Omega) \longmapsto \int_\Omega f d\mu$.

3.2.4.1. *Positive distributions*

EXAMPLE 3.10 (Positive distribution).– A distribution $T \in \mathcal{D}'(\Omega)$ is said to be positive when: $\forall \varphi \in \mathcal{D}(\Omega), \varphi \geq 0 \Longrightarrow \langle T, \varphi \rangle \geq 0$.

A positive distribution is a zeroth order distribution.

Furthermore, let K a compact of Ω and let $\chi \in \mathcal{D}(\Omega), \chi = 1$ on K and $0 \leq \chi \leq 1$.

If $\varphi \in \mathcal{D}(\Omega)$, $\operatorname{supp} \varphi \subset K$ and φ is real, then the functions

$$\psi_\pm : x \longmapsto \left(\chi(x) \max_{x \in K} |\varphi(x)|\right) \pm \varphi(x)$$

are in $\mathcal{D}(\Omega)$ and are positive. Then, $\langle T, \psi_\pm \rangle \geq 0$, from which

$$|\langle T, \varphi \rangle| \leq |\langle T, \chi \rangle| \max_{x \in K} |\varphi(x)| = C_K \max_{x \in K} |\varphi(x)|.$$

This means that T is at most of zeroth order.

3.2.4.2. *Infinite order distribution*

EXAMPLE 3.11 (Infinite order distribution).– Let T be the linear form on $\mathcal{D}(\mathbb{R})$ defined by $\forall\varphi \in \mathcal{D}(\mathbb{R}), \langle T, \varphi\rangle = \sum_{n=0}^{+\infty} \varphi^{(n)}(n)$. T is a distribution. In fact, since a compact of $\mathbb{R}$ contains only a finite number of integers $n \in \mathbb{N}$, the above sum is a finite sum of non-zero terms.

Let K be a compact of $\mathbb{R}$ and $N = \max(\mathbb{N} \cap K)$.

Then, for any $\varphi \in \mathcal{D}(\mathbb{R})$ such that $\operatorname{supp}\varphi \subset K$, we have

$$\begin{aligned}
|\langle T, \varphi\rangle| &= \left|\sum_{n=0}^{N} \varphi^{(n)}(n)\right| \\
&\leq \sum_{n=0}^{N} \max_{x\in K} \left|\varphi^{(n)}(x)\right| \\
&\leq (N+1) \max_{x\in K, 0\leq n\leq N} \left|\varphi^{(n)}(x)\right| \\
&\leq (N+1)\, P_{N,K}(\varphi).
\end{aligned}$$

Hence T is a distribution.

We can also show that T is a distribution of infinite order (see exercise 3.8).

Let $N \in \mathbb{N}$ be fixed, and let K be a compact of $\mathbb{R}$ and $\varphi \in \mathcal{D}(\mathbb{R})$ such that $\operatorname{supp}\varphi \subset K$. We set $\langle T_N, \varphi\rangle = \sum_{n=0}^{N} \varphi^{(n)}(n)$. Therefore, we can see that T_N is linear. On the other hand, since $\operatorname{supp}\varphi^{(n)} \subset \operatorname{supp}\varphi$, then

$$|\langle T_N, \varphi\rangle| \leq \sum_{n=0}^{N} \max_{x\in K}|\varphi^{(n)}(x)| \leq (N+1) \max_{x\in K, n\leq N}|\varphi^{(n)}(x)| = (N+1)P_{N,K}(\varphi).$$

Therefore, T_N is at most a Nth-order distribution.

It can be observed that $\langle T, \varphi\rangle = \lim_{N\to+\infty} \langle T_N, \varphi\rangle$.

REMARK 3.3.–

1) Implicitly, we can see that $\mathcal{D}'(\Omega)$ is a $\mathbb{K}-$vector space with $\mathbb{K} = \mathbb{R}$ or $\mathbb{C}$.

2) The above example leads to defining the notion of convergence in the space of distributions $\mathcal{D}'(\Omega)$.

3.3. Convergence of sequences of distributions

As sequences of continuous linear mappings, distribution sequences behave in a very "simple" way. This is mainly due to the Banach–Steinhaus theorem, which is a result of the uniformization of bounds on families of continuous linear forms on a Banach space. We start by giving the definition of convergence in $\mathcal{D}'(\Omega)$.

3.3.1. *Definition and examples*

In the space of distributions $\mathcal{D}'(\Omega)$, the weak topology is the topology generated by the family of semi-norms defined by

$$P_\varphi(T) = |\langle T, \varphi\rangle|, \varphi \in \mathcal{D}(\Omega), T \in \mathcal{D}'(\Omega),$$

making $\mathcal{D}'(\Omega)$ a locally convex space. The convergence for this topology is the simple convergence. More precisely, we have the following definition.

DEFINITION 3.6.– *We say that the sequence of distributions $(T_n)_n$ of $\mathcal{D}'(\Omega)$ converges to the distribution T if $\lim\limits_{n\to\infty} \langle T_n, \varphi\rangle = \langle T, \varphi\rangle$ for any test function φ in $\mathcal{D}(\Omega)$.*

REMARK 3.4.– *If this limiting distribution T exists, it is unique.*

EXAMPLE 3.12.– Let $\left(T_n\right)_{n\geq 1}$ be a sequence of $\mathcal{D}'(\mathbb{R})$ defined by: $T_n = T_{e^{inx}}$. We have $T_{e^{inx}} \underset{\mathcal{D}'(\mathbb{R})}{\longmapsto} 0$.

As a matter of fact, let $\varphi \in \mathcal{D}(\mathbb{R})$. There exists $A > 0$ such that $\operatorname{supp}\varphi \Subset [-A, A]$. We have

$$\begin{aligned}
\langle T_n, \varphi\rangle &= \int_{-\infty}^{+\infty} e^{inx}\varphi(x)\mathrm{d}x \\
&= \int_{-A}^{A} e^{inx}\varphi(x)\mathrm{d}x \\
&= \left[\frac{1}{in}e^{inx}\varphi(x)\right]_{-A}^{A} - \frac{1}{in}\int_{-A}^{A} e^{inx}\varphi'(x)\mathrm{d}x.
\end{aligned}$$

Therefore, $|\langle T_n, \varphi\rangle| \leq \dfrac{1}{n}\displaystyle\int_{-A}^{A} \varphi'(x) \leq \dfrac{2}{n}A\|\varphi'\|_\infty \underset{n\longrightarrow+\infty}{\longrightarrow} 0$. Consequently, $\forall\varphi \in \mathcal{D}(\mathbb{R})$, $\langle T_n, \varphi\rangle \underset{n\longrightarrow+\infty}{\longrightarrow} 0$. Then, $T_{e^{inx}} \overset{\mathcal{D}'(\mathbb{R})}{\longmapsto} 0$.

REMARK 3.5.–

1) As a result, we have $T_{\cos nx} \overset{\mathcal{D}'(\mathbb{R})}{\longmapsto} 0$ *and* $T_{\sin nx} \overset{\mathcal{D}'(\mathbb{R})}{\longmapsto} 0$.

2) We have just given an example of a sequence of functions $(\cos nx)_n$*, which does not converge as a sequence of functions but which converges as a sequence of distributions.*

EXAMPLE 3.13.– Let us show that $\delta_n \overset{\mathcal{D}'(\mathbb{R})}{\longrightarrow} 0$.

Indeed, we have $\langle\delta_n, \varphi\rangle = \varphi(n)$ and since φ has compact support, then

$$\lim_{n\longrightarrow+\infty} \langle\delta_n, \varphi\rangle = \lim_{n\longrightarrow+\infty} \varphi(n) = 0.$$

EXAMPLE 3.14 (Sequences of functions converging to δ).– Let (f_n) be a sequence of integrable positive functions such that $\displaystyle\int_{\mathbb{R}} f_n(x)\mathrm{d}x = 1$ and that $f_n(x)$ be zero outside an interval $[-\varepsilon_n, +\varepsilon_n]$ with $\lim \varepsilon_n = 0$. Then, the sequence T_{f_n} of regular distributions defined by f_n tends to δ.

Indeed, let $\varphi \in \mathcal{D}(\mathbb{R})$. We have

$$\int_{\mathbb{R}} f_n(x)\varphi(x)\mathrm{d}x = \int_{\mathbb{R}} f_n(x)\left(\varphi(x) - \varphi(0)\right)\mathrm{d}x + \varphi(0)\int_{\mathbb{R}} f_n(x)\mathrm{d}x.$$

According to the finite increment theorem, since f_n is zero outside an interval $[-\varepsilon_n, +\varepsilon_n]$, we have

$$|\varphi(x) - \varphi(0)| \leq |x| \sup_{|x|\leq\varepsilon_n} |\varphi'(x)| \leq \varepsilon_n \sup_{|x|\leq\varepsilon_n} |\varphi'(x)|.$$

We have $\lim \varepsilon_n = 0$, thus there exists $n_0 \in \mathbb{N}$ such that $\forall n \geq n_0$, $[-\varepsilon_n, +\varepsilon_n] \Subset [-1, 1]$. Hence, for all $n \geq n_0$, we have $\sup\limits_{|x|\leq\varepsilon_n} |\varphi'(x)| \leq \sup\limits_{|x|\leq 1} |\varphi'(x)|$. Therefore, $\forall n \geq n_0$, $\forall x \in [-\varepsilon_n, +\varepsilon_n]$, we have $|\varphi(x) - \varphi(0)| \leq \varepsilon_n C$ where

$C = \sup_{|x|\leq 1} |\varphi'(x)|$ is a constant independent of n. Since (f_n) is a sequence of positive functions, we then have

$$\left|\int_{\mathbb{R}} f_n(x)\left(\varphi(x)-\varphi(0)\right)\mathrm{d}x\right| \leq \int_{[-\varepsilon_n,+\varepsilon_n]} f_n(x)\left|\varphi(x)-\varphi(0)\right|\mathrm{d}x$$
$$\leq C\varepsilon_n \int_{\mathbb{R}} f_n(x)\mathrm{d}x = C\varepsilon_n \underset{n\longrightarrow+\infty}{\longrightarrow} 0.$$

Similarly, by hypothesis on the sequence (f_n), we have $\varphi(0)\displaystyle\int_{\mathbb{R}} f_n(x)\mathrm{d}x = \varphi(0)$.

It is deduced that $\displaystyle\int_{\mathbb{R}} f_n(x)\varphi(x)\mathrm{d}x \underset{n\longrightarrow+\infty}{\longrightarrow} \varphi(0) = \langle\delta,\varphi\rangle$. Hence the desired result.

Using the same demonstration, the result of this example can be generalized as follows.

PROPOSITION 3.5.– If (f_n) is a sequence of functions such that

i) $f_n(x) \geq 0$ for $|x| \leq C, C > 0$ a fixed constant,

ii) f_n uniformly converges to 0 in any interval not containing the origin,

iii) for all $a > 0$ (fixed), $\displaystyle\lim_{n\longrightarrow+\infty}\int_{-a}^{a} f_n(x)\mathrm{d}x = 1$,

then the sequence (f_n) converges to δ the Dirac distribution.

NOTE 3.1.– *A sequence of regular distributions can converge to a regular distribution as it can converge to a singular distribution as can be seen in example 3.14.*

EXAMPLE 3.15 (Sequences of functions converging to δ).– Let f be an integrable function verifying $\displaystyle\int_{\mathbb{R}} f(x)\mathrm{d}x = 1$. Consider, for $\varepsilon > 0$, the regular distribution T_{f_ε} defined by the function $f_\varepsilon = \frac{1}{\varepsilon}f(\frac{x}{\varepsilon})$. Then T_{f_ε} tends to δ when $\varepsilon \longrightarrow 0$.

EXAMPLE 3.16 (Dipole of moment $\mu = 1$).– Let $(T_n)_{n\geq 1}$ be a sequence of $\mathcal{D}'(\mathbb{R})$ defined by: $\forall n \in \mathbb{N}^*, \quad T_n = n\left(\delta_{\frac{1}{n}} - \delta_{\frac{-1}{n}}\right)$.

We have $T_n \underset{\mathcal{D}'(\mathbb{R})}{\longmapsto} T$ with T defined by $\langle T,\varphi\rangle = 2\varphi'(0)$.

Indeed, let $\varphi \in \mathcal{D}(\mathbb{R})$. We have

$$\langle T_n, \varphi \rangle = n \left\langle \delta_{\frac{1}{n}} - \delta_{\frac{-1}{n}}, \varphi \right\rangle$$
$$= n\big(\varphi(\frac{1}{n}) - \varphi(\frac{-1}{n})\big).$$

We recall that $\varphi(x) = \varphi(0) + x\psi(x)$ where $\psi(x) = \displaystyle\int_0^1 \varphi'(tx)\mathrm{d}t$. Therefore,

$$\varphi(\frac{1}{n}) = \varphi(0) + \frac{1}{n}\psi(\frac{1}{n}) \quad \text{and} \quad \varphi(\frac{-1}{n}) = \varphi(0) - \frac{1}{n}\psi(\frac{-1}{n}).$$

Consequently,

$$n(\varphi(\frac{1}{n}) - \varphi(\frac{-1}{n})) = \psi(\frac{1}{n}) + \psi(\frac{-1}{n}) \longrightarrow 2\psi(0) = 2\varphi'(0).$$

Then, $\langle T_n, \varphi \rangle \longrightarrow \langle T, \varphi \rangle$. Hence, $T_n \overset{\mathcal{D}'(\mathbb{R})}{\longmapsto} T$.

We will see in Chapter 4 that T is exactly $T = -2\delta_0'$.

Analogously to the sequences of distributions, the convergence of a series of distributions can be defined.

DEFINITION 3.7.– *We consider the sequence of distributions $\left(T_n\right)_{n\geq 1}$. It is said that the series of distributions $\sum_{n\geq 1} T_n$ defines a distribution T of $\mathcal{D}'(\Omega)$, if the sequence of partial sums $S_N = \sum_{n=1}^{N} T_n$ converges in the sense of distributions to T.*

In practice, we very often find series similar to $\sum_{n=-\infty}^{\infty} T_n$. In this case, the definition of convergence is quite similar to the previous case: in fact, we merely have to consider the convergence of the sequence of symmetric partial sums, $S_N = \sum_{n=-N}^{N} T_n$.

EXAMPLE 3.17.– The sequence $\Delta_N = \sum_{n=-N}^{N} \delta_n$ converges in $\mathcal{D}'(\mathbb{R})$ to a distribution denoted $\sum_{n\in\mathbb{Z}} \delta_n$, which is called Cha and written III (a letter from the Cyrillic alphabet). In physics, it is also called “Dirac comb”.

3.3.2. *Other convergence results*

When a sequence of distributions (T_{f_n}) is defined by locally integrable functions f_n, we have to clearly distinguish the difference between convergence in the classical sense (simple convergence) and convergence in the sense of distributions. To prove this point, we only have to look at example 3.12. Theorem 3.3 characterizes the sequences of functions for which the simple limits coincide with the limits in $D'(\Omega)$.

THEOREM 3.3.– *Let (f_n) be a sequence of locally integrable functions in Ω that converges to f almost everywhere in Ω, and let $g \in L^1_{loc}(\Omega)$ such that $\forall n \in \mathbb{N}, |f_n| \leq g$, then f_n converges to f in $\mathcal{D}'(\Omega)$, that is, $T_{f_n} \overset{\mathcal{D}'(\Omega)}{\longrightarrow} T_f$.*

Proof. Let $K \subset \Omega$ be a compact and let $\varphi \in \mathcal{D}(\Omega)$ be such that $\operatorname{supp}\varphi \subset K$. We have $\langle T_{f_n}, \varphi\rangle = \displaystyle\int_\Omega f_n(x)\varphi(x)\mathrm{d}x$.

Since f_n converges to f almost everywhere in Ω, $f_n\varphi$ therefore converges to $f\varphi$ almost everywhere in Ω. Moreover, we have $|f_n\varphi| \leq |g\varphi|$. Then, by application of Lebesgue's dominated convergence theorem, we have

$$\begin{aligned}
\lim_{n\longrightarrow+\infty} \langle T_{f_n}, \varphi\rangle &= \lim_{n\longrightarrow+\infty} \int_\Omega f_n(x)\varphi(x)\mathrm{d}x \\
&= \int_\Omega \lim_{n\longrightarrow+\infty} (f_n(x))\varphi(x)\mathrm{d}x \\
&= \int_\Omega f(x)\varphi(x)\mathrm{d}x \\
&= \langle T, \varphi\rangle .
\end{aligned}$$

Consequently, $T_{f_n} \overset{\mathcal{D}'(\Omega)}{\longrightarrow} T_f$. □

NOTE 3.2.– *The convergence **almost everywhere** on Ω **does not imply** the convergence in $\mathcal{D}'(\Omega)$.*

Counterexample: We consider the sequence $(f_n)_{n\geq 1}$ of $L^1_{loc}(\Omega)$ defined by

$$f_n : x \longmapsto \sqrt{n}e^{-nx^2}.$$

We have for all $x \neq 0$ $f_n(x) \longmapsto 0$, that is $f_n \overset{a.e.}{\longmapsto} 0$, but the sequence $(T_{f_n})_{n\geq 1}$ converges in $\mathcal{D}'(\Omega)$ to $\sqrt{\pi}\delta_0$ and not to the null distribution. Indeed, let $\varphi \in \mathcal{D}(\Omega)$.

We have

$$\langle T_{f_n}, \varphi \rangle = \sqrt{n} \int_{\mathbb{R}} e^{-nx^2} \varphi(x) \mathrm{d}x$$
$$= \int_{\mathbb{R}} e^{-y^2} \varphi(\frac{y}{\sqrt{n}}) \mathrm{d}y.$$

Based on the dominated convergence theorem, we obtain

$$\int_{\mathbb{R}} e^{-y^2} \varphi(\frac{y}{\sqrt{n}}) \mathrm{d}y \longmapsto \sqrt{\pi}\varphi(0).$$

Hence, $\left(T_{f_n}\right)_{n \geq 1}$ converges in $\mathcal{D}'(\Omega)$ to $\sqrt{\pi}\delta_0$.

Theorem 3.3 generalizes to the spaces $L^p_{loc}(\Omega), 1 \leq p \leq +\infty$ as follows.

THEOREM 3.4.– *The convergence in* $L^p_{loc}(\Omega), 1 \leq p \leq +\infty$ *implies the convergence in* $\mathcal{D}'(\Omega)$. *In other words,*

$$f_n \overset{L^p_{loc}(\Omega)}{\longrightarrow} f \Rightarrow T_{f_n} \overset{\mathcal{D}'(\Omega)}{\longrightarrow} T_f.$$

Proof. Let $\left(f_n\right)$ be a sequence of functions in $L^p_{loc}(\Omega)$ that converges in $L^p_{loc}(\Omega)$ to a function f of $L^p_{loc}(\Omega)$.

Let $K \subset \Omega$ denote a compact subset and $\varphi \in \mathcal{D}(\Omega)$ such that $\operatorname{supp} \varphi \subset K$. We have

$$|\langle T_{f_n}, \varphi \rangle - \langle T_f, \varphi \rangle| = |\langle T_{f_n} - T_f, \varphi \rangle|$$
$$= \left| \int_{\mathbb{R}^n} (f_n - f)(x)\varphi(x) \mathrm{d}x \right|$$
$$= \left| \int_K (f_n - f)(x)\varphi(x) \mathrm{d}x \right|$$
$$\leq \left[\int_K |f_n(x) - f(x)|^p \mathrm{d}x \right]^{\frac{1}{p}} \|\varphi(x)\|_{L^q},$$

where q is such that $\frac{1}{p} + \frac{1}{q} = 1$.

It should be noted that $\varphi \in L^q(\Omega)$ because it is in $\mathcal{D}(\Omega)$.

Since $f_n \overset{L^p_{loc}(\Omega)}{\longrightarrow} f$, then $\left[\int_K |f_n(x) - f(x)|^p \mathrm{d}x\right]^{\frac{1}{p}} \underset{n \longrightarrow +\infty}{\longmapsto} 0.$

Moreover, we have $\|\varphi\|_{L^q} < \infty$, then $T_{f_n} \overset{\mathcal{D}'(\Omega)}{\longrightarrow} T_f$. □

THEOREM 3.5 (Uniform boundedness principle).– *Let Ω an open subset of $\mathbb{R}^d$ and $K \subset \Omega$ a compact subset. Let $(T_n)_{n \geq 0}$ a sequence distributions on Ω such that for all $\varphi \in \mathcal{D}(\Omega)$ with support in K, the sequence $(\langle T_n, \varphi\rangle)_{n \in \mathbb{N}}$ admits a limit in $\mathbb{K} = \mathbb{R}$ or $\mathbb{C}$. Then $\exists m \in \mathbb{N}$, $\exists c > 0$, such that*

$$\forall \varphi \in \mathcal{D}(\Omega) \text{ s.t. } \operatorname{supp} \varphi \subset K, \quad \sup_{n \in \mathbb{N}} |\langle T_n, \varphi\rangle| \leq c P_{m,K}(\varphi).$$

REMARK 3.6.–

1) The proof is based on the adaptation of the Banach–Steinhaus theorem.

2) The constant c and the integer m are independent of n.

COROLLARY 3.1.– *Let $(T_n)_{n \geq 0}$ be a sequence of distributions on Ω such that for all $\varphi \in \mathcal{D}(\Omega)$ the sequence $(\langle T_n, \varphi\rangle)_{n \in \mathbb{N}}$ admits limits in $\mathbb{K} = \mathbb{R}$ or $\mathbb{C}$. Then the linear form T defined on $\mathcal{D}(\Omega)$ by*

$$\forall \varphi \in \mathcal{D}(\Omega), \quad \langle T, \varphi\rangle = \lim_{n \longrightarrow +\infty} \langle T_n, \varphi\rangle$$

is a distribution.

Proof. Therefore, we simply have to verify that this linear form T is continuous on $\mathcal{D}(\Omega)$. Let $K \subset \Omega$ be a compact subset. According to the principle of uniform boundedness, there exists an integer $m \in \mathbb{N}$, and a constant $c_K > 0$ such that

$$\forall \varphi \in \mathcal{D}(\Omega) \text{s.t. } \operatorname{supp} \varphi \subset K, \quad \sup_{n \in \mathbb{N}} |\langle T_n, \varphi\rangle| \leq c_K P_{m,K}(\varphi).$$

By passing to the limit in the above inequality, it can be seen that

$$|\langle T, \varphi\rangle| \leq c_K P_{m,K}(\varphi)$$

for any function $\varphi \in \mathcal{D}(\Omega)$ s.t. $\operatorname{supp} \varphi \subset K$, which ensures that $T \in \mathcal{D}'(\Omega)$. □

COROLLARY 3.2 (Sequential continuity of the duality bracket).– *Let* $(T_n)_{n\geq 0}$ *be a sequence of distributions on* Ω *and* $(\varphi_n)_{n\geq 0}$ *be a sequence of elements of* $\mathcal{D}(\Omega)$.

$$\text{If } T_n \underset{n\longrightarrow+\infty}{\overset{\mathcal{D}'(\Omega)}{\longrightarrow}} T \text{ and } \varphi_n \underset{n\longrightarrow+\infty}{\overset{\mathcal{D}(\Omega)}{\longrightarrow}} \varphi, \text{ then } \langle T_n, \varphi_n\rangle \underset{n\longrightarrow+\infty}{\longrightarrow} \langle T, \varphi\rangle .$$

Proof. We have

$$\begin{aligned}|\langle T_n, \varphi_n\rangle - \langle T, \varphi\rangle| &= |\langle T_n, \varphi_n\rangle - \langle T_n, \varphi\rangle + \langle T_n, \varphi\rangle - \langle T, \varphi\rangle| \\ &= |\langle T_n, \varphi_n - \varphi\rangle + \langle T_n - T, \varphi\rangle| \\ &\leq |\langle T_n, \varphi_n - \varphi\rangle| + |\langle T_n - T, \varphi\rangle| .\end{aligned}$$

We have

$$\begin{aligned}|\langle T_n, \varphi_n - \varphi\rangle| &\leq \sup_{n\in\mathbb{N}} |\langle T_n, \varphi_n - \varphi\rangle| \\ &\leq cP_m(\varphi_n - \varphi) \underset{n\longrightarrow+\infty}{\longrightarrow} 0.\end{aligned}$$

On the other hand, $|\langle T_n - T, \varphi\rangle| \underset{n\longrightarrow+\infty}{\longrightarrow} 0$. Consequently,

$$\langle T_n, \varphi_n\rangle \underset{n\longrightarrow+\infty}{\longrightarrow} \langle T, \varphi\rangle .$$

□

3.4. Exercises with solutions

EXERCISE 3.1.– Let $\varphi \in \mathcal{D}(\mathbb{R})$. Show that the following expressions define distributions whose order will be determined:

1) $\displaystyle\int_{\mathbb{R}} \varphi(x^2)\mathrm{d}x;$

2) $\displaystyle\int_{\mathbb{R}} \varphi'(x)\mathrm{e}^{x^2}\mathrm{d}x.$

SOLUTION 3.1.–

1) Let $[a, b] \subset \mathbb{R}$. Let $\varphi \in \mathcal{D}(\mathbb{R})$ such that $\operatorname{supp}\varphi \Subset [a, b]$.

The mapping $T : \varphi \mapsto \displaystyle\int_{\mathbb{R}} \varphi(x^2)\mathrm{d}x$ is a linear form on $\mathcal{D}(\mathbb{R})$; in addition,

$$\begin{aligned}\left|\int_{\mathbb{R}} \varphi(x^2)\mathrm{d}x\right| &= \left|\int_{\left\{x\in\mathbb{R}: x^2\in[a,\, b]\right\}} \varphi(x^2)\mathrm{d}x\right| \\ &\leq \int_{\left\{x\in\mathbb{R}: x^2\in[a,\, b]\right\}} \left|\varphi(x^2)\right| \mathrm{d}x \\ &\leq \|\varphi\|_\infty \int_{\left\{x\in\mathbb{R}:\, x^2\in[a,\, b]\right\}} \mathrm{d}x \\ &\leq C\|\varphi\|_\infty.\end{aligned}$$

Therefore, T is at most a zeroth-order distribution.

2) The mapping $T : \varphi \mapsto \displaystyle\int_{\mathbb{R}} \varphi'(x)e^{x^2}\mathrm{d}x$ is a linear form on $\mathcal{D}(\mathbb{R})$.

Let $\varphi \in \mathcal{D}(\mathbb{R})$ and $[a, b] \subset \mathbb{R}$ such that $\operatorname{supp}\varphi \Subset [a, b]$. Then,

$$\begin{aligned}\left|\int_{\mathbb{R}} \varphi'(x)e^{x^2}\mathrm{d}x\right| &= \left|\left[\varphi(x)e^{x^2}\right]_a^b - \int_a^b \varphi(x)2xe^{x^2}\mathrm{d}x\right| \\ &\leq \int_a^b |\varphi(x)| \left|2xe^{x^2}\right| \mathrm{d}x \\ &\leq \|\varphi\|_\infty \left(e^{b^2} - e^{a^2}\right) = C\|\varphi\|_\infty.\end{aligned}$$

So, T is at most a zeroth-order distribution.

EXERCISE 3.2.–

1) Let $\varphi \in \mathcal{D}(\mathbb{R})$. Show that the expression $\displaystyle\int_0^\infty \varphi'(x)\ln(x)\mathrm{d}x$ defines at most a first-order distribution T.

2) Let φ_n denote a plateau function equal to 1 on $\left[\frac{1}{n}, 1\right]$ and whose support is included in $\left[\frac{1}{2n}, 2\right]$.

a) Apply a lower bound on $\langle T, \varphi_n\rangle$.

b) Deduce therefrom that T is exactly a first-order distribution.

SOLUTION 3.2.–

1) The mapping $T : \varphi \mapsto \displaystyle\int_0^{\infty} \varphi'(x)\ln(x)\mathrm{d}x$ is a linear form on $\mathcal{D}(\mathbb{R})$.

Let $[a,\, b] \subset \mathbb{R}$ and $\varphi \in \mathcal{D}(\mathbb{R})$ such that $\operatorname{supp}\varphi \Subset [a,\, b]$. Then

$$|\langle T, \varphi\rangle| = \left|\int_0^{\infty} \varphi'(x)\ln(x)\mathrm{d}x\right| = \left|\int_{[a,\, b]\cap\mathbb{R}+} \varphi'(x)\ln(x)\mathrm{d}x\right|$$
$$\leq \|\varphi'\|_{\infty}\left|\int_{[a,\, b]\cap\mathbb{R}+} |\ln(x)|\mathrm{d}x\right| .$$

Therefore, T is at most a first-order distribution.

2) a) With φ_n as in the text, we have

$$|\langle T, \varphi_n\rangle| = \left|\int_{\frac{1}{2n}}^{2} \varphi_n'(x)\ln(x)\mathrm{d}x\right| = \left|[\varphi_n(x)\ln(x)]_{\frac{1}{2n}}^{2} - \int_{\frac{1}{2n}}^{2} \frac{\varphi_n(x)}{x}\mathrm{d}x\right|$$
$$\geq \int_{\frac{1}{2n}}^{1} \frac{\varphi_n(x)}{x}\mathrm{d}x \geq \int_{\frac{1}{n}}^{1} \frac{\mathrm{d}x}{\mathrm{d}x} = \ln(n).$$

So, $\forall n \geq 1, |\ \langle T, \varphi_n\rangle\ | \geq \ln(n) \underset{n\to\infty}{\longrightarrow} +\infty.$

b) We have $\|\varphi_n\|_{\infty} = 1$ for all $n \geq 1$, thus, with $[a,\, b] = [0,\, 2]$, there exists $[a,\, b] \subset \mathbb{R}$, such that for all $c > 0$ and all $n \geq 1$, there exists $\varphi_n \in \mathcal{D}(\mathbb{R})$, $\operatorname{supp}\varphi_n \subset [0,\, 2]$ and

$$|\langle T, \varphi_n\rangle| \geq c\|\varphi_n\|_{\infty} = c$$

(because $\ln(n) \underset{n\to\infty}{\longrightarrow} +\infty$).

Therefore, T is not of zeroth order; it is then of first order.

EXERCISE 3.3.– Let $I \subset \mathbb{R}$ be an open interval and let $x_0 \in$ I. Show that there is no function $f \in \mathrm{L}^1_{loc}(I)$ such that $T_f = \delta_{x_0}$.

SOLUTION 3.3.– Using a reasoning by contradiction, we assume that there exists $f \in L^1_{loc}(I)$ such that $\delta_{x_0} = T_f$. Then

$$\forall\varphi \in \mathcal{D}(I), \int_I f(x)\varphi(x)\mathrm{d}x = \varphi(x_0). \qquad [3.1]$$

In particular, for any $\varphi \in \mathcal{D}(I)$, if $x_0 \notin \operatorname{supp}\varphi$, then $\displaystyle\int_I f(x)\varphi(x)\mathrm{d}x = 0$.

Let $J = I \cap \{x \in \mathbb{R} : x < x_0\}$ and let us assume that $\operatorname{supp}\varphi \subset J$; then, $\displaystyle\int_J f\varphi = 0$.

We have, for all $\varphi \in \mathcal{D}(J)$, $\displaystyle\int_J f\varphi = 0$. From exercise 2.5, we have $f = 0$ on J. Similarly, $f = 0$ on $J' = I \cap \{x \in \mathbb{R} : x > x_0\}$.

Since $I = J \cup \{x_0\} \cup J'$, then $f = 0$ almost everywhere I and

$$\forall \varphi \in \mathcal{D}(I), \quad \int_I f(x)\varphi(x)\mathrm{d}x = 0. \qquad [3.2]$$

However, if $\varphi \in \mathcal{D}(I)$ such that $\varphi(x_0) \neq 0$, then equality [3.2] contradicts equality [3.1]. Therefore, a function $f \in L^1_{loc}(I)$ such that $\delta_{x_0} = T_f$ does not exist.

EXERCISE 3.4.– Let $\varphi \in \mathcal{D}(\mathbb{R})$.

1) Show that there exists $\psi \in \mathcal{C}^\infty(\mathbb{R})$ such that, $\forall x \in \mathbb{R}, \varphi(x) = \varphi(0) + x\psi(x)$.

2) Show that the limit $\displaystyle\lim_{\varepsilon \to 0} \int_{|x|>\varepsilon} \frac{\varphi(x)}{x} dx$ exists.

3) Show that this expression defines at most a first-order distribution. This distribution is called principal value of $\dfrac{1}{x}$ and is denoted by $\mathrm{Vp}(\dfrac{1}{x})$.

4) Considering φ_n as in exercise 3.2, show that $\mathrm{Vp}(\dfrac{1}{x})$ is exactly of first order.

SOLUTION 3.4.–

1) We have $\forall x \in \mathbb{R}, \varphi(x) = \varphi(0) + \displaystyle\int_0^x \varphi'(t)\mathrm{d}t$. We set $t = xu$, then

$$\forall x \in \mathbb{R}, \varphi(x) = \varphi(0) + x\int_0^1 \varphi'(xu)\mathrm{d}u.$$

We set $\psi(x) = \int_0^1 \varphi'(xu)\mathrm{d}u$, for all $x \in \mathbb{R}$. Then, as $\varphi \in \mathcal{D}(\mathbb{R})$ and $[0,1]$ is compact, by the differentiation theorem under the integral sign we have $\psi \in \mathcal{C}^\infty(\mathbb{R})$ and $\forall x \in \mathbb{R}, \varphi(x) = \varphi(0) + x\psi(x)$.

2) Let $\epsilon > 0$. Let $[-a, a] \subset \mathbb{R}$ and $\varphi \in \mathcal{D}(\mathbb{R})$ such that $\operatorname{supp}(\varphi) \Subset [-a, a]$. We have

$$\int_{|x|>\epsilon} \frac{\varphi(x)}{x}\mathrm{d}x = \int_{a\geqslant|x|>\epsilon} \frac{\varphi(x)}{x}\mathrm{d}x = \int_{a\geqslant|x|>\epsilon} \frac{\varphi(0)}{x} + \psi(x)\mathrm{d}x$$

$$= \underbrace{\int_{a\geqslant|x|>\epsilon} \frac{\varphi(0)}{x}\mathrm{d}x}_{=0 \text{ by imparity}} + \int_{a\geqslant|x|>\epsilon} \psi(x)\mathrm{d}x) \underset{\epsilon\to 0}{\longrightarrow} \int_{a\geqslant|x|\geqslant 0} \psi(x)\mathrm{d}x$$

(because $\psi \in \mathcal{D}(\mathbb{R})$).

Therefore, the limit exists and is $\displaystyle\int_{a\geqslant|x|\geqslant 0} \psi(x)\mathrm{d}x$.

3) Let $\varphi \in \mathcal{D}(\mathbb{R})$ such that $\operatorname{supp}(\varphi) \Subset [-a, a]$. Then, the mapping $\varphi \to \displaystyle\lim_{\epsilon\to 0+} \int_{|x|>\epsilon} \frac{\varphi(x)}{x}\mathrm{d}x$ is a linear form on $\mathcal{D}(\mathbb{R})$ and

$$\begin{aligned}
\left|\langle \mathrm{Vp}\left(\frac{1}{x}\right), \varphi\rangle\right| &= \left|\int_{-a}^{a} \psi(x)\mathrm{d}x\right| \\
&= \left|\int_{-a}^{a} \left(\int_0^1 \varphi'(xu)\mathrm{d}u\right)\mathrm{d}x\right| \\
&\leqslant \|\varphi'\|_\infty \left|\int_{-a}^{a} \left(\int_0^1 \mathrm{d}u\right)\mathrm{d}x\right| \\
&\leq 2a\|\varphi'\|_\infty.
\end{aligned}$$

Therefore, $\mathrm{Vp}(\frac{1}{x})$ is at most a first-order distribution.

4) Let $n \geqslant 1$ be fixed and $\epsilon > 0$ such that $\epsilon \leqslant \frac{1}{2n}$. We have

$$\begin{aligned}
\left|\langle \mathrm{Vp}(\frac{1}{x}), \varphi_n\rangle\right| &\geqslant \int_\epsilon^2 \frac{\varphi_n(x)}{x}\mathrm{d}x \\
&\geq \int_{\frac{1}{2n}}^2 \frac{\varphi_n(x)}{x}\mathrm{d}x \\
&\geqslant \int_{\frac{1}{n}}^1 \frac{\varphi_n(x)}{x}\mathrm{d}x \\
&\geq \int_{\frac{1}{n}}^1 \frac{\mathrm{d}x}{x} = \ln(n).
\end{aligned}$$

Therefore, $\forall n \geqslant 1, \left|\langle \text{Vp}(\frac{1}{x}), \varphi_n \rangle\right| \geq \ln(n)$. Then by making n tend to infinity, we obtain $\left|\langle \text{Vp}(\frac{1}{x}), \varphi_n \rangle\right| \underset{n\to\infty}{\longrightarrow} +\infty$; moreover, $\|\varphi_n\|_\infty = 1$, for all $n \geqslant 1$. Therefore, $\text{Vp}(\frac{1}{x})$ is not of zeroth-order, it is thus a first-order distribution.

EXERCISE 3.5.– Let T be the linear form on $\mathcal{D}(\mathbb{R})$ defined by $\langle T, \varphi \rangle = \varphi'(0)$.

1) Show that T is a distribution of order less than or equal to 1.

2) We want to show that T is not a zeroth-order distribution.

a) For any $n \geq 1$, give an example of a function f_n of class $\mathcal{C}^\infty$ such that $f_n'(0) = n$ and $\|f_n\|_\infty = 1$.

b) Deduce, for all $n \geq 1$, the existence of a function $\varphi_n \in \mathcal{D}(\mathbb{R})$, with support in $[-1, 1]$ and such that $\varphi_n'(0) = n$ and $\|\varphi_n\|_\infty \leq 1$.

c) Conclude.

SOLUTION 3.5.–

1) Let K be a compact of $\mathbb{R}$ and $\varphi \in \mathcal{D}_K(\mathbb{R})$. We have

$$|\langle T, \varphi \rangle| = |\varphi'(0)| \leq \max_{x\in K} |\varphi'(x)|,$$

which shows that T is indeed a distribution of order less than or equal to 1.

2) a) The function $f_n(x) = \sin(nx)$ is appropriate.

b) Let $\theta \in \mathcal{C}^\infty(\mathbb{R})$ be a plateau function equal to 1 on $[-\frac{1}{2}, \frac{1}{2}]$ and $\text{supp}(\theta) \Subset [-1, 1]$. Then, $\theta(0) = 1$ and $\theta'(0) = 0$. Let $\varphi_n = \theta f_n$. It is clearly visible that the support of φ_n is contained in that of θ and thus in $[-1, 1]$ and that $\|\varphi_n\|_\infty \leq \|f_n\|_\infty = 1$. Moreover,

$$\varphi_n'(0) = \theta'(0) f_n(0) + \theta(0) f_n'(0) = n.$$

c) We assume that T is of zeroth-order. Then, for the compact $[-1, 1]$, there exists a constant C, such that any function $\varphi \in \mathcal{D}(\mathbb{R})$ with support in $[-1, 1]$ verifies

$$|\langle T, \varphi \rangle| \leq C \max_{x\in[-1,1]} |\varphi(x)|.$$

If this inequality is applied to the function φ_n defined just before, we will have, for all $n \geq 1, n = \varphi_n'(0) \leq C$. This is a contradiction and the distribution is not of zeroth order.

EXERCISE 3.6.– Let T be the linear form on $\mathcal{D}(\mathbb{R})$ defined by

$$\langle T, \varphi \rangle = \sum_{n=1}^{+\infty} \varphi^{(n)}(\frac{1}{n}).$$

1) Show that the following formula $\langle S, \varphi \rangle = \sum_{n=1}^{+\infty} \varphi^{(n)}(\frac{1}{n})$ defines an element of $\mathcal{D}'(\mathbb{R}^*)$.

2) We want to show that there is no distribution $T \in \mathcal{D}'(\mathbb{R})$ such that, for any $\varphi \in \mathcal{D}(\mathbb{R}^*)$, $\langle T, \varphi \rangle = \langle S, \varphi \rangle$. We reason by contradiction and assume that there is such a distribution T.

a) Show that for any sequence $(a_k)_k$ of complex numbers, there exists a function $f \in \mathcal{D}(\mathbb{R})$ of support included in $\left]\frac{3}{4}, \frac{5}{4}\right[$ and such that $f^{(j)}(1) = a_j$ for all $j \in \mathbb{N}^*$.

b) We set $f_k(x) = f\left(k^2 x - k + 1\right)$. Show that the supports of the functions f_k are pairwise disjoint.

c) Compute $f_p^{(n)}\left(\frac{1}{n}\right)$.

d) By taking $K = \left[0, \frac{5}{4}\right], \varphi_m = \sum_{p=1}^{m} f_p$ and $a_k = 1$, show that T cannot be continuous on $\mathcal{D}(\mathbb{R})$.

SOLUTION 3.6.–

1) The linear form on $\mathcal{D}(\mathbb{R})$ defined by $\langle S_N, \varphi \rangle = \sum_{n=1}^{N} \varphi^{(n)}(\frac{1}{n})$ is clearly a distribution. For each $\varphi \in \mathcal{D}(\mathbb{R}^*)$, the previous sequence converges to $\langle S, \varphi \rangle$ (the sum is in fact finite). According to the theorem on distribution sequences, $S \in \mathcal{D}'(\mathbb{R}^*)$.

2) a) According to Borel's theorem (see exercise 2.7), there exists a function g of class $\mathcal{C}^\infty$ on $\mathbb{R}$ and such that $g^{(n)}(0) = a_n$ for each n. Let us take θ a plateau function with support in $\left]\frac{3}{4}, \frac{5}{4}\right[$ and equal to 1 in the vicinity of 1; the function $f(x) = \theta(x) g(x-1)$ can be used.

b) If x is in the support of f_k, we have

$$k^2 x - k + 1 \in \left]\frac{3}{4}, \frac{5}{4}\right[\Leftrightarrow x \in \left]\frac{-1}{4k^2} + \frac{1}{k}, \frac{1}{4k^2} + \frac{1}{k}\right[.$$

If $k \geq 1$, we have

$$\frac{-1}{4k^2} + \frac{1}{k} - (\frac{1}{4(k+1)^2} + \frac{1}{k+1}) = \frac{2k^2 + 2k - 1}{4k^2(k+1)^2} > 0.$$

The supports are indeed disjoints.

c) If $p = n$, then we have $f_n^{(n)}\left(\frac{1}{n}\right) = n^{2n} f^{(n)}(1) = n^{2n} a_n$. Since the supports are disjoint, then $f_p^{(n)}(\frac{1}{n}) = 0$ if $p \neq n$.

d) For $K = [0, \frac{5}{4}]$, there exists an integer j and a constant C such that, if φ has support in K,

$$|\langle T, \varphi \rangle| \leq C \max_{\alpha \leq j, x \in K} \left| \varphi^{(\alpha)}(x) \right|.$$

Let us take the functions φ_m as shown. φ_m has a compact support contained in both $\mathbb{R}^*$ and in K. We thus have

$$\langle T, \varphi_m \rangle = \sum_{p=1}^{m} \sum_{n=1}^{+\infty} f_p^{(n)}(\frac{1}{n}) = \sum_{p=1}^{m} p^{2p}.$$

On the other hand, since the functions f_p have disjoint supports, for $\alpha \leq j$, we have

$$\max_{x \in \mathbb{R}} \left| \varphi_m^{(\alpha)}(x) \right| \leq \sum_{p=1}^{m} \max_{x \in \mathbb{R}} \left| f_p^{(\alpha)}(x) \right| \leq c \sum_{p=1}^{m} p^{2j},$$

where c is a constant that depends only on f. Putting this together, we get

$$m^{2m} \leq \sum_{p=1}^{m} p^{2p} \leq \sum_{p=1}^{m} p^{2j} \leq cm^{2j+1}.$$

This is impossible when m becomes large.

EXERCISE 3.7.– Let $\varphi \in \mathcal{D}(\mathbb{R}^2)$. Show that the expression

$$\langle T, \varphi \rangle = \int_0^{\infty} \left(\varphi\left(\frac{1}{t^2}, \sin t\right) - \varphi(0, \sin t) \right) \mathrm{d}t$$

defines a distribution T of first order at most.

SOLUTION 3.7.– T is a linear form on $\mathcal{D}(\mathbb{R}^2)$.

Let $[-a, a]^2 \subset \mathbb{R}^2$, $a > 0$, be a compact and $\varphi \in \mathcal{D}(\mathbb{R}^2)$ such that $\operatorname{supp}\varphi \Subset [-a, a]^2$. Therefore,

$$\begin{aligned}
|\langle T, \varphi\rangle| &= \left|\int_0^\infty \left(\varphi\left(\frac{1}{t^2}, \sin t\right) - \varphi(0, \sin t)\right) dt\right| \\
&= \left|\int_0^\infty \int_0^1 \partial_x\varphi\left(\frac{s}{t^2}, \sin t\right) \frac{ds}{t^2} dt\right| \\
&\leq \int_{\frac{1}{\sqrt{a}}}^\infty \int_0^1 \left|\partial_x\varphi\left(\frac{s}{t^2}, \sin t\right)\right| ds \frac{dt}{t^2} \\
&\leq \|\partial_x\varphi\|_\infty \int_{\frac{1}{\sqrt{a}}}^\infty \frac{dt}{t^2} \\
&\leq \sqrt{a}\|\partial_x\varphi\|_\infty \leq \sqrt{a}P_1(\varphi).
\end{aligned}$$

We deduce that T is at most a first-order distribution.

EXERCISE 3.8.– Let $\varphi \in \mathcal{D}(\mathbb{R})$. Show that the expression $\langle T, \varphi\rangle = \sum_{p=0}^{+\infty} \varphi^{(p)}(p)$ defines an infinite-order distribution on $\mathbb{R}$.

SOLUTION 3.8.– Let T be defined by: $\langle T, \varphi\rangle = \sum_{p=0}^{+\infty} \varphi^{(p)}(p)$.

It is easy to see that T is a linear form on $\mathcal{D}(\mathbb{R})$.

Let us study the continuity of T.

Let $K \subset \mathbb{R}$ be a compact. We have $K \subset [-N, N]$ where $N \in \mathbb{N}^*$.

For any $\varphi \in \mathcal{D}(\mathbb{R})$ such that $\operatorname{supp}\varphi \subset K$, we have

$$|\langle T, \varphi\rangle| \leq \sum_{j=0}^{N} \left|\varphi^{(j)}(j)\right| \leq \sum_{j=0}^{N} \max_{x\in K} \left|\varphi^{(j)}(x)\right| \leq (N+1)P_N(\varphi).$$

Therefore, T is a distribution.

Let us show that T is of infinite order. To this end, we follow on reasoning by contradiction.

If T were of finite order, there would exist $k \in \mathbb{N}$ such that for all compact K of $\mathbb{R}$ we have

$$|\langle T, \varphi \rangle| \leq C \sum_{j=0}^{k} \max_{x \in K} \left| \varphi^{(j)}(x) \right|. \qquad [3.3]$$

Let $\psi_0 \in \mathcal{D}(\mathbb{R})$, such that

$$\begin{cases} \psi_0(x) = 1, & \text{if } |x| \leq \frac{1}{4} \\ \psi_0(x) = 0, & \text{if } |x| \geq \frac{1}{2} \\ 0 \leq \psi_0(x) \leq 1. \end{cases}$$

Let us set $\psi(x) = \frac{x^{k+1}}{(k+1)!}\psi_0(x)$. The Leibniz formula shows that $\psi^{(k+1)}(0) = 1$.

Let $\varphi(x) = \psi\left(\lambda(x - (k+1))\right)$ where λ is meant to tend to $+\infty$. We have

$$\operatorname{supp} \varphi \subset \left\{ x \in \mathbb{R}, |x - (k+1)| \leq \frac{1}{2\lambda} \right\} \subset \left[k + \frac{1}{2}, k + \frac{3}{2} \right].$$

It can be deduced that $\varphi^{(j)}(j) = 0$ if $j \neq k+1$ and $\varphi^{(k+1)}(k+1) = \lambda^{k+1}\psi^{(k+1)}(0) = \lambda^{k+1}$. On the other hand, for $j \leq k$,

$$\left| \varphi^{(j)}(x) \right| \leq \lambda^{j} \max_{y \in \mathbb{R}} \left| \psi^{(j)}(y) \right| \leq \lambda^{k} \max_{\mathbb{R}} \left| \psi^{(j)}(x) \right|.$$

Inequality [3.3] applied to φ would thus imply that

$$\lambda^{k+1} \leq C\lambda^{k} \sum_{j=0}^{k} \max_{\mathbb{R}} \left| \psi^{(j)}(x) \right|.$$

This is absurd if λ is made to tend to $+\infty$.

EXERCISE 3.9.– Let $\varphi \in \mathcal{D}(\mathbb{R})$ and $\varepsilon > 0$.

1) For $\alpha \in]-2, -1[$, show that $\displaystyle\int_0^{\infty} x^{\alpha}\varphi(x)\mathrm{d}x = A_{\varphi}\varepsilon^{\alpha+1} + R_{\varepsilon}$, where $A_{\varphi} \in \mathbb{R}$ does not depend on ε and where R_{ε} tends to a limit when ε tends to 0^{+}.

2) We set $\langle \mathrm{Pf}(x^{\alpha}), \varphi \rangle = \lim\limits_{\varepsilon \to 0^{+}} R_{\varepsilon}$. Show that $\mathrm{Pf}(x^{\alpha})$ is at most a first-order distribution.

SOLUTION 3.9.–

1) Let $[-a, a] \subset \mathbb{R}, a > 0$. Let $\varphi \in \mathcal{D}(\mathbb{R})$ with $\operatorname{supp} \varphi \Subset [-a, a]$.

For $\alpha \in]-2, -1[$ and $\epsilon > 0$, we have

$$\int_\epsilon^\infty x^\alpha \varphi(x)\mathrm{d}x = \int_\epsilon^a x^\alpha \varphi(x)\mathrm{d}x = \int_\epsilon^a \left(x^\alpha \varphi(0) + x^{\alpha+1}\psi(x)\right) \mathrm{d}x,$$

with $\psi(x) = \displaystyle\int_0^1 \varphi'(xu)\mathrm{d}u$ and thus $\varphi(x) = \varphi(0) + x\psi(x)$. Then,

$$\int_\epsilon^{+\infty} x^\alpha \varphi(x)\mathrm{d}x = \int_\epsilon^a x^\alpha \varphi(0)\mathrm{d}x + \int_\epsilon^a x^{\alpha+1}\psi(x)\mathrm{d}x.$$

On the other hand, we have

$$\int_\epsilon^a x^\alpha \varphi(0)\mathrm{d}x = \left[\frac{x^{\alpha+1}}{\alpha+1}\varphi(0)\right]_\epsilon^a = \frac{a^{\alpha+1}}{\alpha+1}\varphi(0) - \frac{\epsilon^{\alpha+1}}{\alpha+1}\varphi(0).$$

We set $A_\varphi = -\dfrac{\varphi(0)}{\alpha+1}$ and $R_\epsilon = \displaystyle\int_\epsilon^a x^{\alpha+1}\psi(x)\mathrm{d}x + \frac{\varphi(0)}{\alpha+1}a^{\alpha+1}$.

Since $\alpha \in]-2, -1[, x \mapsto x^{\alpha+1}$ is integrable in the neighborhood of 0, thus as ψ belongs to $\mathcal{D}(\mathbb{R})$ and bounded, $x \mapsto x^{\alpha+1}\psi(x)$ is also integrable in the neighborhood of 0.

Therefore, R_ϵ has a limit when ϵ tends to 0^+ given by

$$\int_0^a x^{\alpha+1}\psi(x)\mathrm{d}x + \frac{\varphi(0)}{\alpha+1}a^{\alpha+1}.$$

We have indeed $\displaystyle\int_\epsilon^\infty x^\alpha \varphi(x)\mathrm{d}x = A_\varphi \epsilon^{\alpha+1} + R_\epsilon$ with R_ϵ having a limit in 0^+.

2) We set $\langle \mathrm{Pf}(x^\alpha), \varphi\rangle = \dfrac{\varphi(0)}{\alpha+1}a^{\alpha+1} + \displaystyle\int_0^a x^{\alpha+1}\Big(\int_0^1 \varphi'(xu)\mathrm{d}u\Big)\mathrm{d}x$. Then,

$$|\langle \mathrm{Pf}(x^\alpha), \varphi\rangle| \le C_0\|\varphi\| + C_1\|\varphi'\| \le CP_1(\varphi),$$

where $C = 2\max(C_0, C_1)$. Consequently, $\mathrm{Pf}(x^\alpha)$ is a first-order distribution at most.

EXERCISE 3.10.– For all $n \in \mathbb{N}$, we denote by T_n the distribution associated with the locally integrable function $f_n : t \longrightarrow \frac{\sin(nt)}{\pi t}$. Show that the sequence $(T_n)_{n\in\mathbb{N}}$ converges in $\mathcal{D}'(\mathbb{R})$ to the distribution δ_0.

Hint: The identity $\displaystyle\int_0^{\infty} \frac{\sin(t)}{t} = \frac{\pi}{2}$ can be used.

SOLUTION 3.10.– Let $\varphi \in \mathcal{D}(\mathbb{R})$, such that $\text{supp}(\varphi) \Subset [-A, A]$. For all $n \in \mathbb{N}$, we have

$$\langle T_n, \varphi\rangle = \int_{-A}^{A} \frac{\sin(nt)}{\pi t} \mathrm{d}t.$$

φ can be decomposed into $\varphi(t) = \varphi(0) + t\psi(t)$ with $\psi \in \mathcal{C}^{\infty}(\mathbb{R})$.

Then $\forall n \in \mathbb{N}$, $\displaystyle\langle T_n, \varphi\rangle = \varphi(0) \int_{-A}^{A} \frac{\sin(nt)}{\pi t} \mathrm{d}t + \frac{1}{\pi}\int_{-A}^{A} \psi(t)\sin(nt)\mathrm{d}t.$

According to the Riemann–Lebesgue lemma, the second integral tends to 0 when n tends to infinity. Then, using the change in variable $u = nt$ in the first integral, we obtain

$$\int_{-nA}^{nA} \frac{\sin(u)}{\pi u} \mathrm{d}u \longrightarrow \frac{1}{\pi}\int_{-\infty}^{+\infty} \frac{\sin(u)}{u} \mathrm{d}u = 1.$$

Therefore, $\displaystyle\lim_{n\longrightarrow\infty} \langle T_n, \varphi\rangle = \varphi(0) = \langle \delta_0, \varphi\rangle$, thus T_n converges to δ_0 in $\mathcal{D}'(\mathbb{R})$.

EXERCISE 3.11.– Compute the limits, in $\mathcal{D}'(\mathbb{R})$, of the following distribution sequences:

1) $A_n = \sin(nx)$;

2) $B_n = n\, g(nx)$ where $g \in L^1(\mathbb{R})$;

3) $C_n = \displaystyle\frac{1}{n}\sum_{p=0}^{n-1} \delta_{\frac{p}{n}}$;

4) $E_n = e^{inx}\mathrm{Vp}\left(\dfrac{1}{x}\right)$;

5) $\mathbf{D}_n(x) = \displaystyle\sum_{k=-n}^{k=+n} e^{ikx}$.

SOLUTION 3.11.–

1) Let $\varphi \in \mathcal{D}(\mathbb{R})$ such that $\text{supp}(\varphi) \Subset [-A, A]$. Then for all $n \in \mathbb{N}$, it follows that

$$\begin{aligned}\langle A_n, \varphi\rangle &= \int_{-A}^{A} \sin(nx)\varphi(x)\mathrm{d}x \\ &= \int_{-A}^{A} \frac{\cos(nx)}{n}\varphi'(x)\mathrm{d}x.\end{aligned}$$

Therefore,

$$\mid \langle A_n, \varphi\rangle \mid \leq \frac{1}{n}\int_{-A}^{A} \mid \varphi'(x) \mid \mathrm{d}x.$$

Then, A_n tends to 0 when n tends to infinity.

REMARK 3.7.– *Using the same method, it can also be shown that* $\cos(nx)$ *and* e^{inx} *tend to* 0 *when* n *tends to infinity.*

2) Let $\varphi \in \mathcal{D}(\mathbb{R})$ such that $\text{supp}(\varphi) \Subset [-A, A]$. Then for all $n \in \mathbb{N}$, it follows that

$$\begin{aligned}\langle B_n, \varphi\rangle &= \int_{-A}^{A} n.g(nx)\varphi(x)\mathrm{d}x \\ &= \int_{-nA}^{nA} g(u)\varphi(\frac{u}{n})\mathrm{d}u \\ &= \int_{\mathbb{R}} g(u)\varphi(\frac{u}{n})\mathbb{1}_{[-nA,nA]}\mathrm{d}u\end{aligned}$$

For all $u \in \mathbb{R}$, we have

$$g(u)\varphi(\frac{u}{n})\mathbb{1}_{[-nA,nA]} \longrightarrow g(u)\varphi(0)$$

and

$$\left|g(u)\varphi(\frac{u}{n})\mathbb{1}_{[-nA,nA]}\right| \leq |g(u)| \parallel \varphi \parallel_\infty .$$

So, according to the dominated convergence theorem, we obtain that B_n converges to $\left(\int_{\mathbb{R}} g(u)\mathrm{d}u\right)\delta_0$ in $\mathcal{D}'(\mathbb{R})$.

3) Let $\varphi \in \mathcal{D}(\mathbb{R})$. Then,

$$\langle C_n, \varphi\rangle = \frac{1}{n}\sum_{p=0}^{n-1} \varphi(\frac{p}{n}) \longrightarrow \int_0^1 \varphi(x)\mathrm{d}x = \int_{\mathbb{R}} \varphi(x)\mathbb{1}_{[0,1]}\mathrm{d}x.$$

Therefore, C_n converges to the distribution associated with $\mathbb{1}_{[0,1]}$.

4) Let $\varphi \in \mathcal{D}(\mathbb{R})$ be such that $\mathrm{supp}(\varphi) \Subset [-A, A]$. Then for all n $\in \mathbb{N}$, we have

$$\begin{aligned}
\langle E_n, \varphi\rangle &= \left\langle e^{inx}\mathrm{Vp}(\frac{1}{x}), \varphi\right\rangle \\
&= \left\langle \mathrm{Vp}(\frac{1}{x}), e^{inx}\varphi(x)\right\rangle \\
&= \lim_{\epsilon\longrightarrow 0}\int_{|x|>\epsilon} \frac{e^{inx}\varphi(x)}{x}\mathrm{d}x \\
&= \lim_{\epsilon\longrightarrow 0}\left[\int_{A\geq|x|>\epsilon} \cos(nx)\frac{\varphi(x)}{x}\mathrm{d}x + i\int_{A\geq|x|>\epsilon} \sin(nx)\frac{\varphi(x)}{x}\mathrm{d}x\right].
\end{aligned}$$

For the real part: It can be written that $\varphi(x) = \varphi(0) + x\psi(x)$ with $\psi \in \mathcal{C}^\infty(\mathbb{R})$. We have

$$\int_{A\geq|x|>\epsilon} \cos(nx)\frac{\varphi(x)}{x}\mathrm{d}x = \varphi(0)\int_{A\geq|x|>\epsilon} \frac{\cos(nx)}{x}\mathrm{d}x + \int_{A\geq|x|>\epsilon} \cos(nx)\psi(x)\mathrm{d}x.$$

On the other hand, $\varphi(0)\displaystyle\int_{A\geq|x|>\epsilon} \frac{\cos(nx)}{x}\mathrm{d}x = 0$ (by imparity).

For the imaginary part:

$$\int_{A\geq|x|>\epsilon} \sin(nx)\frac{\varphi(x)}{x}\mathrm{d}x = \varphi(0)\int_{A\geq|x|>\epsilon} \frac{\sin(nx)}{x}\mathrm{d}x + \int_{A\geq|x|>\epsilon} \sin(nx)\psi(x)\mathrm{d}x.$$

Therefore,

$$\lim_{\epsilon\longrightarrow 0}\int_{A\geq|x|>\epsilon} \sin(nx)\frac{\varphi(x)}{x}\mathrm{d}x = \varphi(0)\int_{-A}^{A} \frac{\sin(nx)}{x}\mathrm{d}x + \int_{-A}^{A} \sin(nx)\psi(x)\mathrm{d}x.$$

On the other hand, $\displaystyle\int_{-A}^{A} \frac{\sin(nx)}{x}\mathrm{d}x + \int_{-A}^{A} \sin(nx)\psi(x)\mathrm{d}x = \pi.$

Since $\displaystyle\int_{-A}^{A} \sin(nx)\phi(x)\mathrm{d}x = 0$, then, E_n converges to $i\pi\delta_0$ in $\mathcal{D}'(\mathbb{R})$.

5) A simple calculation makes it possible to see that

$$\mathbf{D}_n(x) = \begin{cases} \frac{\sin\left(\left(n+\frac{1}{2}\right)x\right)}{\sin\left(\frac{x}{2}\right)}, & x \notin 2\pi\mathbb{Z}, \\ 2n+1, & \text{elsewhere.} \end{cases}$$

Let $\varphi \in \mathcal{D}(\mathbb{R})$ be a test function such that $\mathrm{supp}(\varphi) \Subset [a,b] \subset [-(2n+1)\pi, (2n+1)\pi]$ (for n large enough). We have

$$\begin{aligned} \langle \mathbf{D}_n, \varphi \rangle &= \int_{-\infty}^{+\infty} \frac{\sin\left(\left(n+\frac{1}{2}\right)x\right)}{\sin\left(\frac{x}{2}\right)} \varphi(x) \mathrm{d}x \\ &= \sum_{k=-n}^{k=n} \int_{(2k-1)\pi}^{(2k+1)\pi} \frac{\sin\left(\left(n+\frac{1}{2}\right)x\right)}{\sin\left(\frac{x}{2}\right)} \varphi(x) \mathrm{d}x \\ &= \sum_{k=-n}^{k=n} \int_{-\pi}^{\pi} \frac{\sin\left(\left(n+\frac{1}{2}\right)x\right)}{\sin\left(\frac{x}{2}\right)} \varphi(x+2k\pi) \mathrm{d}x \\ &= \int_{-\pi}^{\pi} \frac{\sin\left(\left(n+\frac{1}{2}\right)x\right)}{\sin\left(\frac{x}{2}\right)} \psi(x) \mathrm{d}x, \end{aligned}$$

with $\psi(x) = \sum_{k=-n}^{k=n} \varphi(x+2k\pi)$. It can also be written that

$$\begin{aligned} \langle \mathbf{D}_n, \varphi \rangle &= \int_{-\pi}^{\pi} \frac{\sin\left(\left(n+\frac{1}{2}\right)x\right)}{\sin\left(\frac{x}{2}\right)} \psi(0) \mathrm{d}x + \int_{-\pi}^{\pi} \frac{\sin\left(\left(n+\frac{1}{2}\right)x\right)}{\sin\left(\frac{x}{2}\right)} \left(\psi(x) - \psi(0)\right) \mathrm{d}x \\ &= \psi(0) \int_{-\pi}^{\pi} \frac{\sin\left(\left(n+\frac{1}{2}\right)x\right)}{\sin\left(\frac{x}{2}\right)} \mathrm{d}x \\ &+ \int_{-\pi}^{\pi} \sin\left(\left(n+\frac{1}{2}\right)x\right) \frac{x}{\sin\left(\frac{x}{2}\right)} \frac{\left(\psi(x) - \psi(0)\right)}{x} \mathrm{d}x \\ &= 2\pi \sum_{k=-n}^{k=n} \varphi(2k\pi) + \int_{-\pi}^{\pi} \sin\left(\left(n+\frac{1}{2}\right)x\right) \frac{x}{\sin\left(\frac{x}{2}\right)} \frac{\left(\psi(x) - \psi(0)\right)}{x} \mathrm{d}x. \end{aligned}$$

This last integral tends to 0 by virtue of the Riemann–Lesbegue lemma. Then $\langle \mathbf{D}_n, \varphi \rangle \underset{n \longrightarrow +\infty}{\longrightarrow} 2\pi \sum_{k=-\infty}^{+\infty} \varphi(2k\pi)$. It should be recalled that the sum has a finite support, since φ has a compact support.

Conclusion: The sequence $(\mathbf{D}_n)$ tends to the Dirac comb $2\pi \sum_{k=-\infty}^{+\infty} \delta_{2k\pi}$.

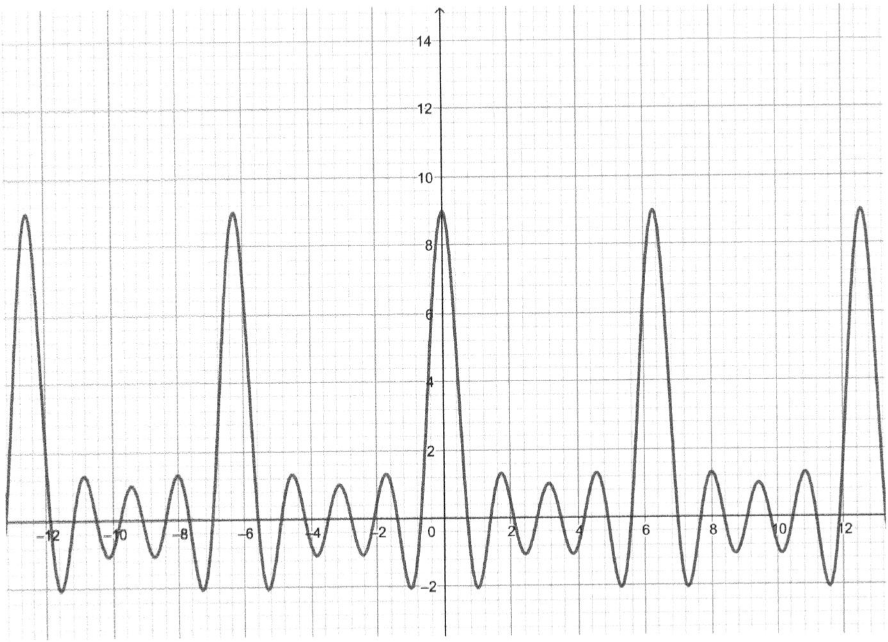

Figure 3.1. D_4

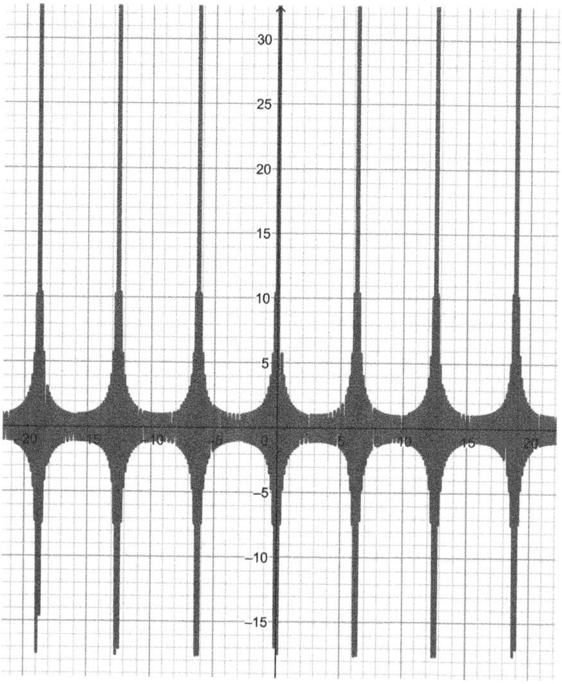

Figure 3.2. D_{40}

EXERCISE 3.12.– Show that in the sense of the distributions, we have

1) $\lim\limits_{\varepsilon \longmapsto 0^+} \dfrac{1}{\pi}\dfrac{\varepsilon}{x^2+\varepsilon^2} = \delta;$

2) $\lim\limits_{\varepsilon \longmapsto 0^+} \dfrac{1}{2\sqrt{\pi t}} e^{-\frac{x^2}{4t}} = \delta;$

3) $\lim\limits_{\lambda \longmapsto +\infty} \lambda e^{-\lambda x} H(x) = \delta.$

SOLUTION 3.12.–

1) Let $\varphi \in \mathcal{D}(\mathbb{R})$. We have

$$\begin{aligned}\left\langle \frac{1}{\pi}\frac{\varepsilon}{x^2+\varepsilon^2}, \varphi \right\rangle &= \frac{1}{\pi}\int_{-\infty}^{+\infty} \frac{\varepsilon}{x^2+\varepsilon^2}\varphi(x)\mathrm{d}x \\ &= \frac{1}{\pi}\int_{-\infty}^{+\infty} \frac{1}{t^2+1}\varphi(\varepsilon t)\mathrm{d}t.\end{aligned}$$

Since $\frac{1}{t^2+1}\varphi(\varepsilon t) \underset{\varepsilon\longmapsto 0+}{\longrightarrow} \frac{1}{t^2+1}\varphi(0)$, then by Lebesgue's dominated convergence theorem (φ is bounded), we obtain

$$\frac{1}{\pi}\int_{-\infty}^{+\infty}\frac{1}{t^2+1}\varphi(\varepsilon t)\mathrm{d}t \underset{\varepsilon\longmapsto 0+}{\longrightarrow} \frac{1}{\pi}\int_{-\infty}^{+\infty}\frac{1}{t^2+1}\varphi(0)\mathrm{d}t = \varphi(0).$$

Therefore, $\left\langle \frac{1}{\pi}\frac{\varepsilon}{x^2+\varepsilon^2}, \varphi\right\rangle \underset{\varepsilon\longmapsto 0+}{\longrightarrow} \langle\delta,\varphi\rangle$. Consequently, $\lim\limits_{\varepsilon\longmapsto 0+}\frac{1}{\pi}\frac{\varepsilon}{x^2+\varepsilon^2} = \delta$.

2) Let $\varphi \in \mathcal{D}(\mathbb{R})$. We have

$$\begin{aligned}\left\langle \frac{1}{2\sqrt{\pi t}}e^{-\frac{x^2}{4t}}, \varphi\right\rangle &= \frac{1}{2\sqrt{\pi t}}\int_{-\infty}^{+\infty} e^{-\frac{x^2}{4t}}\varphi(x)\mathrm{d}x \\ &= \frac{1}{\sqrt{\pi}}\int_{-\infty}^{+\infty} e^{-u^2}\varphi(2u\sqrt{t})\mathrm{d}u.\end{aligned}$$

Since $e^{-u^2}\varphi(2u\sqrt{t}) \underset{t\longmapsto 0+}{\longrightarrow} e^{-u^2}\varphi(0)$, then by Lebesgue's dominated convergence theorem (φ is bounded), we obtain

$$\frac{1}{\sqrt{\pi}}\int_{-\infty}^{+\infty} e^{-u^2}\varphi(2u\sqrt{t})\mathrm{d}u \underset{t\longmapsto 0+}{\longrightarrow} \frac{1}{\sqrt{\pi}}\int_{-\infty}^{+\infty} e^{-u^2}\varphi(0)\mathrm{d}u = \varphi(0).$$

Consequently,

$$\left\langle \frac{1}{2\sqrt{\pi t}}e^{-\frac{x^2}{4t}}, \varphi\right\rangle \underset{t\longmapsto 0+}{\longrightarrow} \langle\delta,\varphi\rangle.$$

Therefore, $\lim\limits_{t\longmapsto 0+}\frac{1}{2\sqrt{\pi t}}e^{-\frac{x^2}{4t}} = \delta$.

3) Let $\varphi \in \mathcal{D}(\mathbb{R})$. We have

$$\begin{aligned}\langle\lambda e^{-\lambda x}H(x), \varphi\rangle &= \int_{-\infty}^{+\infty}\lambda e^{-\lambda x}H(x)\varphi(x)\mathrm{d}x \\ &= \int_{0}^{+\infty}\lambda e^{-\lambda x}\varphi(x)\mathrm{d}x \\ &= \int_{0}^{+\infty} e^{-u}\varphi(\frac{u}{\lambda})\mathrm{d}u.\end{aligned}$$

By Lebesgue's dominated convergence theorem, we obtain

$$\int_0^{+\infty} e^{-u}\varphi(\frac{u}{\lambda})\mathrm{d}u \underset{\lambda\longmapsto+\infty}{\longrightarrow} \int_0^{+\infty} e^{-u}\varphi(0)\mathrm{d}u = \varphi(0).$$

Therefore, $\left\langle \lambda e^{-\lambda x}H(x), \varphi\right\rangle \underset{t\longmapsto 0^+}{\longrightarrow} \langle \delta, \varphi\rangle$. Consequently,

$$\lim_{\lambda\longmapsto+\infty} \lambda e^{-\lambda x}H(x) = \delta.$$

EXERCISE 3.13.– Show that in the sense of distributions, we have

1) $\lim\limits_{\varepsilon\longmapsto 0^+} \dfrac{x}{x^2+\varepsilon^2} = \mathrm{Vp}\left(\dfrac{1}{x}\right)$;

2) $\lim\limits_{\varepsilon\longmapsto 0^+} \dfrac{1}{x+i\varepsilon} = \mathrm{Vp}\left(\dfrac{1}{x}\right) - i\pi\delta.$

SOLUTION 3.13.–

1) Let $\varphi \in \mathcal{D}(\mathbb{R})$. We have

$$\begin{aligned}\left\langle \frac{x}{x^2+\varepsilon^2}, \varphi\right\rangle &= \int_{-\infty}^{+\infty} \frac{x}{x^2+\varepsilon^2}\varphi(x)\mathrm{d}x\\ &= \int_0^{+\infty} \frac{x^2}{x^2+\varepsilon^2}\frac{\varphi(x)-\varphi(-x)}{x}\mathrm{d}x.\end{aligned}$$

However, $\left|\frac{x^2}{x^2+\varepsilon^2}\frac{\varphi(x)-\varphi(-x)}{x}\right| \leq \left|\frac{\varphi(x)-\varphi(-x)}{x}\right|$ and since the major function is integrable, then by Lebesgue's dominated convergence theorem, we get

$$\int_0^{+\infty} \frac{x^2}{x^2+\varepsilon^2}\frac{\varphi(x)-\varphi(-x)}{x}\mathrm{d}x \underset{\varepsilon\longmapsto 0^+}{\longrightarrow} \int_0^{+\infty} \frac{\varphi(x)-\varphi(-x)}{x}\mathrm{d}x = \left\langle \mathrm{Vp}\left(\frac{1}{x}\right), \varphi\right\rangle.$$

Therefore, $\left\langle \frac{x}{x^2+\varepsilon^2}, \varphi\right\rangle \underset{t\longmapsto 0^+}{\longrightarrow} \langle \mathrm{Vp}\left(\frac{1}{x}\right), \varphi\rangle$. Consequently,

$$\lim_{\varepsilon\longmapsto 0^+} \frac{x}{x^2+\varepsilon^2} = \mathrm{Vp}\left(\frac{1}{x}\right).$$

2) Let $\varphi \in \mathcal{D}(\mathbb{R})$. We have

$$\begin{aligned}\left\langle \frac{1}{x+i\varepsilon}, \varphi \right\rangle &= \int_{-\infty}^{+\infty} \frac{x-i\varepsilon}{x^2+\varepsilon^2}\varphi(x)\mathrm{d}x \\ &= \int_{-\infty}^{+\infty} \frac{x}{x^2+\varepsilon^2}\varphi(x)\mathrm{d}x - i\int_{-\infty}^{+\infty} \frac{\varepsilon}{x^2+\varepsilon^2}\varphi(x)\mathrm{d}x \\ &= \left\langle \frac{x}{x^2+\varepsilon^2}, \varphi \right\rangle - i\left\langle \frac{\varepsilon}{x^2+\varepsilon^2}, \varphi \right\rangle.\end{aligned}$$

The first question gives $\left\langle \frac{x}{x^2+\varepsilon^2}, \varphi \right\rangle \underset{\varepsilon \longmapsto 0+}{\longrightarrow} \left\langle \mathrm{Vp}\left(\frac{1}{x}\right), \varphi \right\rangle$. From exercise 3.13, it follows that $\left\langle \frac{\varepsilon}{x^2+\varepsilon^2}, \varphi \right\rangle \underset{\varepsilon \longmapsto 0+}{\longrightarrow} \langle \pi\delta, \varphi \rangle$.

Therefore, $\left\langle \frac{1}{x+i\varepsilon}, \varphi \right\rangle \underset{\varepsilon \longmapsto 0+}{\longrightarrow} \left\langle \mathrm{Vp}\left(\frac{1}{x}\right) - \pi\delta, \varphi \right\rangle$. Then,

$$\lim_{\varepsilon \longmapsto 0+} \frac{1}{x+i\varepsilon} = \mathrm{Vp}\left(\frac{1}{x}\right) - i\pi\delta.$$

4

Operations on Distributions

The multiplication of two distributions does not define an internal law in the set of distributions. However, if a is an indefinitely differentiable function and T an arbitrary distribution, the product aT of a by T can be defined as a distribution. Multiplication by an indefinitely differentiable function is then one of the important operations of distributions and will be the subject of section 4.1. Second, the notion of differentiation in the sense of distributions is presented and it is shown that the distributions are indefinitely differentiable and that their successive derivatives are also distributions. The chapter concludes with some properties concerning translation, scaling, parity and homogeneity of a distribution. Throughout this chapter, Ω denotes an open set of $\mathbb{R}^d$.

4.1. Multiplication by a $\mathcal{C}^\infty$ function

Multiplication or division of two distributions is something that is not always valid and can create problems, it is in general not defined; as can be seen in this example for $f(x) = g(x) = \frac{1}{\sqrt{x}} \in \mathcal{D}'(\mathbb{R})$ since they are locally integrable but their product $f(x)g(x) = \frac{1}{x} \notin \mathcal{D}'(\mathbb{R})$. On the other hand, if $f \in \mathcal{D}'(\mathbb{R})$ and $\psi \in \mathcal{C}^\infty(\mathbb{R})$, the product ψf can have a meaning and defines a distribution on $\mathcal{D}(\mathbb{R})$ as can be seen in section 4.1.1.

4.1.1. *Definition and some properties*

PROPOSITION 4.1 (and definition).– Let T be a distribution of $\mathcal{D}'(\Omega)$ and a a $\mathcal{C}^\infty(\Omega)$ function. The linear form S defined on $\mathcal{D}(\Omega)$ by

$$\langle S, \varphi\rangle = \langle T, a\varphi\rangle, \forall\varphi \in \mathcal{D}(\Omega)$$

is a distribution and is called the "product of a and T", and is denoted by $S = aT$.

Proof. First of all, we actually have $a\varphi \in \mathcal{D}(\Omega)$. Therefore, the right-hand side member is well defined.

Let $K \subset \Omega$ be a compact and let $\varphi \in \mathcal{D}(\Omega)$ such that $\operatorname{supp}\varphi \Subset K$. Since $T \in \mathcal{D}'(\Omega)$, there exists $m \in \mathbb{N}$ and $C > 0$ such that, $|\langle T, \varphi\rangle| \leq CP_m(\varphi)$. However, $a\varphi \in \mathcal{D}(\Omega)$ and $\operatorname{supp}(a\varphi) \subset \operatorname{supp}\varphi \subset K$, then: $|\langle S, \varphi\rangle| = |\langle T, a\varphi\rangle| \leq CP_m(a\varphi)$ By Leibniz's formula, we have $\partial^\alpha(a\varphi) = \sum\limits_{0\leq\beta\leq\alpha} \binom{\alpha}{\beta} \partial^\beta a \partial^{\alpha-\beta}\varphi$. Then,

$$\max_{x\in K}|\partial^\alpha(a\varphi)(x)| \leq 2^{|\alpha|} \max_{x\in K, |\beta_1|\leq|\alpha|}|\partial^{\beta_1}a(x)| \max_{x\in K, |\beta_2|\leq|\alpha|}|\partial^{\beta_2}\varphi(x)|.$$

Hence, for any α such that $|\alpha| \leq m$, we have

$$\max_{x\in K}|\partial^\alpha(a\varphi)(x)| \leq \tilde{C}P_m(\varphi) \quad \text{with } \tilde{C} = 2^m \max_{x\in K, |\beta_1|\leq m}|\partial^{\beta_1}a(x)|.$$

Then $P_m(a\varphi) \leq \tilde{C}P_m(\varphi)$. We finally have

$$|\langle S, \varphi\rangle| = |\langle T, a\varphi\rangle| \leq C\tilde{C}P_m(\varphi) \leq C_S\, P_m(\varphi),$$

which shows that S is a distribution on Ω. □

The following properties can be verified quite easily.

PROPOSITION 4.2.– For all and $T, S \in \mathcal{D}'(\Omega)$, we have

i) $(a + b)T = aT + bT$;

ii) $(ab)T = a(bT)$;

iii) $a(T + S) = aT + aS$.

4.1.2. *Examples*

EXAMPLE 4.1.– Let $a \in \mathcal{C}^\infty(\Omega)$ and $f \in L^1_{loc}(\Omega)$. We have

$$aT_f = T_{af}.$$

EXAMPLE 4.2.– Let $\alpha \in \mathcal{C}^\infty(\Omega)$ and $x_0 \in \Omega$. We have

$$\alpha\delta_{x_0} = \alpha(x_0)\delta_{x_0}.$$

In particular, in $\mathbb{R}$ we have $\alpha\delta = \alpha(0)\delta$ and $x\delta_0 = 0$. This relation is often interpreted in physics by stating that δ is an eigen distribution of the multiplication operator by a function α, with the eigenvalue $\alpha(0)$.

EXAMPLE 4.3.– We have $x\mathrm{Vp}\left(\frac{1}{x}\right) = 1$.

As it happens, if $\varphi \in \mathcal{D}(\Omega)$, then

$$\begin{aligned}\left\langle x\mathrm{Vp}\left(\frac{1}{x}\right), \varphi\right\rangle &= \left\langle \mathrm{Vp}\left(\frac{1}{x}\right), x\varphi\right\rangle \\ &= \lim_{\varepsilon\to 0+}\int_{|x|>\varepsilon}\frac{x\varphi(x)}{x}\,\mathrm{d}x \\ &= \lim_{\varepsilon\to 0+}\int_{|x|>\varepsilon}\varphi(x)\,\mathrm{d}x \\ &= \int_{\mathbb{R}}\varphi(x)\,\mathrm{d}x \\ &= \langle 1, \varphi\rangle .\end{aligned}$$

Consequently, $x\mathrm{Vp}\left(\frac{1}{x}\right) = 1$.

REMARK 4.1.– *The relation $x\mathrm{Vp}\left(\frac{1}{x}\right) = 1$ is extremely interesting, because it says that the inverse of x in the sense of distributions is nothing else than the distribution $\mathrm{Vp}\left(\frac{1}{x}\right)$ and not $\frac{1}{x}$, which obviously cannot define a regular distribution.*

4.1.3. *Convergence properties*

Another interesting result is that multiplication by a function $\mathcal{C}^\infty$ is a continuous operation. More precisely, we have the following theorem.

THEOREM 4.1.– *Let $T \in \mathcal{D}'(\Omega)$ and $a \in \mathcal{C}^\infty(\Omega)$. Let $\left(a_n\right)_{n\geq 0}$ be a sequence which converges to a in $\mathcal{C}^\infty(\Omega)$ and let $\left(T_n\right)_{n\geq 0}$ be a sequence which converges to T in $\mathcal{D}'(\Omega)$. We then have*

i) $a_n T \underset{n\longrightarrow+\infty}{\longrightarrow} aT$;

ii) $aT_n \underset{n\longrightarrow+\infty}{\longrightarrow} aT$;

iii) $a_n T_n \underset{n\longrightarrow+\infty}{\longrightarrow} aT$.

Convergence is in $\mathcal{D}'(\Omega)$.

Proof. It is obvious that point (iii) has to be demonstrated.

Let $\varphi \in \mathcal{D}(\Omega)$. We set for all n, $\psi_n = a_n\varphi$. By Leibniz's formula, we have $\psi_n \underset{n\longrightarrow+\infty}{\longrightarrow} a\varphi$. From corollary 3.2, we have

$$\langle a_n T_n, \varphi\rangle = \langle T_n, \psi_n\rangle \underset{n\longrightarrow+\infty}{\longrightarrow} \langle T, a\varphi\rangle = \langle aT, \varphi\rangle\,. \qquad \square$$

REMARK 4.2.– *Resulting from this theorem, the mapping*

$$(a, T) \in \mathcal{C}^\infty(\Omega) \times \mathcal{D}'(\Omega) \longrightarrow aT \in \mathcal{D}'(\Omega)$$

is continuous.

4.1.4. *Solution of the equations $xT = 0, xT = 1$ and $xT = S$*

These three equations will be studied in the one-dimensional case $d = 1$. A more general result will then be given in any dimension d. We start with a technical lemma that will be used later.

LEMMA 4.1.– *We have the following equivalence*

$$\psi = x\theta, \theta \in \mathcal{D}(\mathbb{R}) \iff \psi \in \mathcal{D}(\mathbb{R}) \text{ and } \psi(0) = 0.$$

Proof. First of all, if $\psi = x\theta$, with $\theta \in \mathcal{D}(\mathbb{R})$, it is clear that $\psi \in \mathcal{D}(\mathbb{R})$ and that $\psi(0) = 0$.

Conversely, let $\psi \in \mathcal{D}(\mathbb{R})$ such that $\psi(0) = 0$. Let us assume that $\operatorname{supp} \psi \in]-A, A[$. By the Taylor formula with integral remainder at first order, it can be written that

$$\psi(x) = \psi(0) + x \int_0^1 \psi'(tx)\mathrm{d}t = x\theta(x),$$

where $\theta(x) = \int_0^1 \psi'(tx)\mathrm{d}t$. Then, $\theta \in \mathcal{C}^\infty(\mathbb{R})$, and if $|x| > A$, we have $\psi(x) = 0$ from which $\theta(x) = 0$. Therefore, $\theta \in \mathcal{D}(\mathbb{R})$. □

PROPOSITION 4.3.– Let $T \in \mathcal{D}'(\mathbb{R})$ denote a distribution. The following two assertions are equivalent:

1) $xT = 0$;

2) $T = c\delta_0$ where c is a constant.

Proof. We have already seen that if $T = c\delta_0$, then $xT = cx\delta_0 = c0 = 0$. Hence a first implication.

For the other sense, suppose that $xT = 0$. For $\theta \in \mathcal{D}(\mathbb{R})$, we have

$$0 = \langle xT, \theta \rangle = \langle T, x\theta \rangle .$$

Therefore, T is null with all functions of the form $x\theta, \theta \in \mathcal{D}(\mathbb{R})$. According to lemma 4.1, the distribution T cancels out on all the functions $\psi \in \mathcal{D}(\mathbb{R})$ such that $\psi(0) = 0$.

We set $\chi \in \mathcal{D}(\mathbb{R})$ such that $\chi(x) = 1$ for $|x| \leq 1$. Let $\varphi \in \mathcal{D}(\mathbb{R})$. Let us set $\psi = \varphi - \varphi(0)\chi$. Then $\psi \in \mathcal{D}(\mathbb{R})$ and $\psi(0) = 0$. Therefore, $\langle T, \psi \rangle = 0$, or still:

$$\langle T, \varphi \rangle = \varphi(0) \langle T, \chi \rangle = c \langle \delta_0, \varphi \rangle \text{ with } c = \langle T, \chi \rangle .$$

Thereafter, $T = c\delta_0$. Hence the other implication. □

A similar equation with a second member can then be studied. We start by looking at the equation $xT = 1$.

PROPOSITION 4.4.– Let $T \in \mathcal{D}'(\mathbb{R})$ denote a distribution. The following two assertions are equivalent:

1) $xT = 1$;

2) $T = c\delta_0 + \text{Vp}(\frac{1}{x})$ where c is a constant.

Proof. We already saw, in example 4.3, that $x\text{Vp}(\frac{1}{x}) = 1$.

So, if T is a solution of the equation $xT = 1$, we must have

$$x\left(T - \text{Vp}(\frac{1}{x})\right) = 0.$$

By the previous proposition,

$$T - \text{Vp}(\frac{1}{x}) = c\delta_0,$$

with c as a constant.

A simple verification shows the opposite implication. □

Here, we come once more across the general principle of solving linear equations: the set of solutions is an affine space directed by the kernel of the linear application that defines the equation under consideration (i.e. the set of solutions of the associated homogeneous equation) and involving a particular solution of the equation. The equation $xT = S$ can in fact be solved for any second member $S \in \mathcal{D}'(\mathbb{R})$.

PROPOSITION 4.5.– Let $S \in \mathcal{D}'(\mathbb{R})$ be a distribution. Then the equation $xT = S$ admits a solution $T \in \mathcal{D}'(\mathbb{R})$.

Proof. Let $\varphi \in \mathcal{D}(\mathbb{R})$ and $\chi \in \mathcal{D}(\mathbb{R})$ be such that $\chi(x) = 1$ for $|x| \leq 1$. Let us set $\tilde{\varphi} = \varphi - \varphi(0)\chi$. Then $\tilde{\varphi}(0) = 0$. There thus exists $\theta \in \mathcal{D}(\mathbb{R})$ such that $\tilde{\varphi} = x\theta$. We define the distribution T by

$$\langle T, \varphi \rangle = \langle S, \theta \rangle .$$

Then, $\langle xT, \varphi \rangle = \langle T, x\varphi \rangle = \langle S, \varphi \rangle$, from which $xT = S$. Actually,

$$\tilde{x\varphi} = x\varphi - (x\varphi)(0)\chi = x\varphi,$$

and in this case $\theta = \varphi$.

All there is to do now is verify that the formula $\langle T, \varphi \rangle = \langle S, \theta \rangle$ defines correctly a distribution $T \in \mathcal{D}'(\mathbb{R})$. First of all, if $\varphi \in \mathcal{D}(\mathbb{R})$, we have $\operatorname{supp} \theta \subset \operatorname{supp} \varphi \cup \operatorname{supp} \chi$, therefore $\theta \in \mathcal{C}^\infty(\mathbb{R})$. Next, since $S \in \mathcal{D}'(\mathbb{R})$, then $|\langle S, \theta \rangle| \leq C_S P_m(\theta)$.

However, $\theta = \displaystyle\int_0^1 (\varphi'(tx) - \varphi(0)\chi'(tx))\,\mathrm{d}t$, thus $P_m(\theta) \leq C P_{m+1}(\varphi)$. As such,

$$|\langle T, \varphi \rangle| \leq C_S C P_{m+1}(\varphi) = C' \, P_{m+1}(\varphi).$$

Consequently, $T \in \mathcal{D}'(\mathbb{R})$. □

EXAMPLE 4.4.– We have the following two identities:

$$x\mathrm{Pf}\left(\frac{1}{x^2}\right) = \mathrm{Vp}\left(\frac{1}{x}\right) \quad \text{and} \quad x^2\mathrm{Pf}\left(\frac{1}{x^2}\right) = 1.$$

Indeed, let $\varphi \in \mathcal{D}(\mathbb{R})$. We have

$$\left\langle \mathrm{Pf}\left(\frac{1}{x^2}\right), \varphi \right\rangle = \lim_{\varepsilon \longrightarrow 0} \left(\int_{|x|>\varepsilon} \frac{\varphi(x)}{x^2}\mathrm{d}x - 2\frac{\varphi(0)}{\varepsilon} \right).$$

Then

$$\left\langle x\mathrm{Pf}\left(\frac{1}{x^2}\right), \varphi \right\rangle = \left\langle \mathrm{Pf}\left(\frac{1}{x^2}\right), x\varphi \right\rangle = \lim_{\varepsilon \longrightarrow 0} \int_{|x|>\varepsilon} \frac{\varphi(x)}{x}\mathrm{d}x = \left\langle \mathrm{Vp}\left(\frac{1}{x}\right), \varphi \right\rangle.$$

Consequently,

$$x\mathrm{Pf}\left(\frac{1}{x^2}\right) = \mathrm{Vp}\left(\frac{1}{x}\right) \quad \text{and} \quad x^2\mathrm{Pf}\left(\frac{1}{x^2}\right) = x\mathrm{Vp}\left(\frac{1}{x}\right) = 1.$$

4.2. Differentiation of a distribution

In this section, we introduce the notion of derivation for distributions. A generalization of the notion of derivative of functions for distributions is given but keeping the same classical notion for ordinary functions. By a particular choice of a few examples, the difference between classical differentiation (i.e. differentiation of functions) and differentiation in the sense of distributions is discussed.

4.2.1. *Definition and examples*

4.2.1.1. *Definition*

Let $f \in \mathcal{C}^1(\mathbb{R})$ and $\varphi \in \mathcal{D}(\mathbb{R})$. Then, an integration by parts gives

$$\langle f', \varphi \rangle = \int_{\mathbb{R}} f'(x)\varphi(x)\, \mathrm{d}x = -\int_{\mathbb{R}} f(x)\varphi'(x)\, \mathrm{d}x = -\langle f, \varphi' \rangle,$$

which motivates the following definition.

DEFINITION 4.1.– *We call **derivative** of a distribution $T \in \mathcal{D}'(\mathbb{R})$ the distribution $T' \in \mathcal{D}'(\mathbb{R})$ defined by $\langle T', \varphi \rangle = -\langle T, \varphi' \rangle, \quad \forall \varphi \in \mathcal{D}(\mathbb{R})$.*

PROPOSITION 4.6.– Any distribution $T \in \mathcal{D}'(\mathbb{R})$ has derivatives of any order which are also distributions and we have, for all $m \in \mathbb{N}$,

$$\left\langle T^{(m)}, \varphi \right\rangle = (-1)^m \left\langle T, \varphi^{(m)} \right\rangle, \quad \forall \varphi \in \mathcal{D}(\mathbb{R}).$$

Proof. Let $m \in \mathbb{N}$ be fixed and $\varphi \in \mathcal{D}(\mathbb{R})$.

By reiterating the definition for successive derivatives of the distribution T, it can be seen that

$$\left\langle T^{(m)}, \varphi \right\rangle = (-1)^m \left\langle T, \varphi^{(m)} \right\rangle.$$

Since $\varphi \in \mathcal{D}(\mathbb{R})$, then $\forall j \in \mathbb{N}, \quad \varphi^{(j)} \in \mathcal{D}(\mathbb{R})$, which gives a meaning to the right-hand member of the above identity. We now show that, thus defined, the functional $T^{(m)}$ is indeed a distribution. The linearity is trivial, all that remains is to show continuity.

Let K a compact of $\mathbb{R}$ and let $\varphi \in \mathcal{D}(\mathbb{R})$ be such that $\operatorname{supp}\varphi \subset K$. Since T is a distribution, there exists $C_K > 0$ and an integer $n \in \mathbb{N}$ such that

$$|\langle T, \varphi \rangle| \leq C_K P_{n,K}(\varphi).$$

But $\varphi^{(m)} \in \mathcal{D}(\mathbb{R})$, therefore

$$\left|\left\langle T, \varphi^{(m)}\right\rangle\right| \leq C_K P_{n,K}(\varphi^{(m)}) \leq C_K P_{n+m,K}(\varphi).$$

It can then be deduced that

$$\left|\left\langle T^{(m)}, \varphi\right\rangle\right| \leq C_K P_{n,K}(\varphi^{(m)}) \leq C_K P_{n+m,K}(\varphi).$$

Therefore, $T^{(m)}$ is indeed a distribution. □

REMARK 4.3.– *The proof given above leads to deducing that*

1) if T is a distribution of order p, then $T^{(m)}$ is at most of order $p + m$;

2) any distribution is indefinitely differentiable, that is of $\mathcal{C}^\infty$ class.

By reasoning about partial derivatives, the notion of differentiation of distributions can be generalized to any dimension and order.

PROPOSITION 4.7 (and definition).– Let $T \in \mathcal{D}'(\Omega)$ and $\alpha \in \mathbb{N}^d$ be a multi-index. The αth-order derivative $\partial^\alpha T$ of T is the distribution defined by

$$\langle \partial^\alpha T, \varphi\rangle = (-1)^{|\alpha|} \langle T, \partial^\alpha \varphi\rangle, \quad \forall \varphi \in \mathcal{D}(\Omega).$$

4.2.1.2. *Examples*

We give a series of examples of differentiations in the sense of distributions.

EXAMPLE 4.5.– The derivative of a distribution T_f with $f \in \mathcal{C}^1(\mathbb{R})$ is the distribution $T_{f'}$. This is the subject of the calculation carried out at the beginning of this section.

EXAMPLE 4.6.–

1) Let H be the Heaviside function on $\mathbb{R}$ defined by

$$\begin{cases} H(x) = 1, & x > 0 \\ H(x) = 0, & x < 0. \end{cases}$$

Then, $H' = \delta$. Indeed, $\forall \varphi \in \mathcal{D}(\mathbb{R})$,

$$\begin{aligned}
\langle H', \varphi \rangle &= -\langle H, \varphi' \rangle \\
&= -\int_{\mathbb{R}} H(x)\varphi'(x)\mathrm{d}x \\
&= -\int_0^{+\infty} \varphi'(x)\mathrm{d}x \\
&= \varphi(0) \\
&= \langle \delta, \varphi \rangle.
\end{aligned}$$

2) $(\mathrm{sign}(x))' = 2\delta$, where $\mathrm{sign}(x) = H(x) - H(-x)$.

EXAMPLE 4.7.– The pth derivative of δ is the distribution $\delta^{(p)}$ defined by

$$\left\langle \delta^{(p)}, \varphi \right\rangle = (-1)^p \varphi^{(p)}(0)$$

for any test function φ.

The distribution $\delta^{(p)}$ is of order p. In fact, we simply have to take the sequence of functions $\varphi_n(x) = x^p \psi(nx)$ with $\psi \in \mathcal{D}(]-1,1[)$ such that $\psi = 1$ close to 0. We have

$$\left| \left\langle \delta^{(p)}, \varphi_n \right\rangle \right| = p!,$$

whereas for $m < p$,

$$\lim_{n \longrightarrow +\infty} \sup_{x \in [-1,1]} (\varphi_n)^{(m)}(x) = \lim_{n \longrightarrow +\infty} \sup_{x \in [-1,1]} (x^p \psi(nx))^{(m)} = 0.$$

EXAMPLE 4.8.– The two distributions δ and δ' are *linearly independent*, that is, if a and b are two arbitrary complex numbers, then we have

$$a\delta + b\delta' = 0 \iff a = b = 0.$$

The proof follows from the fact that the order of δ is 0, while that of δ' is 1.

EXAMPLE 4.9 (Differentiation of a product aT).– If T is a distribution and a is a function of $\mathcal{C}^\infty$ class, then we have $(aT)' = a'T + aT'$.

Indeed, for $\varphi \in \mathcal{D}(\Omega)$, we have

$$\begin{aligned}\langle (aT)', \varphi\rangle &= -\langle aT, \varphi'\rangle \\ &= -\langle T, a\varphi'\rangle \\ &= -\langle T, (a\varphi)' - a'\varphi\rangle \\ &= \langle T', (a\varphi)\rangle + \langle T, a'\varphi\rangle \\ &= \langle aT', \varphi\rangle + \langle a'T, \varphi\rangle \\ &= \langle aT' + a'T, \varphi\rangle .\end{aligned}$$

Generally speaking, Leibniz's formula for the product of a distribution by a function of class $\mathcal{C}^\infty$ can be demonstrated and we have

$$\partial^\alpha(aT) = \sum_{0\le\beta\le\alpha} \binom{\alpha}{\beta} \partial^\beta a \partial^{\alpha-\beta} T$$

for $T \in \mathcal{D}'(\Omega)$ and $a \in \mathcal{C}^\infty(\Omega)$.

Application: For $a \in \mathcal{C}^\infty(\Omega)$, we have: $a\delta' = a(0)\delta - a'(0)\delta$.

Special case: $x\delta' = -\delta$ and $\forall k > 1,\ x^k\delta' = 0$.

EXAMPLE 4.10.– Let f be the function defined on $\mathbb{R}$ by $\begin{cases} f(x) = \ln(|x|), & x \neq 0, \\ f(x) = 0, & x = 0. \end{cases}$

Since $f \in L^1_{loc}(\mathbb{R})$, then it can be associated with a distribution $T_f \in \mathcal{D}(\mathbb{R})$. We then have $(T_f)' = \mathrm{Vp}(\frac{1}{x})$. To see this, the calculation made in example 3.5 should be revisited.

EXAMPLE 4.11.– Let $u \in L^1_{loc}(\mathbb{R})$ and set, for $x \in \mathbb{R}$, $v(x) = \displaystyle\int_0^x u(t)dt$.

Then v is a continuous function on $\mathbb{R}$ and $v' = u$ in the sense of distributions.

We begin by showing the continuity of v. Let $b \in \mathbb{R}$ and let (x_n) be a sequence that converges toward b. We have $\forall n \geq 0, v(x_n) = \displaystyle\int_0^{x_n} u(t)dt = \int_{\mathbb{R}} \mathbb{1}_{(0,x_n)} u(t)dt$. By

Lebesgue's dominated convergence theorem, the sequence $(v(x_n))_n$ then converges to $\int_{\mathbb{R}} \mathbb{1}_{(0,b)} u(t)\mathrm{d}t = v(b)$, hence the continuity of v in b, thus on $\mathbb{R}$.

Let $\varphi \in \mathcal{D}(\mathbb{R})$ and assume that $\operatorname{supp} \varphi \Subset [-A, A]$. We have

$$
\begin{aligned}
\langle v', \varphi \rangle &= - \langle v, \varphi' \rangle \\
&= - \int_{\mathbb{R}} v(x)\varphi'(x)\mathrm{d}x \\
&= - \int_{-A}^{A} v(x)\varphi'(x)\mathrm{d}x \\
&= - \int_{-A}^{A} \Big(\int_0^x u(t)\mathrm{d}t \Big) \varphi'(x)\mathrm{d}x \\
&= \underbrace{- \int_0^A \int_0^x u(t)\varphi'(x)\mathrm{d}t\mathrm{d}x}_{I_1} + \underbrace{\int_{-A}^0 \int_x^0 u(t)\varphi'(x)\mathrm{d}t\mathrm{d}x}_{I_2}.
\end{aligned}
$$

We use Fubini's theorem to calculate the integral I_1. We have

$$
\begin{cases} 0 \leq t \leq x \\ 0 \leq x \leq A \end{cases} \iff \begin{cases} 0 \leq t \leq A \\ t \leq x \leq A. \end{cases}
$$

Then,

$$
\begin{aligned}
I_1 &= - \int_0^A u(t) \Big(\int_t^A \varphi'(x)\mathrm{d}x \Big) \mathrm{d}t \\
&= - \int_0^A u(t)(\varphi(A) - \varphi(t))\mathrm{d}t \\
&= \int_0^A u(t)\varphi(t)\mathrm{d}t \quad \text{(because } \varphi(A) = 0\text{)}.
\end{aligned}
$$

In the same way, we show that $I_2 = \int_{-A}^{0} u(t)\varphi(t)\mathrm{d}t$. Consequently,

$$\begin{aligned}\langle v', \varphi\rangle &= \int_0^A u(t)\varphi(t)\mathrm{d}t + \int_{-A}^{0} u(t)\varphi(t)\mathrm{d}t \\ &= \int_{-A}^{A} u(t)\varphi(t)\mathrm{d}t \\ &= \int_{\mathbb{R}} u(t)\varphi(t)\mathrm{d}t \\ &= \langle u, \varphi\rangle .\end{aligned}$$

Indeed, we have $T_{v'} = T_u$, otherwise $v' = u$ in $\mathcal{D}'(\mathbb{R})$.

4.2.2. *Continuity of the differentiation operator*

In this section, we give a simple but interesting result of the continuity of the derivative operator on the space of distributions.

THEOREM 4.2.– *Let (T_n) be a sequence of distributions such that $T_n \longrightarrow T$ in $\mathcal{D}'(\Omega)$. Then, for every fixed multi-index $\alpha \in \mathbb{N}^d$, we have $\partial^\alpha T_n \longrightarrow \partial^\alpha T$ in $\mathcal{D}'(\Omega)$.*

Proof. Let $\varphi \in \mathcal{D}(\mathbb{R})$. We have $\langle \partial^\alpha T_n, \varphi\rangle = (-1)^{|\alpha|} \langle T_n, \partial^\alpha \varphi\rangle$.

However, $T_n \longrightarrow T$, then $\langle T_n, \partial^\alpha\varphi\rangle \longrightarrow \langle T, \partial^\alpha\varphi\rangle$. Consequently,

$$\langle \partial^\alpha T_n, \varphi\rangle = (-1)^{|\alpha|} \langle T_n, \partial^\alpha \varphi\rangle \longrightarrow (-1)^{|\alpha|} \langle T, \partial^\alpha \varphi\rangle = \langle \partial^\alpha T, \varphi\rangle .$$

Hence, $\partial^\alpha T_n \longrightarrow \partial^\alpha T$ in $\mathcal{D}'(\Omega)$. □

4.2.3. *Solution of the equations* $T' = 0$ *and* $\partial_{x_i} T = 0$

We start with a technical lemma that will be used to characterize a constant distribution.

LEMMA 4.2.– *We have the following equivalence:*

$$\psi = \varphi', \varphi \in \mathcal{D}(\mathbb{R}) \iff \psi \in \mathcal{D}(\mathbb{R}) \text{ and } \int_{\mathbb{R}} \psi(x)dx = 0.$$

Proof.

$\Longrightarrow$) : This direct direction is obvious since φ has a compact support.

$\Longleftarrow$) : We assume that $\psi \in \mathcal{D}(\mathbb{R})$ and $\int_{\mathbb{R}} \psi(x)\mathrm{d}x = 0$.

We consider φ defined by $\varphi(x) = \int_{-\infty}^{x} \psi(t)\mathrm{d}t$ and $\operatorname{supp}\psi \Subset [-M, M]$.

It is clear that $\varphi \in \mathcal{C}^\infty(\mathbb{R})$ and that $\varphi'(x) = \psi(x), \forall x \in \mathbb{R}$.

Let us show that $\operatorname{supp}\varphi \Subset [-M, M]$.

If $x > M$, then

$$\begin{aligned}
\varphi(x) &= \int_{-\infty}^{x} \psi(t)\mathrm{d}t \\
&= \int_{-\infty}^{M} \psi(t)\mathrm{d}t + \underbrace{\int_{M}^{x} \psi(t)\mathrm{d}t}_{=0} \\
&= \int_{-\infty}^{M} \psi(t)\mathrm{d}t \\
&= \int_{-M}^{M} \psi(t)\mathrm{d}t \\
&= \int_{\mathbb{R}} \psi(t)\mathrm{d}t \\
&= 0.
\end{aligned}$$

Hence, $x > M \Longrightarrow \varphi(x) = 0$.

Similarly, we show that $x < -M \Longrightarrow \varphi(x) = 0$.

Then, $\operatorname{supp}\varphi \Subset [-M, M]$. Therefore, $\varphi \in \mathcal{D}(\mathbb{R})$.

Consequently, there exists $\varphi \in \mathcal{D}(\mathbb{R})$ such that $\psi = \varphi'$. It is concluded that

$$\psi = \varphi', \varphi \in \mathcal{D}(\mathbb{R}) \iff \psi \in \mathcal{D}(\mathbb{R}) \text{ and } \int_{\mathbb{R}} \psi(x)\mathrm{d}x = 0. \qquad \square$$

PROPOSITION 4.8.– Let $T \in \mathcal{D}'(\mathbb{R})$ denote a distribution on $\mathbb{R}$. We have

$$T' = 0 \iff T \text{ is constant.}$$

Proof. $\Longleftarrow$) : It is assumed that $T \equiv c$.

Let $\varphi \in \mathcal{D}(\mathbb{R})$ such that $\operatorname{supp}\varphi \Subset [-A, A]$. We have

$$\begin{aligned}
\langle T', \varphi \rangle &= -\langle T, \varphi' \rangle \\
&= -\int_{\mathbb{R}} c\varphi'(t)\mathrm{d}t \\
&= -\int_{-A}^{A} c\varphi'(t)\mathrm{d}t \\
&= -c\int_{-A}^{A} \varphi'(t)\mathrm{d}t \\
&= -c\big(\varphi(A) - \varphi(-A)\big) \\
&= 0.
\end{aligned}$$

Therefore, $\langle T', \varphi \rangle = 0, \forall \varphi \in \mathcal{D}(\mathbb{R})$. Then $T' = 0$.

$\Rightarrow$) : For the direct direction, we assume that $T' = 0$. Then,

$$\forall \varphi \in \mathcal{D}(\mathbb{R}), \langle T', \varphi \rangle = -\langle T, \varphi' \rangle = 0.$$

Therefore, T cancels out on all the functions $\psi \in \mathcal{D}(\mathbb{R})$ such as $\psi = \varphi'$ where $\varphi \in \mathcal{D}(\mathbb{R})$. According to lemma 4.2, we have

$$\psi = \varphi', \varphi \in \mathcal{D}(\mathbb{R}) \iff \psi \in \mathcal{D}(\mathbb{R}) \text{ and } \int_{\mathbb{R}} \psi(x)\mathrm{d}x = 0.$$

From this equivalence, it is inferred that for all $\psi \in \mathcal{D}(\mathbb{R})$ such that $\int_{\mathbb{R}} \psi(x)\mathrm{d}x = 0$, we have $\langle T, \psi \rangle = 0$.

Let $\chi \in \mathcal{D}(\mathbb{R})$ be fixed such that $\int_{\mathbb{R}} \chi(x)\mathrm{d}x = 1$. Let $\varphi \in \mathcal{D}(\mathbb{R})$. We set:

$$\forall x \in \mathbb{R}, \psi(x) = \varphi(x) - \Big(\int_{\mathbb{R}} \varphi(x)\mathrm{d}x\Big)\chi(x).$$

Then $\psi \in \mathcal{D}(\mathbb{R})$ and $\int_{\mathbb{R}} \psi(x)\mathrm{d}x = 0$. Consequently, $\langle T, \psi \rangle = 0$. Then, by linearity of T, we obtain:

$$\langle T, \varphi \rangle = \langle T, \chi \rangle \int_{\mathbb{R}} \varphi(x)\mathrm{d}x. = c \langle 1, \varphi \rangle = \langle c, \varphi \rangle \text{ with } c = \langle T, \chi \rangle .$$

Therefore, T is constant and $T = c$.

Conclusion: $T' = 0 \iff T \in \mathrm{vect}(1)$. □

The result holds in higher dimension, but its proof is more difficult. A more general result will be admitted from which the desired result will be immediately deduced.

THEOREM 4.3.– *Let $T \in \mathcal{D}'(\Omega)$ be a distribution on Ω. It is assumed that there exists $f_1, f_2, \cdots, f_d$ of the continuous functions on Ω such that $\forall i \in \{1, \cdots, d\}, \partial_{x_i} T = f_i$.*

Then there exists $f \in \mathcal{C}^1(\Omega)$, such that $T = f$.

COROLLARY 4.1.– *Let $T \in \mathcal{D}'(\Omega)$ be a distribution on Ω a connected open set of $\mathbb{R}^d$.*

It is assumed that $\forall i \in \{1, \cdots, d\}, \partial_{x_i} T = 0$. Then, T is constant.

Proof. We simply have to apply theorem 4.3 and then the classical result on functions of class $\mathcal{C}^1$ with zero derivatives on a connected open (result based on finite increases). □

4.2.4. *Jump formula in dimension 1*

4.2.4.1. *Piecewise C^k-class function*

DEFINITION 4.2.– *A function $f : [a, b] \longrightarrow \mathbb{R}$ is piecewise of class C^k if there exists a subdivision $a < a_1 < a_2 < \ldots < a_n = b$ of $[a, b]$ such that for all $i = 0, \cdots, n-1$, if we set $g_i = f|_{]a_i, a_{i+1}[}$, then g_i extends into a C^k-class function, to the whole interval $[a_i, a_{i+1}]$.*

REMARK 4.4.– *If f is a piecewise C^k-class function, then*

1) the function f is of class C^k on $]a_i, a_{i+1}[$ for all $i = 0, \cdots, n-1$;

2) for any integer $j \leq k$ and at each a_i ; $i = 1, \cdots, n$, the differentiated function $f^{(k)}$ has a left limit $f^{(k)}\left(a_i^-\right)$ and a right limit $f^{(k)}\left(a_i^+\right)$;

3) the derivatives $f^{(j)}$ for $j \leq k$ exists in the ordinary sense everywhere except at points a_i; $i = 0, \cdots, n$;

4) f is locally integrable on $[a, b]$, so defines a distribution T_f;

5) if now f is defined on an interval I of $\mathbb{R}$ which is not necessarily a segment anymore, then it is said that f is piecewise of class C^k on I if it is piecewise of class C^k on any segment included in I.

4.2.4.2. *Jump formula*

We will consider here the jump formula in dimension one only. To this end, we start by considering a piecewise C^1-class function.

PROPOSITION 4.9 (Jump formula).– Let f be a piecewise C^1-class function. The derivative of the distribution T_f determined by f is given by

$$(T_f)' = T_{f'} + \sum_{i=1}^{n} \left(f\left(a_i^+\right) - f(a_i^-)\right) \delta_{a_i}. \tag{4.1}$$

This formula is called the jump formula.

Proof. Let $\varphi \in \mathcal{D}(\mathbb{R})$, such that $\operatorname{supp}\varphi \Subset [-A, A]$, then for a discontinuity point a, we have

$$\begin{aligned}
\langle (T_f)', \varphi \rangle &= -\langle T_f, \varphi' \rangle \\
&= -\int_{\mathbb{R}} f(x)\varphi'(x)\mathrm{d}x \\
&= -\int_{-\infty}^{a} f(x)\varphi'(x)\mathrm{d}x - \int_{a}^{+\infty} f(x)\varphi'(x)\mathrm{d}x \\
&= -\int_{-A}^{a} f(x)\varphi'(x)\mathrm{d}x - \int_{a}^{A} f(x)\varphi'(x)\mathrm{d}x \\
&= -f(a^-)\varphi(a) + \int_{-A}^{a} f'(x)\varphi(x)\mathrm{d}x + f(a^+)\varphi(a) + \int_{a}^{A} f'(x)\varphi(x)\mathrm{d}x \\
&= \big(f(a^+) - f(a^-)\big)\varphi(a) + \int_{-A}^{a} f'(x)\varphi(x)\mathrm{d}x + \int_{a}^{A} f'(x)\varphi(x)\mathrm{d}x
\end{aligned}$$

$$\begin{aligned}
\langle (T_f)', \varphi \rangle &= \big(f(a^+) - f(a^-)\big)\varphi(a) + \int_{-A}^{A} f'(x)\varphi(x)\mathrm{d}x \\
&= \big(f(a^+) - f(a^-)\big)\varphi(a) + \int_{\mathbb{R}} f'(x)\varphi(x)\mathrm{d}x \\
&= \big\langle \big(f(a^+) - f(a^-)\big)\delta_a, \varphi \big\rangle + \langle T_{f'}, \varphi \rangle \\
&= \big\langle \big(f(a^+) - f(a^-)\big)\delta_a + T_{f'}, \varphi \big\rangle .
\end{aligned}$$

Consequently, $\big\langle (T_f)', \varphi \big\rangle = \big\langle T_{f'} + \big(f(a^+) - f(a^-)\big)\delta_a, \varphi \big\rangle, \forall \varphi \in \mathcal{D}(\mathbb{R})$.

Conclusion: $(T_f)' = T_{f'} + \left(f\left(a^+\right) - f(a^-)\right)\delta_a = T_{f'} + \omega(a)\delta_a$.

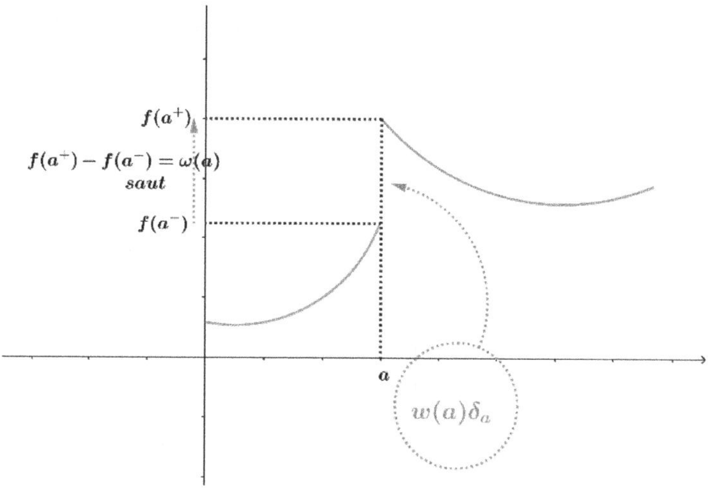

Figure 4.1. *Jump function. For a color version of this figure, see www.iste.co.uk/aitbenhassi/distribution.zip*

We now replace a by a_i and carry out the sum $i = 1, 2, \dots, n$.

If we assume that $a_1 < a_2 < \cdots < a_n$, then for all $\varphi \in \mathcal{D}(\mathbb{R})$, we have

$$\begin{aligned}\langle (T_f)', \varphi \rangle &= -\langle T_f, \varphi \rangle \\ &= -\int_{-\infty}^{a_1} f(x)\varphi'(x)\mathrm{d}x - \sum_{i=1}^{n-1} \int_{a_i}^{a_{i+1}} f(x)\varphi'(x)\mathrm{d}x - \int_{a_n}^{+\infty} f(x)\varphi'(x)\mathrm{d}x.\end{aligned}$$

By integrations by parts, it can be easily seen that

$$\langle (T_f)', \varphi \rangle = \sum_{i=1}^{n} \left(f\left(a_i^+\right) - f\left(a_i^-\right) \right) \varphi(a_i) + \int_{-\infty}^{+\infty} f'(x)\varphi(x)\mathrm{d}x, \quad \forall \varphi \in \mathcal{D}(\mathbb{R}).$$

Consequently, $(T_f)' = T_{f'} + \sum_{i=1}^{n} \left(f\left(a_i^+\right) - f\left(a_i^-\right) \right) \delta_{a_i} = T_{f'} + \sum_{i=1}^{n} \omega(a_i)\delta_{a_i}$. □

REMARK 4.5.– *The same thing can be obtained even with a countable number of discontinuity points* $(a_i)_{i=1}^{+\infty}$*:*

$$\textit{that is, } (T_f)' = T_{f'} + \sum_{i=1}^{+\infty} \left(f\left(a_i^+\right) - f\left(a_i^-\right) \right) \delta_{a_i} = T_{f'} + \sum_{i=1}^{+\infty} \omega(a_i)\delta_{a_i}.$$

EXAMPLE 4.12.– Using the jump formula, the following results can be found:

1) $H' = \delta_0$

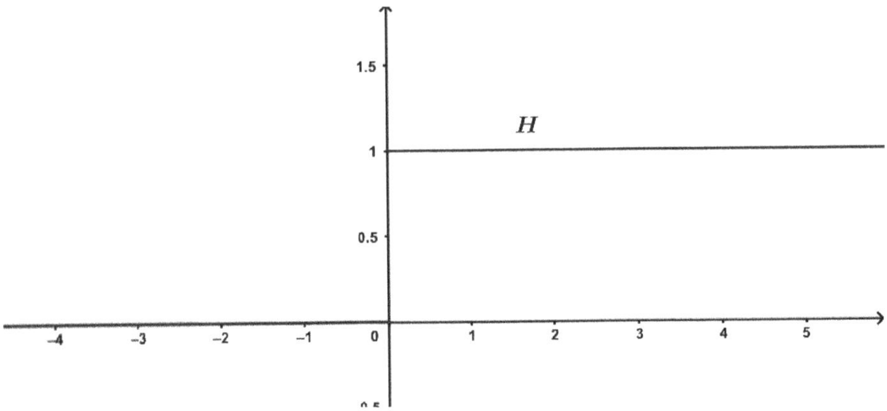

Figure 4.2. *Heaviside function H. For a color version of this figure, see www.iste.co.uk/aitbenhassi/distribution.zip*

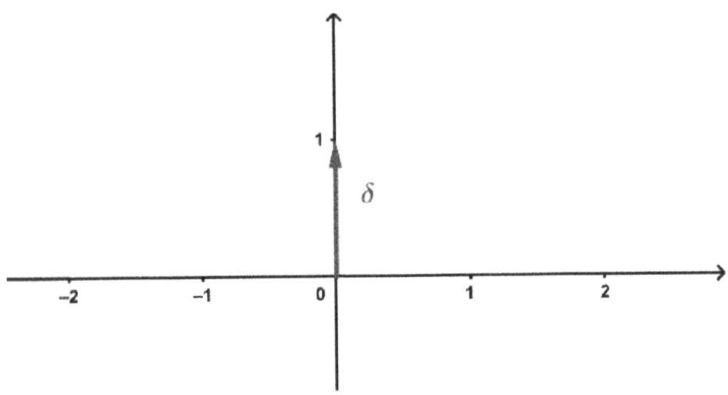

Figure 4.3. *Derivative of H. For a color version of this figure, see www.iste.co.uk/aitbenhassi/distribution.zip*

2) $\text{sign}(x)' = 2\delta_0$

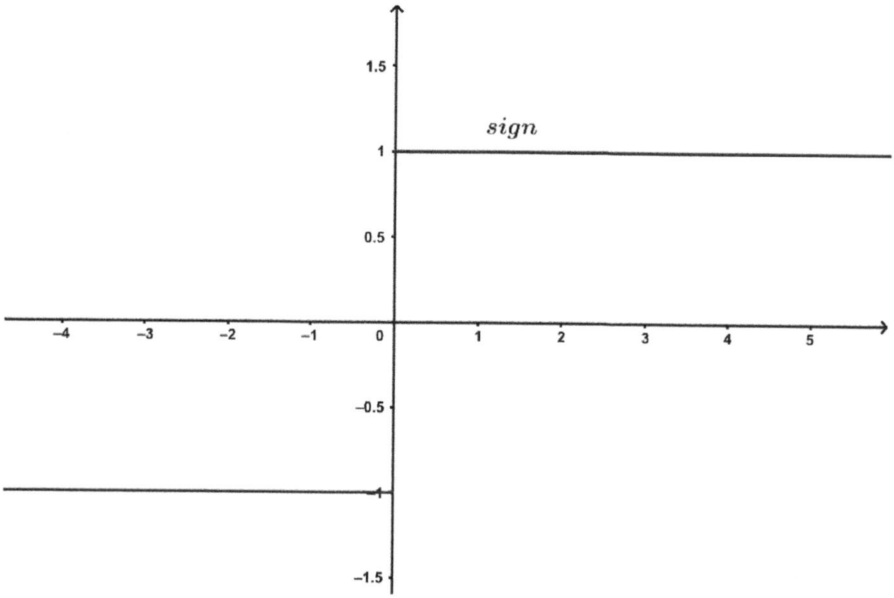

Figure 4.4. *Function sign$_x$. For a color version of this figure, see www.iste.co.uk/aitbenhassi/distribution.zip*

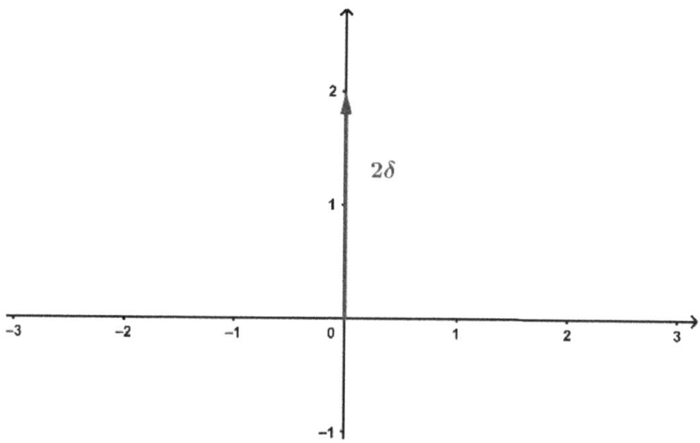

Figure 4.5. *Derivative of sign$_x$. For a color version of this figure, see www.iste.co.uk/aitbenhassi/distribution.zip*

EXAMPLE 4.13.– Let $E(x)$ be the integer part of the real x.

The discontinuity points of this function are the points $x = m$ with $m \in \mathbb{Z}$. On each interval $[m, m+1[$, we have $E(x) = m$, thus $E'(x) = 0$.

The jump formula gives

$$E' = \sum_{m=-\infty}^{+\infty} \delta_m.$$

The distribution $\sum_{m=-\infty}^{+\infty} \delta_m$ is denoted by $\triangle_1$ and called Dirac comb.

EXAMPLE 4.14.– Let f denote the $2\pi-$periodic function defined by

$$f(x) = \frac{1}{2}\left(1 - \frac{x}{\pi}\right), \quad x \in [0, 2\pi].$$

The jump at discontinuity points $x = \pm 2m\pi, m \in \mathbb{N}$, is $\omega(x) = 1$. Then by applying the jump formula, we obtain

$$(T_f)' = -\frac{1}{2\pi} + \sum_{m=-\infty}^{+\infty} \delta_{2m\pi}.$$

EXAMPLE 4.15 (General case of a periodic function).– Let f denote the $a-$periodic function ($a > 0$) defined by

$$f(x) = \frac{x}{a}.$$

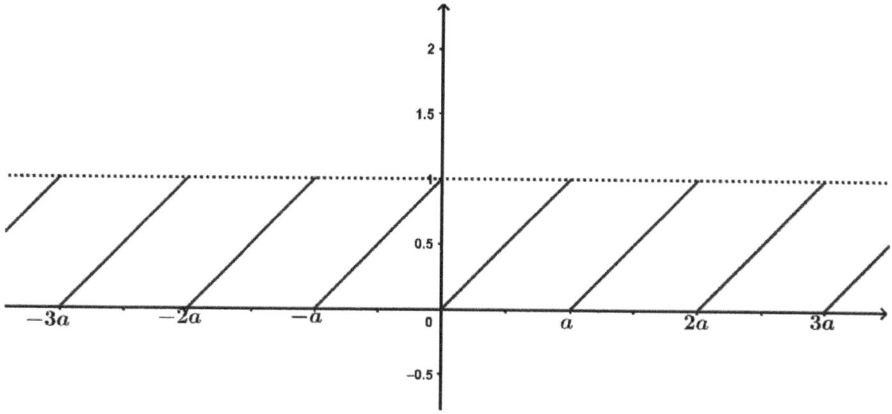

Figure 4.6. *Function* $f(x) = \frac{x}{a}$

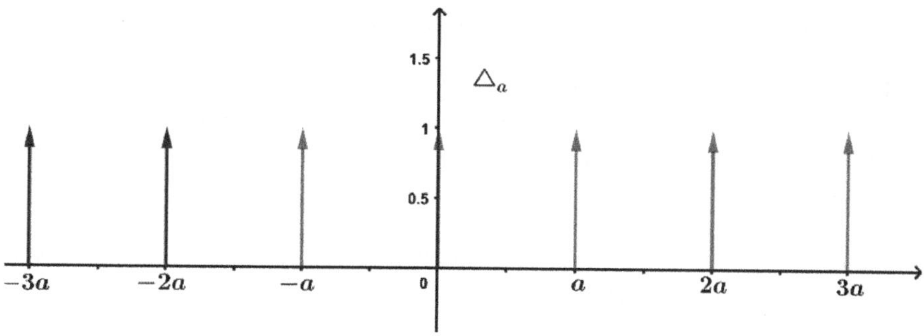

Figure 4.7. *The Dirac comb* $\triangle_a$

The jump at the discontinuity points $x_k = \pm ka, k \in \mathbb{Z}$, is $\omega(x_k) = 1$. Then, applying the jump formula, we get

$$f' = \frac{1}{a} - \sum_{k=-\infty}^{+\infty} \delta_{ka}. \qquad [4.2]$$

For $a > 0$, the distribution defined by $\triangle_a = \sum\limits_{m=-\infty}^{+\infty} \delta_{ma}$ is called the Dirac comb.

The result from the jump formula can be extended to the successive derivatives by considering the jumps of f and those of its derivatives.

PROPOSITION 4.10 (Jump formula with $\mathcal{C}^k$).– Let f denote a piecewise $\mathcal{C}^k$-class function. The successive derivatives of the distribution T_f determined by f are given by

$$(T_f)^{(m)} = T_{f^{(m)}} + \sum_{i=1}^{n} (f(a_i^+) - f(a_i^-))\delta_{a_i}^{(m-1)}$$
$$+ \sum_{i=1}^{n} (f'(a_i^+) - f'(a_i^-))\delta_{a_i}^{(m-2)} + \dots$$
$$+ \sum_{i=1}^{n} (f^{(m-1)}(a_i^+) - f^{(m-1)}(a_i^-))\delta_{a_i}$$

for all $m \in \mathbb{N}^*$ such that $m \leq k$.

4.2.5. *Differentiation/integration under the duality bracket*

The following two statements generalize, on the one hand, the differentiation theorem under the summation sign and, on the other hand, Fubini's theorem. They will play an important part in differentiating the convolution product of two distributions.

In the following two statements, the notation $\langle T, \varphi(., y)\rangle$ designates $\langle T, \varphi_y\rangle$ where $\varphi_y \in \mathcal{D}(\Omega)$ is the function defined by $\varphi_y(x) := \varphi(x, y)$.

PROPOSITION 4.11 (Differentiation under the bracket).– Let $T \in \mathcal{D}'(\Omega)$ be a distribution and φ a function of $\mathcal{D}(\Omega \times \mathbb{R}^d)$. Then the function $F : y \longmapsto \langle T, \varphi(., y)\rangle$ is of class $\mathcal{C}^\infty$ on $\mathbb{R}^d$ and we have for all $\alpha \in \mathbb{N}^d$, $\partial_y^\alpha F(y) = \langle T, \partial_y^\alpha \varphi(., y)\rangle$.

Proof. We begin by demonstrating the result for $|\alpha| = 1$.

Let $y_0 \in \mathbb{R}^d$, and we will show that the differential of F at y_0 is the mapping

$$h \longmapsto \langle T, \nabla_y \varphi(., y_0).h\rangle .$$

Let $(h_n)_{n\in\mathbb{N}}$ a sequence of $\mathbb{R}^d$ converging to 0 and let us show that the sequence

$$\frac{1}{|h_n|} (F(y_0 + h_n) - F(y_0) - \langle T, \nabla_y \varphi(., y_0).h_n\rangle)$$

converges to 0 when n tends to infinity. Simply, we observe that the sequence of functions

$$\varphi_n(x) := \frac{1}{|h_n|}\left(\varphi(x, y_0 + h_n) - \varphi(x, y_0) - \nabla_y \varphi(., y_0).h_n\right)$$

converges to the null function in $\mathcal{D}(\Omega)$. So, we have

$$\langle T, \varphi_n \rangle \underset{n \longrightarrow +\infty}{\longrightarrow} 0,$$

so F is differentiable on $\mathbb{R}^d$ and its partial derivatives are

$$\partial_{y_j} F(y) = \left\langle T, \partial_{y_j} \varphi(., y) \right\rangle$$

for all $y \in \mathbb{R}^d$. The argument is iterated by induction on the differentiation order. □

REMARK 4.6.– *In the above text, the hypothesis $\varphi \in \mathcal{D}(\Omega \times \mathbb{R}^d)$ can be weakened by assuming only supp $\varphi \subset K \times \mathbb{R}^d$ with K compact, that is, the support of φ_y is included in K for all $y \in \mathbb{R}^d$.*

PROPOSITION 4.12 (Integration under the bracket).– Let $T \in \mathcal{D}'(\Omega)$ denote a distribution and φ a function of $\mathcal{D}(\Omega \times \mathbb{R}^d)$. We have

$$\int_{\mathbb{R}^d} \langle T, \varphi(., y) \rangle \, \mathrm{d}y = \left\langle T, \int_{\mathbb{R}^d} \varphi(., y) \mathrm{d}y \right\rangle .$$

Proof. We consider the case $d = 1$, the general case is then obtained by induction on d using Fubini's theorem. Let R be a positive real such that φ have support in $[-R, R]^2$. Let ξ denote a $\mathcal{C}^\infty$-class function on $\mathbb{R}$ of integral 1, supported in $[-R, R]$. Let ψ be the function defined by

$$\psi(x, y) := \varphi(x, y) - \xi(y) \int_{\mathbb{R}} \varphi(x, u) \mathrm{d}u.$$

This function is of class $\mathcal{C}^\infty$ on $\Omega \times \mathbb{R}$ and has support in $[-R, R]^2$. Moreover, $\displaystyle\int_{\mathbb{R}} \psi(x, y) \mathrm{d}y = 0$, for all $x \in \Omega$. Then the function $\Phi(x, y) := \displaystyle\int_{-\infty}^{y} \psi(x, u) \mathrm{d}u$ is also of class $\mathcal{C}^\infty$ on $\Omega \times \mathbb{R}$ with support in $[-R, R]^2$. By proposition 4.12 on the differentiation under the bracket, we have

$$\frac{\mathrm{d}}{\mathrm{d}y} \langle T, \Phi(., y) \rangle = \langle T, \partial_y \Phi(., y) \rangle = \langle T, \psi(., y) \rangle .$$

It is deduced that the mapping $y \longmapsto \langle T, \Phi(x,y)\rangle - \displaystyle\int_{-\infty}^{y} \langle T, \psi(x,u)\rangle \,\mathrm{d}u$ is constant. However, $y \longmapsto \langle T, \Phi(x,y)\rangle$ and $y \longmapsto \displaystyle\int_{-\infty}^{y} \langle T, \psi(x,u)\rangle \,\mathrm{d}u$ are identically null for $y < -R$, therefore these two functions are equal. In particular, $\displaystyle\int_{\mathbb{R}} \langle T, \psi(x,y)\rangle \,\mathrm{d}y = 0$. Then,

$$\begin{aligned}
0 &= \int_{\mathbb{R}} \langle T, \psi(x,y)\rangle \,\mathrm{d}y \\
&= \int_{\mathbb{R}} \langle T, \varphi(x,y)\rangle \,\mathrm{d}y - \left\langle T, \int_{-\infty}^{y} \varphi(x,u)\mathrm{d}u \right\rangle \int_{\mathbb{R}} \xi(y)\mathrm{d}y \\
&= \int_{\mathbb{R}} \langle T, \varphi(x,y)\rangle \,\mathrm{d}y - \left\langle T, \int_{-\infty}^{y} \varphi(x,u)\mathrm{d}u \right\rangle,
\end{aligned}$$

hence the result. □

4.3. Transformations of distributions

This section presents a study on the effect of translation and dilation on distributions.

4.3.1. *Distribution translation*

We first note that for a function f locally integrable in $\mathbb{R}^d$ and $a \in \mathbb{R}^d$; by change of variable, we have

$$\int_{\mathbb{R}^d} f(x-a)\varphi(x)\mathrm{d}x = \int_{\mathbb{R}^d} f(x)\varphi(x+a)\mathrm{d}x.$$

Denoting by $\tau_a g$ the translation by the vector a of the function g and which is defined by

$$\tau_a g(x) = g(x-a), \forall x \in \mathbb{R}^d,$$

we get $\displaystyle\int_{\mathbb{R}^d} \tau_a f(x)\varphi(x)\mathrm{d}x = \int_{\mathbb{R}^d} f(x)\tau_{-a}\varphi(x)\mathrm{d}x$. Otherwise,

$$\langle \tau_a f, \varphi\rangle = \langle f, \tau_{-a}\varphi\rangle .$$

The result can be generalized to any distribution.

DEFINITION 4.3.– *Let $T \in \mathcal{D}'(\mathbb{R}^d)$ be a distribution and let $a \in \mathbb{R}^d$.*

The translation of T by the vector a is the distribution denoted $\tau_a T$ and defined by

$$\langle \tau_a T, \varphi \rangle = \langle T, \tau_{-a}\varphi \rangle, \quad \forall \varphi \in \mathcal{D}(\mathbb{R}).$$

EXAMPLE 4.16.– We have $\tau_a \delta = \delta_a$. As such, for any $\varphi \in \mathcal{D}(\mathbb{R})$, we have

$$\begin{aligned}\langle \tau_a \delta, \varphi \rangle &= \langle \delta, \tau_{-a}\varphi \rangle \\ &= \tau_{-a}\varphi(0) \\ &= \varphi(a) \\ &= \langle \delta_a, \varphi \rangle .\end{aligned}$$

Then, $\tau_a \delta = \delta_a$.

DEFINITION 4.4.– *Let $T \in \mathcal{D}'(\mathbb{R})$ and $a \in \mathbb{R}_+^*$. The distribution T is said to be periodic of period a, if $\tau_a T = T$.*

EXAMPLE 4.17.– Let $\varphi \in \mathcal{D}(\mathbb{R})$. We have

$$\begin{aligned}\langle \tau_{2\pi} T_{\sin}, \varphi \rangle &= \langle T_{\sin}, \tau_{-2\pi}\varphi \rangle \\ &= \int_{\mathbb{R}} \sin x \varphi(x + 2\pi) \mathrm{d}x \\ &= \int_{\mathbb{R}} \sin(x - 2\pi)\varphi(x) \mathrm{d}x \\ &= \int_{\mathbb{R}} \sin(x)\varphi(x) \mathrm{d}x \\ &= \langle T_{\sin}, \varphi \rangle .\end{aligned}$$

Then, $\tau_{2\pi} T_{\sin} = T_{\sin}$. Consequently, $T_{\sin}$ is 2π–periodic.

EXAMPLE 4.18.– The distribution T defined by $\langle T, \varphi \rangle = \sum_{-\infty}^{+\infty} \varphi(na), a > 0$ is periodic of period a.

EXAMPLE 4.19.– Let α be an indefinitely differentiable function, and let T a distribution of $\mathcal{D}'(\mathbb{R})$. For any real a, we have: $\tau_a(\alpha T) = (\tau_a \alpha)(\tau_a T)$.

EXAMPLE 4.20.– For any real a, we have

$$(x-a)T = 0 \iff \exists k \in \mathbb{R}, T = k\delta_a.$$

4.3.2. *Distribution dilation*

For $\lambda \in \mathbb{R} \setminus \{0\}$, we recall that the dilation $d_\lambda g$ of a function g defined on $\mathbb{R}^d$ the function defined by

$$d_\lambda g(x) = g(\frac{x}{\lambda}), \forall x \in \mathbb{R}^d.$$

DEFINITION 4.5.– *Let $\lambda \in \mathbb{R} \setminus \{0\}$, and let T be a distribution of $\mathcal{D}'(\mathbb{R})$. The dilation $d_\lambda T$ of the distribution T is the distribution defined by*

$$\langle d_\lambda T, \varphi \rangle = \left\langle T, |\lambda| d_{\frac{1}{\lambda}} \varphi \right\rangle, \forall \varphi \in \mathcal{D}(\mathbb{R}).$$

EXAMPLE 4.21.– We have $d_\lambda \delta = |\lambda| \delta$ and $d_\lambda \delta_{x_0} = |\lambda| \delta_{\lambda x_0}$.

4.3.3. *Distribution parity*

DEFINITION 4.6.– *Let T denote a distribution of $\mathcal{D}'(\mathbb{R})$. The distribution denoted by $\check{T} := d_{-1}T$ and defined by $\forall \varphi \in \mathcal{D}(\mathbb{R}), \langle \check{T}, \varphi \rangle = \langle T, \check{\varphi} \rangle$ where $\check{\varphi}(x) = \varphi(-x)$, is called the symmetric distribution of the distribution T.*

DEFINITION 4.7.– *It is said that T is an even distribution if $\check{T} = T$ and odd if $\check{T} = -T$.*

EXAMPLE 4.22.– The distribution of δ is even, as is $\delta_{x_0} + \delta_{-x_0}$.

EXAMPLE 4.23.– The distribution $\text{Vp}\left(\frac{1}{x}\right)$ is odd, as is $\text{sign}_x = 2H - 1$.

EXAMPLE 4.24.– The distribution H has no well-defined parity, as is δ_{x_0} for $x_0 \neq 0$.

APPLICATION EXERCISE.– Any distribution can be uniquely written as a sum of an even and an odd distribution:

$$T = \frac{T + \check{T}}{2} + \frac{T - \check{T}}{2}.$$

Apply this formula to the distributions H and δ_{x_0}.

4.3.4. *Distribution homogeneity*

The purpose of this section is to study the effect of a change of variable on a distribution.

Consider the homothety h_λ defined on $\mathbb{R}^d$ by: $h_\lambda(x) = \lambda x, \forall x \in \mathbb{R}^d$. Let T denote a distribution of $\mathcal{D}'(\mathbb{R})$. This distribution T is associated with a distribution denoted by $T \circ h_\lambda$ and defined by $\langle T \circ h_\lambda, \varphi \rangle = \frac{1}{|\lambda|} \langle T, \varphi \circ h_\lambda \rangle, \forall \varphi \in \mathcal{D}(\mathbb{R})$.

REMARK 4.7.– *If $T \circ h_\lambda = T$, we say that the distribution T is invariant by h_λ.*

DEFINITION 4.8.– *We say that T is a homogeneous distribution of degree $p \in \mathbb{N}$ if*

$$T \circ h_\lambda = \lambda^p T, \quad \forall \lambda > 0.$$

EXAMPLE 4.25.–

1) The distribution $|x|$ is homogeneous of degree 1.

2) The distribution sign_x is homogeneous of degree 0.

3) The distribution $\text{Vp}(\frac{1}{x})$ is homogeneous of degree -1.

4.4. Exercises with solutions

EXERCISE 4.1.– Let $T \in \mathcal{D}'(\mathbb{R})$. Show that $\big(\sin x\big)T = 0$ if and only if there exists a sequence $\big(c_n\big)_{n\in\mathbb{Z}}$ of complex numbers such that $T = \sum\limits_{n\in\mathbb{Z}} c_n \delta_{n\pi}$.

SOLUTION 4.1.–

i) Let $O_n = \left[-\dfrac{3\pi}{4} + n\pi, \dfrac{3\pi}{4} + n\pi\right]$. Then $\big(O_n\big)_{n\in\mathbb{Z}}$ is a locally finite covering of $\mathbb{R}$. Let $(\chi_n)_n$ be a unit partition associated with $\big(O_n\big)_{n\in\mathbb{N}}$. Let $\varphi \in \mathcal{C}_c^\infty(\mathbb{R})$. We decompose φ into: $\varphi = \sum\limits_{n\in\mathbb{Z}} \varphi\chi_n := \sum\limits_{n\in\mathbb{Z}} \varphi_n$. According to the Taylor formula with integral remainder, each φ_n is written as

$$\forall x \in \mathbb{R}, \varphi_n = \varphi(n\pi) + (x - n\pi)\psi_n(x) \quad \text{with} \quad \psi_n \in \mathcal{C}^\infty(\mathbb{R}).$$

Moreover, it can be seen that $x \longrightarrow \dfrac{x - n\pi}{\sin(x)}$ is of class $\mathcal{C}^\infty$ and has compact support in O_n. Finally, if ν_n is a plateau function equal to 1 on O_n and with support in $]-\pi + n\pi, \pi + n\pi[$, it can then be written that, for all $x \in \mathbb{R}$,

$$\begin{aligned}\varphi_n(x) &= \varphi(n\pi)\nu_n(x) + \sin(x)\frac{x - n\pi}{\sin(x)}\psi_n(x)\nu_n(x) \\ &= \varphi(n\pi)\nu_n(x) + \sin(x)\theta_n(x).\end{aligned}$$

So,

$$\begin{aligned}\langle T, \varphi\rangle &= \sum_{n\in\mathbb{Z}} \left[\langle T, \nu_n\rangle \varphi(n\pi) + \langle(\sin(x)T), \theta_n\rangle\right] \\ &= \sum_{n\in\mathbb{Z}} \langle T, \nu_n\rangle \varphi(n\pi).\end{aligned}$$

If we set $c_n = \langle T, \nu_n\rangle$, we correctly have: $T = \sum_{n\in\mathbb{Z}} c_n\delta_{n\pi}$.

ii) Conversely, let us assume that $T = \sum_{n\in\mathbb{Z}} c_n\delta_{n\pi}$. Since $\sin(n\pi) = 0$, we have indeed $\sin(x)T = 0$.

Conclusion: $\big(\sin x\big)T = 0$ if and only if there exists a sequence $\left(c_n\right)_{n\in\mathbb{Z}}$ of complex numbers such that $T = \sum_{n\in\mathbb{Z}} c_n\delta_{n\pi}$.

EXERCISE 4.2.– For $a \neq b$, solve, in $\mathcal{D}'(\mathbb{R})$, the following equations:

1) $(x - a)(x - b)T = 0$;

2) $(x - a)(x - b)T = 1$.

SOLUTION 4.2.–

1) We recall that $f\,\delta_a = f(a)\delta_a$ for any function f indefinitely differentiable. Therefore, $(x - a)\delta_a = 0$. Let $\varphi \in \mathcal{D}(\mathbb{R})$. According to the Taylor formula with integral remainder, there exists a function ψ indefinitely differentiable on $\mathbb{R}$ such that $\forall x \in \mathbb{R}, \quad \varphi(x) = \varphi(a) + (x - a)\psi(x)$. We then have

$$\begin{aligned}\langle T, \varphi\rangle &= \varphi(a)\,\langle T, \mathbf{1}\rangle + \langle T, (x - a)\psi\rangle \\ &= \varphi(a)\,\langle T, \mathbf{1}\rangle + \langle(x - a)T, \psi\rangle \\ &= \langle T, \mathbf{1}\rangle\,\langle\delta_a, \varphi\rangle + \langle(x - a)T, \psi\rangle\,.\end{aligned}$$

Therefore, $(x-a)T=0 \iff T \in \mathbb{R}\delta_a$. Returning to our equation, we get

$$(x-a)(x-b)T=0 \iff (x-b)T \in \mathbb{R}\delta_a.$$

We then solve the equation $(x-b)T=\delta_a$.

It is known that $(x-b)\frac{\delta_a}{a-b}=\delta_a$, thus $(x-b)\left(T-\frac{\delta_a}{a-b}\right)=0$. Then $T=\frac{\delta_a}{a-b}+c\delta_b$ where c is a real constant. It is deduced that

$$(x-a)(x-b)T=0 \Longrightarrow T \in \text{Vect}(\delta_a,\delta_b).$$

Since the reciprocal is trivial, we obtain the following equivalence:

$$(x-a)(x-b)T=0 \iff T \in \text{Vect}(\delta_a,\delta_b),$$

for $a=b$.

We have $(x-a)^2T=0 \Longrightarrow (x-a)T \in \mathbb{R}\delta_a$ and we are looking for a particular solution of $(x-a)T=\delta_a$. To this end, we should observe that $(x-a)\delta_a'=\delta_a$. Therefore, $T-\delta_a' \in \mathbb{R}\delta_a$. As such,

$$(x-a)^2T=0 \Longrightarrow T \in \text{Vect}(\delta_a,\delta_a').$$

Since the reciprocal is trivial, the following is deduced:

$$(x-a)^2T=0 \iff T \in \text{Vect}(\delta_a,\delta_a').$$

2) If we recall Question 1, solving the equation $(x-a)(x-b)T=1$ is equivalent to finding a particular solution to this equation. We then simply observe that $(x-a)\text{Vp}\left(\frac{1}{x-a}\right)=1$. A small decomposition into simple elements gives

$$\frac{1}{(x-a)(x-b)}=\frac{1}{(b-a)}\left(\frac{1}{x-b}-\frac{1}{x-a}\right).$$

Let us set

$$T_0=\frac{1}{(b-a)}\left(\text{Vp}\left(\frac{1}{x-b}\right)-\text{Vp}\left(\frac{1}{x-a}\right)\right).$$

We have

$$\begin{aligned}(x-a)(x-b)T_0 &= \frac{1}{(b-a)}\left((x-a)(x-b)\text{Vp}\left(\frac{1}{x-b}\right)\right.\\&\left.-(x-b)(x-a)\text{Vp}\left(\frac{1}{x-a}\right)\right)\\&=\frac{1}{(b-a)}((x-a)-(x-b))\\&=1.\end{aligned}$$

Consequently, $(x-a)(x-b)T = 1 \iff T \in T_0 + \text{Vect}(\delta_a, \delta_b)$.

EXERCISE 4.3.– Solve in $\mathcal{D}'(\mathbb{R})$ the equation: $xT = \text{Vp}\left(\frac{1}{x}\right)$.

SOLUTION 4.3.– The general solution of $xT = 0$ is $T = c\delta$, where c is a constant. A particular solution of the equation $xT = \text{Vp}\left(\frac{1}{x}\right)$ simply has to be obtained. From example 4.4, we have $x\text{Pf}\left(\frac{1}{x^2}\right) = \text{Vp}\left(\frac{1}{x}\right)$. Therefore, the general solution to the proposed equation is $T = c\delta + \text{Pf}\left(\frac{1}{x^2}\right)$, where c is a constant.

EXERCISE 4.4.– Let H be a Heaviside function (i.e. indicative of $\mathbb{R}_+$).

1) a) Show that $H' = \delta_0$ in $\mathcal{D}'(\mathbb{R})$.

b) Show that $(\ln(|\,x\,|))' = \text{Vp}\left(\frac{1}{x}\right)$ in $\mathcal{D}'(\mathbb{R})$.

c) Show that in $\mathcal{D}'(\mathbb{R})$,

$$\text{Vp}\left(\frac{1}{x}\right)' = \lim_{\epsilon\longrightarrow 0}\left(\int_{|x|>\epsilon}\frac{\varphi(x)}{x^2} - \frac{2\varphi(0)}{\epsilon}\right) = \text{Pf}\left(\frac{1}{x^2}\right).$$

2) Let $m \in \mathbb{N}$.

Calculate the successive derivatives in $\mathcal{D}'(\mathbb{R})$ of the function $x \longmapsto \frac{x^m}{m!}H(x)$.

SOLUTION 4.4.–

1) a) Let $\varphi \in \mathcal{D}(\mathbb{R})$. Then,

$$\begin{aligned}\langle H', \varphi\rangle &= -\langle H, \varphi'\rangle \\ &= \varphi(0) \\ &= \langle \delta_0, \varphi\rangle\end{aligned}$$

hence the result is sought for.

b) Let $\varphi \in \mathcal{D}(\mathbb{R})$. Then,

$$\begin{aligned}\left\langle (\ln \mid x \mid)', \varphi\right\rangle &= -\langle \ln \mid x \mid, \varphi'\rangle \\ &= -\lim_{\epsilon \longrightarrow 0} \int_{A \geq |x| \geq \epsilon} \ln \mid x \mid \varphi'(x)\mathrm{d}x \\ &= \lim_{\epsilon \longrightarrow 0} \int_{|x| \geq \epsilon} \frac{\varphi(x)}{x} \\ &= \left\langle \mathrm{Vp}(\frac{1}{x}), \varphi\right\rangle.\end{aligned}$$

The result is then proven.

c) Let $\varphi \in \mathcal{D}(\mathbb{R})$, $\mathrm{supp}(\varphi) \Subset [-A, A]$, $A \geq 0$. Then,

$$\begin{aligned}\left\langle \mathrm{Vp}\left(\frac{1}{x}\right)', \varphi\right\rangle &= -\left\langle \mathrm{Vp}\left(\frac{1}{x}\right), \varphi'\right\rangle \\ &= -\lim_{\epsilon \longrightarrow 0} \int_{A \geq |x| \geq \epsilon} \frac{\varphi'(x)}{x} \\ &= -\lim_{\epsilon \longrightarrow 0} \left(-\frac{\varphi(-\epsilon)}{\epsilon} - \frac{\varphi(\epsilon)}{\epsilon} + \int_{A \geq |x| \geq \epsilon} \frac{\varphi(x)}{x^2}\right).\end{aligned}$$

Now, $\varphi(\epsilon) = \varphi(0) + \epsilon\psi_1(\epsilon)$ with $\psi_1(\epsilon) \underset{\epsilon \longrightarrow 0}{\longrightarrow} 0$.

And $\varphi(-\epsilon) = \varphi(0) - \epsilon\psi_2(\epsilon)$ with $\psi_2(\epsilon) \underset{\epsilon \longrightarrow 0}{\longrightarrow} 0$.

Here, $\varphi(\epsilon) + \varphi(-\epsilon) = 2\varphi(0) + \epsilon\,(\psi_1(\epsilon) - \psi_2(\epsilon))$. Thereafter,

$$-\frac{\varphi(-\epsilon)}{\epsilon} - \frac{\varphi(\epsilon)}{\epsilon} + \frac{2\varphi(0)}{\epsilon} = (\psi_1(\epsilon) - \psi_2(\epsilon)) \underset{\epsilon \longrightarrow 0}{\longrightarrow} 0.$$

Consequently,

$$\text{Vp}\left(\frac{1}{x}\right)' = \lim_{\epsilon \longrightarrow 0}\left(\int_{|x|>\epsilon} \frac{\varphi(x)}{x^2} - \frac{2\varphi(0)}{\epsilon}\right) = \text{Pf}\left(\frac{1}{x^2}\right).$$

2) By integration by successive parts, for $\varphi \in \mathcal{D}(\mathbb{R})$ and for all $k \leq m$, we have

$$\begin{aligned}\left\langle\left(\frac{x^m}{m!}H(x)\right)^{(k)}, \varphi\right\rangle &= (-1)^k\left\langle\frac{x^m}{m!}H(x), \varphi^{(k)}\right\rangle \\ &= (-1)^k \int_0^{+\infty} \frac{x^m}{m!}\varphi^{(k)}(x)\mathrm{d}x \\ &= (-1)^k.(-1)^k \int_0^{+\infty} \frac{x^{m-k}}{(m-k)!}\varphi(x)\mathrm{d}x.\end{aligned}$$

Hence,

– for $k \leq m$, we have

$$\left(\frac{x^m}{m!}H(x)\right)^{(k)} = \frac{x^{m-k}}{(m-k)!}H(x);$$

– for $k = m+1$, we have

$$\left(\frac{x^m}{m!}H(x)\right)^{(m+1)} = \delta_0;$$

– for $k \geq m+2$, we have

$$\left(\frac{x^m}{m!}H(x)\right)^{(k)} = \delta_0^{k-m-1}.$$

EXERCISE 4.5.– Consider the function f defined by $\begin{cases} f(x) = \cos x, & x > 0, \\ f(x) = 0, & x \leq 0. \end{cases}$

Determine the derivative in the sense of the distributions of the function f and express the result using the Heaviside function $H(x)$.

SOLUTION 4.5.– Since f is piecewise of C^1 class, the jump formula is applicable. We then obtain $f' = \delta_0 - \sin x H(x)$.

EXERCISE 4.6.– Show that the sequence of distributions $\left(T_n\right)_{n\in\mathbb{N}^*}$ defined by

$\forall n \in \mathbb{N}^*$, $T_n = n\big(\delta_{1/n} - \delta_{-1/n}\big)$, converges in $\mathcal{D}'(\mathbb{R})$.

Is the order of the limit of a distribution of order m always m?

SOLUTION 4.6.– Let $\varphi \in \mathcal{D}(\mathbb{R})$. We have $\varphi(t) = \varphi(0) + t\psi(t)$ with $\psi(x) = \int_0^1 \varphi'(xu)\mathrm{d}u$.

Then for all $n \in \mathbb{N}$, we have

$$\begin{aligned}\langle T_n, \varphi\rangle &= n\left(\varphi(\frac{1}{n}) - \varphi(\frac{-1}{n})\right)\\ &= n\left(\varphi(0) + \frac{1}{n}\psi(\frac{1}{n}) - \varphi(0) - \frac{1}{n}\psi(\frac{-1}{n})\right)\\ &= \psi(\frac{1}{n}) + \psi\left(\frac{-1}{n}\right) \longrightarrow 2\psi(0) = 2\int_0^1 \varphi'(0)\mathrm{d}t = 2\varphi'(0) = \langle -2\delta_0', \varphi\rangle .\end{aligned}$$

Therefore, T_n converges to $-2\delta_0'$ in $\mathcal{D}'(\mathbb{R})$.

Now for all $n \in \mathbb{N}$, T_n is of zeroth order and δ_0' is of first order; then in general, there is no connection between the order of the elements of a sequence of distributions and the order of the limit distribution.

EXERCISE 4.7.– Determine the limit, when $\alpha \longrightarrow 0$, of the distribution defined by

$$T_a = \frac{1}{2\alpha}\left(\mathrm{Vp}\frac{1}{x-\alpha} - \mathrm{Vp}\frac{1}{x+\alpha}\right).$$

SOLUTION 4.7.– Let $\varphi \in \mathcal{D}(\mathbb{R})$. We have

$$\begin{aligned}\langle T_a, \varphi\rangle &= \left\langle \frac{1}{2\alpha}\left(\mathrm{Vp}\frac{1}{x-\alpha} - \mathrm{Vp}\frac{1}{x+\alpha}\right), \varphi\right\rangle\\ &= \frac{1}{2\alpha}\left\langle \mathrm{Vp}\frac{1}{x-\alpha}, \varphi\right\rangle - \frac{1}{2\alpha}\left\langle \mathrm{Vp}\frac{1}{x+\alpha}, \varphi\right\rangle .\end{aligned}$$

Now, the distribution $\text{Vp}\frac{1}{x\pm\alpha}$ is the translate of $\text{Vp}\left(\frac{1}{x}\right)$ by the translation: $\tau_{\pm\alpha}$, therefore,

$$\begin{aligned}\langle T_a, \varphi\rangle &= \frac{1}{2\alpha}\left\langle \text{Vp}\left(\frac{1}{x}\right), \varphi(x+\alpha)\right\rangle - \frac{1}{2\alpha}\left\langle \text{Vp}\left(\frac{1}{x}\right), \varphi(x-\alpha)\right\rangle \\ &= \left\langle \text{Vp}\left(\frac{1}{x}\right), \frac{\varphi(x+\alpha)-\varphi(x-\alpha)}{2\alpha}\right\rangle \\ &= \left\langle \text{Vp}\left(\frac{1}{x}\right), \Phi_\alpha\right\rangle.\end{aligned}$$

where $\Phi_\alpha = \dfrac{\varphi(x+\alpha)-\varphi(x-\alpha)}{2\alpha} \in \mathcal{D}(\mathbb{R})$. We have

$$\lim_{\alpha\longrightarrow 0}\Phi_\alpha = \lim_{\alpha\longrightarrow 0}\frac{1}{2}\left(\frac{\varphi(x+\alpha)-\varphi(x)}{\alpha} + \frac{\varphi(x-\alpha)-\varphi(x)}{-\alpha}\right) = \varphi'(x).$$

By reiterating the same procedure for the derivatives of Φ_α, we show that Φ_α tends toward φ' in the sense of the convergence in $\mathcal{D}(\mathbb{R})$. Since $\text{Vp}\left(\frac{1}{x}\right)$ is continuous on $\mathcal{D}(\mathbb{R})$, then

$$\begin{aligned}\lim_{\alpha\longrightarrow 0}\langle T_a, \varphi\rangle &= \lim_{\alpha\longrightarrow 0}\left\langle \text{Vp}\left(\frac{1}{x}\right), \Phi_\alpha\right\rangle \\ &= \left\langle \text{Vp}\left(\frac{1}{x}\right), \varphi'\right\rangle \\ &= -\left\langle \left(\text{Vp}\left(\frac{1}{x}\right)\right)', \varphi\right\rangle \\ &= \left\langle \text{Pf}\frac{1}{x^2}, \varphi\right\rangle.\end{aligned}$$

Consequently, $\lim_{\alpha\longrightarrow 0} T_a = \text{Pf}(\frac{1}{x^2})$.

EXERCISE 4.8.– Solve in $\mathcal{D}'(\mathbb{R})$, the differential equation: $xT' + T = 0$.

SOLUTION 4.8.– We have $(xT)' = xT' + T$. Therefore, $xT' + T = 0 \Leftrightarrow (xT)' = 0$. Then, there exists $\lambda \in \mathbb{R}$ such that $xT = \lambda$. However, $\lambda\text{Vp}(\frac{1}{x})$ is a solution of the latter equation, so

$$xT = \lambda \iff x\left(T - \lambda\text{Vp}\left(\frac{1}{x}\right)\right) = 0.$$

Therefore, there exists $\lambda, \alpha \in \mathbb{R}$, such that: $T = \lambda \text{Vp}\left(\frac{1}{x}\right) + \alpha\delta_0$.

EXERCISE 4.9.– Solve in the space of distributions the following differential equations:

1) $u' + xu = \delta$;

2) $u' + u = H$;

3) $u' + e^{-x}u = \delta$.

SOLUTION 4.9.–

1) We have

$$\begin{aligned}
u' + xu = \delta &\iff (u' + xu)\, e^{\frac{x^2}{2}} = e^{\frac{x^2}{2}}\delta \\
&\iff \frac{\mathrm{d}}{\mathrm{d}x}\left(e^{\frac{x^2}{2}}u\right) = e^{\frac{x^2}{2}}\delta \\
&\iff \frac{\mathrm{d}}{\mathrm{d}x}\left(e^{\frac{x^2}{2}}u\right) = \delta \\
&\iff e^{\frac{x^2}{2}}u(x) = H(x) + C \\
&\iff u(x) = e^{\frac{-x^2}{2}}H(x) + Ce^{\frac{-x^2}{2}}
\end{aligned}$$

with C a real constant.

2) We have

$$\begin{aligned}
u' + u = H &\iff (u' + u)\, e^{x} = e^{x}H(x), \forall x \in \mathbb{R} \\
&\iff \frac{\mathrm{d}}{\mathrm{d}x}\left(e^{x}u\right) = e^{x}H(x), \forall x \in \mathbb{R} \\
&\iff e^{x}u = \int_{-\infty}^{x} e^{t}H(t)\mathrm{d}t + C, \forall x \in \mathbb{R} \\
&\iff u(x) = \left(\int_{-\infty}^{x} e^{t}H(t)\mathrm{d}t + C\right)e^{-x}, \forall x \in \mathbb{R}
\end{aligned}$$

with C a real constant.

However, $\begin{cases} \displaystyle\int_{-\infty}^{x} e^t H(t) \mathrm{d}t = 0, & x < 0, \\ \displaystyle\int_{-\infty}^{x} e^t H(t) \mathrm{d}t = e^x - 1, & x \geq 0, \end{cases}$

then $u' + u = \delta \iff u(x) = e^{-x} \left(H(x) \left(e^x - 1 \right) + C \right), \forall x \in \mathbb{R}.$

3) We have

$$\begin{aligned} u' + e^{-x} u = \delta' &\iff (u' + u) \, e^{-e^{-x}} = e^{-e^{-x}} \delta' \\ &\iff \frac{\mathrm{d}}{\mathrm{d}x} \left(e^{-e^{-x}} u \right) = e^{-e^{-x}} \delta'. \end{aligned}$$

Since $a(x)\delta' = -a'(0)\delta + a(0)\delta'$, then $e^{e^{-x}} \delta' = -e^{-1}\delta + e^{-1}\delta'$. Consequently,

$$\begin{aligned} u' + e^{-x} u = \delta' &\iff \frac{\mathrm{d}}{\mathrm{d}x} \left(e^{-e^{-x}} u \right) = -e^{-1}\delta + e^{-1}\delta' \\ &\iff u(x) = e^{e^{-x}} \left(-e^{-1} H(x) + e^{-1}\delta + C \right), \end{aligned}$$

with C a real constant.

EXERCISE 4.10.– In the $\mathbb{R}^2$ plane, we call Heaviside function to the function H defined by: $\begin{cases} H(x,y) = 1, & x > 0 \text{ and } y > 0, \\ H(x,y) = 0, & \text{elsewhere.} \end{cases}$

In the sense of distributions, we have $\dfrac{\partial^2 H}{\partial x \partial y} = \delta_{(0,0)}$.

SOLUTION 4.10.– Let $\varphi \in \mathcal{D}(\mathbb{R}^2)$. We have

$$\begin{aligned}
\left\langle \frac{\partial^2 H}{\partial x \partial y}, \varphi \right\rangle &= \left\langle H, \frac{\partial^2 \varphi}{\partial x \partial y} \right\rangle \\
&= \int_{\mathbb{R}^2} H(x,y) \frac{\partial^2 \varphi}{\partial x \partial y}(x,y) \mathrm{d}x \mathrm{d}y \\
&= \int_0^{+\infty} \int_0^{+\infty} \frac{\partial^2 \varphi}{\partial x \partial y}(x,y) \mathrm{d}x \mathrm{d}y \\
&= \int_0^{+\infty} \left(\int_0^{+\infty} \frac{\partial^2 \varphi}{\partial x \partial y}(x,y) \mathrm{d}x \right) \mathrm{d}y \\
&= -\int_0^{+\infty} \frac{\partial \varphi}{\partial y}(0,y) \mathrm{d}y \\
&= \varphi(0,0) \\
&= \left\langle \delta_{(0,0)}, \varphi \right\rangle .
\end{aligned}$$

Consequently, $\dfrac{\partial^2 H}{\partial x \partial y} = \delta_{(0,0)}$.

EXERCISE 4.11.–

1) Let p and q represent two natural numbers.
Compute the distribution: $T = x^p \delta^{(q)}$ where $\delta^{(i)}$ is the ith derivative of the Dirac measure on $\mathbb{R}$.

2) Solve the equation $xU' = 0$ in $\mathcal{D}'(\mathbb{R})$.

3) Solve the equation $x^m U = 1$ in $\mathcal{D}'(\mathbb{R})$.

SOLUTION 4.11.–

1) Since $x^p \in \mathcal{C}^\infty(\mathbb{R})$, then $x^p \delta^{(q)}$ makes sense. Let $\varphi \in \mathcal{D}(\mathbb{R})$:

$$\left\langle x^p \delta^{(q)}, \varphi \right\rangle = \left\langle (-1)^q \delta, (x^p \varphi)^{(q)} \right\rangle = (-1)^q \left(x^p \varphi \right)^{(q)} (0).$$

According to the Leibniz rule,

$$(x^p \varphi)^{(q)} = \sum_{i=0}^{q} \binom{q}{i} (x^p)^{(i)} \varphi^{(q-i)} = \sum_{i=0}^{q} F_{i,q}(x).$$

Two cases can be distinguished:

– If $P > q$, then $(x^p\varphi)^{(q)}(0) = 0$. Actually,

$$(x^p)^{(i)} = p(p-1)\dots(p-i+1)x^{p-i}$$

and $p - i$ is strictly positive.

– If $P \leq q$, then

$$\sum_{i=0}^{q} F_{i,q}(x) = \sum_{i=0}^{p-1} F_{i,q}(x) + \sum_{i=p}^{q} F_{i,q}(x).$$

The first of the two sums cancels out at the origin for the same reason as above and the second is written as

$$\sum_{i=p}^{q} F_{i,q}(x) = \binom{q}{p} (x^p)^{(p)}\varphi^{(q-p)}$$

because $(x^p)^{(i)}(0) = 0$ for $i > p+1$.

In this case,

$$\left\langle x^p\delta^{(q)}, \varphi\right\rangle = (-1)^q \binom{q}{p} p!\varphi^{(q-p)}(0) = \frac{(-1)^p q!}{(q-p)!}\left\langle \delta^{(q-p)}, \varphi\right\rangle.$$

Conclusion:

$$x^p\delta^{(q)} = \begin{cases} 0, & p > q, \\ \dfrac{(-1)^p q!}{(q-p)!}\delta^{(q-p)}, & p \leq q \end{cases}.$$

2) We must have $u' = c\delta$ with c a constant. Let H denote the Heaviside function, then $H' = \delta$, that is to say, H is a particular solution to the equation $u' = \delta$. The general solution of $xu' = 0$ is $\{cH + d\delta, (c,d) \in \mathbb{R}^2\}$.

3) It is already known that the solutions of the equation $x^m u = 0$ are given by $\sum_{k=0}^{m-1} c_k\delta^{(k)}$. We should simply find a particular solution to the equation: $x^m u = 1$. If $m = 1$, we know that $\mathrm{vp}(\frac{1}{x})$ is appropriate. In general, we can construct a solution

of $x^{m+1}u = 1$ from a solution u_m to $x^m u = 1$. We derive the equation $x^m u_m = 1$, which gives

$$mx^{m-1}u_m + x^m u_m' = 0,$$

then multiply by x, which yields

$$0 = mx^m u_m + x^{m+1}u_m' = m + x^{m+1}u_m'.$$

Then $u_{m+1} = -\frac{1}{m}u_m'$ is a particular solution to the equation $x^{m+1}u = 1$.

In summary, we find as a particular solution

$$u_1 = \text{Vp}(\frac{1}{x})$$

$$u_2 = -\left(\text{Vp}(\frac{1}{x})\right)'$$

$$u_m = \frac{(-1)^{m-1}}{(m-1)!}\left(\text{Vp}(\frac{1}{x})\right)^{(m-1)}.$$

Knowing that $(\ln|x|)' = \text{Vp}(\frac{1}{x})$, we can also write

$$u_m = \frac{(-1)^{m-1}}{(m-1)!}(\ln|x|)^{(m)}.$$

EXERCISE 4.12.– Compute, in the sense of distributions, the following derivatives:

1) $\left(\frac{\mathrm{d}}{\mathrm{d}x} - \lambda\right) H(x).e^{\lambda x}$;

2) $\left(\frac{\mathrm{d}^2}{\mathrm{d}x^2} + \omega^2\right) \frac{H(x)\sin(\omega x)}{\omega}$;

3) $\frac{\mathrm{d}^m}{\mathrm{d}x^m}\frac{H(x)x^{m-1}}{(m-1)!}$, for $m \in \mathbb{N}^*$.

SOLUTION 4.12.–

1) Let $\varphi \in \mathcal{D}(\mathbb{R})$. We have

$$\begin{aligned}\left\langle \frac{\mathrm{d}}{\mathrm{d}x}\left(H(x)e^{\lambda x}\right), \varphi \right\rangle &= -\left\langle \left(H(x)e^{\lambda x}\right), \varphi' \right\rangle \\ &= -\int_{-\infty}^{+\infty} H(x)e^{\lambda x}\varphi'(x)\mathrm{d}x \\ &= -\int_{0}^{+\infty} e^{\lambda x}\varphi'(x)\mathrm{d}x \\ &= \varphi(0) + \lambda \int_{0}^{+\infty} e^{\lambda x}\varphi(x)\mathrm{d}x.\end{aligned}$$

Consequently:

$$\left(\frac{\mathrm{d}}{\mathrm{d}x} - \lambda\right) H(x).e^{\lambda x} = \delta.$$

Otherwise, by simply applying the differentiation formula we obtain

$$\frac{\mathrm{d}}{\mathrm{d}x}\left(H(x).e^{\lambda x}\right) = \frac{\mathrm{d}}{\mathrm{d}x}\left(H(x)\right).e^{\lambda x} + \lambda H(x)e^{\lambda x} = \delta.e^{\lambda x} + \lambda H(x)e^{\lambda x}.$$

2) Let $\varphi \in \mathcal{D}(\mathbb{R})$. We have

$$\begin{aligned}\left\langle \frac{\mathrm{d}^2}{\mathrm{d}x^2}\frac{H(x)\sin(\omega x)}{\omega}, \varphi \right\rangle &= \left\langle \frac{H(x)\sin(\omega x)}{\omega}, \varphi'' \right\rangle \\ &= \frac{1}{\omega}\int_{0}^{+\infty} \sin(\omega x)\varphi''(x)\mathrm{d}x \\ &= \varphi(0) - \omega\int_{0}^{+\infty} \sin(\omega x)\varphi(x)\mathrm{d}x \\ &= \langle \delta, \varphi \rangle - \omega^2 \left\langle \frac{H(x)\sin(\omega x)}{\omega}, \varphi \right\rangle.\end{aligned}$$

Consequently,

$$\left(\frac{\mathrm{d}^2}{\mathrm{d}x^2} + \omega^2\right)\frac{H(x)\sin(\omega x)}{\omega} = \delta.$$

3) For $m = 1$, we have $\dfrac{\mathrm{d}}{\mathrm{d}x}H(x) = \delta$.

For $m \geq 2$, we have

$$\frac{\mathrm{d}}{\mathrm{d}x}\frac{H(x)x^{m-1}}{(m-1)!} = \frac{x^{m-1}}{(m-1)!}\delta + \frac{x^{m-2}}{(m-2)!}H(x) = \frac{x^{m-2}}{(m-2)!}H(x).$$

Set by set, we get

$$\frac{\mathrm{d}^2}{\mathrm{d}x^2}\frac{H(x)x^{m-1}}{(m-1)!} = \frac{x^{m-3}}{(m-3)!}H(x), \ldots, \frac{\mathrm{d}^{m-2}}{\mathrm{d}x^{m-2}}\frac{H(x)x^{m-1}}{(m-1)!} = xH(x).$$

Then,

$$\frac{\mathrm{d}^{m-1}}{\mathrm{d}x^{m-1}}\frac{H(x)x^{m-1}}{(m-1)!} = x\delta + H(x).$$

Consequently,

$$\frac{\mathrm{d}^m}{\mathrm{d}x^m}\frac{H(x)x^{m-1}}{(m-1)!} = \delta.$$

It should be noted that a more general result is obtained in exercise 4.4 using the definition.

EXERCISE 4.13.– Show that the function f defined by

$$\begin{cases} f(x) = 0, \ x < 0, \\ f(x) = \frac{1}{\sqrt{x}}, \ x > 0 \end{cases}$$

is a distribution, then calculate its derivative.

SOLUTION 4.13.– The function f is locally integrable, therefore it defines a distribution. Its quasi-derivative equals 0 on $]-\infty, 0[$, and $-\dfrac{1}{2}x^{-\frac{3}{2}}$ on $]0, +\infty[$, which is not locally integrable. We are thus going to regularize it. We have

$$\begin{aligned} \langle f', \varphi \rangle &= -\langle f, \varphi' \rangle \\ &= -\int_0^{+\infty} \frac{\varphi'(x)}{\sqrt{x}}\mathrm{d}x \\ &= -\lim_{\varepsilon \longrightarrow 0}\int_{\varepsilon}^{+\infty} \frac{\varphi'(x)}{\sqrt{x}}\mathrm{d}x. \end{aligned}$$

If we integrate by parts,

$$-\int_{\varepsilon}^{+\infty}\frac{\varphi'(x)}{\sqrt{x}}\mathrm{d}x=\frac{\varphi(\varepsilon)}{\sqrt{\varepsilon}}+\int_{\varepsilon}^{+\infty}\frac{-\varphi(x)}{2x^{\frac{3}{2}}}\mathrm{d}x.$$

Since $\varphi(\varepsilon)=\varphi(0)+O(\varepsilon)$ when $\varepsilon\longrightarrow 0$, we finally have

$$\langle f',\varphi\rangle=\lim_{\varepsilon\longrightarrow 0}\left(\frac{\varphi(0)}{\sqrt{\varepsilon}}+\int_{\varepsilon}^{+\infty}\frac{-\varphi(x)}{2x^{\frac{3}{2}}}\mathrm{d}x\right).$$

EXERCISE 4.14.– For $\varphi\in\mathcal{D}(\mathbb{R}^2)$, we consider two linear forms on $\mathcal{D}(\mathbb{R}^2)$ defined by: $\langle T,\varphi\rangle=\int_{\mathbb{R}}\varphi(x,x)\mathrm{d}x$ and $\langle T^+,\varphi\rangle=\int_0^{+\infty}\varphi(x,x)\mathrm{d}x$.

1) a) Show that $T\in\mathcal{D}'(\mathbb{R}^2)$ and that $T^+\in\mathcal{D}'(\mathbb{R}^2)$.

b) Calculate $\partial_x T+\partial_t T$ and $\partial_x T^+ + \partial_t T^+$.

2) Let $\mathbb{D}=\left\{(x,t)\in\mathbb{R}^2\ /\ t\geq\mid x\mid\right\}$.

We denote by E the regular distribution associated with $\mathbb{1}_{\mathbb{D}}$ indicative of $\mathbb{D}$.

Calculate $\partial_t E-\partial_x E$; then, deduce therefrom $\partial_t^2 E-\partial_x^2 E$.

SOLUTION 4.14.–

1) Let $A>0$.

a) For any $\varphi\in\mathcal{D}(\mathbb{R}^2)$ such that $\operatorname{supp}\varphi\Subset[-A,A]^2$, we have

$$|\langle T,\varphi\rangle|\leq 2A\|\varphi\|_\infty \quad\text{and}\quad \left|\langle T^+,\varphi\rangle\right|\leq 2A\|\varphi\|_\infty$$

Consequently, $T\in\mathcal{D}'(\mathbb{R}^2)$ and $T^+\in\mathcal{D}'(\mathbb{R}^2)$; moreover, they are of zeroth order.

b) Let $\varphi\in\mathcal{D}(\mathbb{R}^2)$ such that $\operatorname{supp}\varphi\Subset[-A,A]^2$, we have

$$\begin{aligned}\langle\partial_x T,\varphi\rangle&=-\langle T,\partial_x\varphi\rangle\\&=-\int_{-A}^{A}\partial_x\varphi(x,x)\mathrm{d}x\\&=-\int_{-A}^{A}\partial_x\left(\varphi(x,x)\right)\mathrm{d}x+\int_{-A}^{A}\partial_t\varphi(x,x)\mathrm{d}x\\&=0+\langle T,\partial_t\varphi\rangle\\&=-\langle\partial_t T,\varphi\rangle.\end{aligned}$$

Then, $\partial_x T + \partial_t T = 0$.

In the same way, we show that $\partial_x T^+ + \partial_t T^+ = \delta_{(0,0)}$. In reality,

$$\begin{aligned}
\left\langle \partial_x T^+, \varphi \right\rangle &= -\left\langle T^+, \partial_x \varphi \right\rangle \\
&= -\int_0^A \partial_x \varphi(x,x) \mathrm{d}x \\
&= -\int_0^A \partial_x \left(\varphi(x,x) \right) \mathrm{d}x + \int_0^A \partial_t \varphi(x,x) \mathrm{d}x \\
&= \varphi(0,0) + \left\langle T^+, \partial_t \varphi \right\rangle \\
&= \varphi(0,0) - \left\langle \partial_t T^+, \varphi \right\rangle \\
&= \left\langle \delta_{(0,0)} - \partial_t T^+, \varphi \right\rangle .
\end{aligned}$$

The result sought for can be deduced therefrom.

2) Let $A > 0$. For any $\varphi \in \mathcal{D}(\mathbb{R}^2)$ such that $\operatorname{supp} \varphi \Subset [-A, A]^2$, we have

$$\begin{aligned}
\langle \partial_t E - \partial_x E, \varphi \rangle &= -\langle E, \partial_t \varphi - \partial_x \varphi \rangle \\
&= -\int\int_{t \geq |x|} \partial_t \varphi(t,x) - \partial_x \varphi(t,x) \mathrm{d}t \mathrm{d}x \\
&= -\int_{x \in \mathbb{R}} \Big(\int_{|x|}^{+\infty} \partial_t \varphi(t,x) \mathrm{d}t \Big) \mathrm{d}x - \int_{t \geq 0} \Big(\int_{-t}^{t} \partial_x \varphi(t,x) \mathrm{d}x \Big) \mathrm{d}t \\
&= -\int_{x \in \mathbb{R}} \varphi(|x|, x) \mathrm{d}x - \int_0^{+\infty} \left(\varphi(t,t) - \varphi(t,-t) \right) \mathrm{d}t \\
&= 2 \left\langle T^+, \varphi \right\rangle .
\end{aligned}$$

Therefore, $\partial_t E - \partial_x E = 2T^+$.

On the other hand, we have

$$\begin{aligned}
\left\langle \partial_t^2 E - \partial_x^2 E, \varphi \right\rangle &= \langle \partial_t E - \partial_x E, (\partial_t + \partial_x) \varphi \rangle \\
&= -2 \left\langle T^+, (\partial_t + \partial_x) \varphi \right\rangle \\
&= 2 \left\langle (\partial_t + \partial_x) T^+, \varphi \right\rangle \\
&= \left\langle 2\delta_{(0,0)}, \varphi \right\rangle .
\end{aligned}$$

Consequently, $\partial_t^2 E - \partial_x^2 E = 2\delta_{(0,0)}$.

EXERCISE 4.15.–

1) Show that the function f defined by $f(x,y) = \dfrac{1}{x+iy}$ is an element of $\mathcal{D}'(\mathbb{R}^2)$.

2) The Cauchy–Riemann operator is defined by $\tilde{\partial} = \dfrac{1}{2}\left(\dfrac{\partial}{\partial x} + i\dfrac{\partial}{\partial y}\right)$. Show that $\tilde{\partial} f = \pi\delta$.

SOLUTION 4.15.–

1) First of all,

$$|f(x,y)| = \frac{1}{(x^2+y^2)^{\frac{1}{2}}} = \frac{1}{|X|} \quad \text{where} \quad X = (x,y).$$

We know that the function $\dfrac{1}{|X|^\alpha}$ is integrable in the neighborhood of the origin in $\mathbb{R}^d$ if $\alpha < d$. Here, $\alpha = 1$ and $d = 2$ therefore f is $L^1_{loc}(\mathbb{R}^2)$, which implies that f defines an element of $\mathcal{D}'(\mathbb{R}^2)$.

2) Let $\varphi \in \mathcal{D}(\mathbb{R}^2)$. We have

$$\langle \overline{\partial} f, \varphi \rangle = -\langle f, \overline{\partial}\varphi \rangle = -\frac{1}{2}\int_{\mathbb{R}^2} \frac{1}{x+iy}\left(\frac{\partial f}{\partial x} + i\frac{\partial f}{\partial y}\right) \mathrm{d}x\mathrm{d}y.$$

If we shift to polar coordinates: $x = r\cos\theta$ and $y = r\sin\theta$. We have

$$\mathrm{d}x\mathrm{d}y = r\mathrm{d}r\mathrm{d}\theta$$

$$\frac{\partial}{\partial x} = \cos\theta\frac{\partial}{\partial r} - \frac{\sin\theta}{r}\frac{\partial}{\partial\theta}$$

$$\frac{\partial}{\partial y} = \sin\theta\frac{\partial}{\partial r} + \frac{\cos\theta}{r}\frac{\partial}{\partial\theta}.$$

Then

$$\langle \overline{\partial} f, \varphi \rangle = -\frac{1}{2}\int_0^{2\pi}\int_0^{+\infty} \frac{e^{-i\theta}}{r}\left(e^{i\theta}\frac{\partial\tilde{\varphi}}{\partial r} + i\frac{e^{-i\theta}}{r}\frac{\partial\tilde{\varphi}}{\partial\theta}\right) r\mathrm{d}r\mathrm{d}\theta,$$

where $\tilde{\varphi}(r,\theta) = \varphi(r\cos\theta, r\sin\theta)$.

According to Fubini's theorem,

$$\langle \overline{\partial} f, \varphi \rangle = -\frac{1}{2} \int_0^{2\pi} \left(\int_0^{+\infty} \frac{\partial \tilde{\varphi}}{\partial r} \mathrm{d}r \right) \mathrm{d}\theta - \frac{i}{2} \int_0^{+\infty} \frac{1}{r} \int_0^{2\pi} \left(\frac{\partial \tilde{\varphi}}{\partial \theta} \mathrm{d}\theta \right) \mathrm{d}r.$$

Since $\tilde{\varphi}(0,0) = \varphi(0,0)$ and $\theta \longmapsto \tilde{\varphi}(r,\theta)$ is $2\pi-$periodic, we get

$$\langle \overline{\partial} f, \varphi \rangle = -\frac{1}{2} \times 2\pi \times (-\varphi(0,0)) = \pi\varphi(0,0) = \langle \pi\delta, \varphi \rangle .$$

Consequently, $\overline{\partial} f = \pi\delta$.

EXERCISE 4.16.– In the sense of distributions, solve the following equations:

1) $T' - \lambda T = \delta$;

2) $T'' - 3T' + 2T = \delta$;

3) $T'' - w^2 T = \delta$ and $T'' + w^2 T = \delta$.

SOLUTION 4.16.–

1) We get

$$\begin{aligned} T' - \lambda T = \delta &\iff e^{-\lambda x}\left(T' - \lambda T\right) = e^{-\lambda x}\delta, \forall x \in \mathbb{R} \\ &\iff \frac{\mathrm{d}}{\mathrm{d}x}\left(e^{-\lambda x} T\right) = \delta, \forall x \in \mathbb{R} \\ &\iff e^{-\lambda x} T = H + c, \forall x \in \mathbb{R} \\ &\iff T = e^{\lambda x} H + c e^{\lambda x}, \forall x \in \mathbb{R}, \end{aligned}$$

where c is a constant.

2) The operator $D^2 - 3D + 2I = (D - 2I) \circ (D - I)$ prompts us to introduce $S = T' - T$. Then $S' - 2S = \delta$, therefore $S(x) = e^{2x}\left(H(x) + c\right)$. We then obtain

$$\begin{aligned} T' - T = e^{2x}\left(H(x) + c\right) &\iff e^{-x}(T' - T) = e^{x}\left(H(x) + c\right) \\ &\iff \frac{\mathrm{d}}{\mathrm{d}x}(e^{-x} T) = e^{x}\left(H(x) + c\right). \end{aligned}$$

Let us look for a primitive of $e^x H(x) = H(x) + (e^x - 1)H(x)$. A primitive of $H(x)$ is $xH(x)$. A primitive of the continuous function $(e^x - 1)H(x)$ is

$$\int_{-\infty}^{x} (e^t - 1)H(t)dt = 0 \text{ if } x < 0 \text{ and } e^x - x - 1 \text{ if } x \geq 0$$

In summary, a primitive of $e^x H(x)$ is the function f defined by

$$\begin{cases} f(x) = 0, & x < 0, \\ f(x) = e^x - 1, & x \geq 0. \end{cases}$$

Otherwise, $f = (e^x - 1)\, H(x)$.

Therefore, $e^{-x}T = (e^x - 1)H(x) + ce^x + d$. Consequently,

$$T = \left(e^{2x} - e^x\right) H(x) + ce^{2x} + de^x, \forall x \in \mathbb{R},$$

where c and d are constants.

3) These are second-order equations with constant coefficients. Let us look for a particular solution to the equation $T'' - w^2 T = \delta$ in the form $T(x) = H(x)f(x)$ where f is a $\mathcal{C}^2$-class function to be determined.

We have

$$\begin{aligned} T'' - w^2 T = \delta &\iff H\left(f'' - w^2 f\right) + f'(0)\delta + f(0)\delta' = \delta \\ &\iff f'' - w^2 f = 0, f'(0) = 1, f(0) = 0 \end{aligned}$$

Then, $f(x) = \dfrac{\operatorname{sh}(\omega x)}{\omega}$. The thing that remains is to add the general solution of the associated homogeneous equation:

$$T(x) = A\operatorname{ch}(\omega x) + B\operatorname{sh}(\omega x) + H(x)\frac{\operatorname{sh}(\omega x)}{\omega}, \forall x \in \mathbb{R}.$$

The same procedure is followed for the other equation.

5

Distribution Support

We do not know how to make sense of the convolution product for any pair of distributions unless at least we have a compact support, hence the interest in defining the support of a distribution and giving its fundamental properties. In this chapter, Ω denotes an open set of $\mathbb{R}^d$.

5.1. Distribution restriction and extension

5.1.1. *Unit partitions*

We start by recalling a lemma ensuring the existence of a plateau-type function.

LEMMA 5.1.– *Let $a \in \mathbb{R}^d$ and $r > 0$.*

There exits a function $\varphi \in \mathcal{D}(\Omega)$ such that

i) supp $\varphi \subset B(a, 2r)$;

ii) $0 \leq \varphi \leq 1$;

iii) $\varphi|_{B(a,r)} = 1$.

DEFINITION 5.1.– *Let K be a compact such that $K \subset \Omega$ and $K \subset \bigcup_{j=1}^{p} \Omega_j$ with $(\Omega_j)_{1 \leq j \leq p}$ a finite family of open sets included in Ω. A partition of the unit subordinated to the open cover $\bigcup_{j=1}^{p} \Omega_j$ of K is a sequence of functions $(\chi_j)_{1 \leq j \leq p}$ in $\mathcal{D}(\Omega)$ such that* *i)* $0 \leq \chi_j \leq 1, \quad \forall j \in \{1, \cdots, p\}$;

ii) supp $(\chi_j) \subset \Omega_j, \quad \forall j \in \{1, \cdots, p\}$;

iii) there exists a neighborhood V *of* K *such that* $\forall x \in V, \sum_{j=1}^{p} \chi_j(x) = 1$.

We give a technical lemma that will be very useful later on; it is the lemma of unit partitions. This lemma is a tool that allows us shifting from local to global.

LEMMA 5.2.– *For all* K *a compact of* Ω *and for any open cover* $\bigcup_{j=1}^{p} \Omega_j$ *of* K*, there exists a partition of the unit subordinate to the cover* $\bigcup_{j=1}^{p} \Omega_j$.

Proof. First, if S is a compact included in an open set U of $\mathbb{R}^d$, then there exists an open subset $V \subset \overline{V} \subset U$ such that $S \subset V$ and $\overline{V}$ is compact. We actually simply have to construct $V = \{x \in U \big| d(x, S) < \delta/2\}$ where $\delta = \min_{x \in S} d(x, U^c)$.

Let $S_1 = K \backslash (\Omega_2 \cup \cdots \cup \Omega_p)$. Then, S_1 is closed, bounded and therefore compact. Moreover, $S_1 \subset \Omega_1$. Therefore, there exists an open subset V_1 of compact adherence such that $S_1 \subset V_1$ and $\overline{V_1} \subset \Omega$. Consequently, $K \subset S_1 \cup \Omega_2 \cup \cdots \cup \Omega_p$, from which $K \subset V_1 \cup \Omega_2 \cup \cdots \cup \Omega_p$. We set $S_2 = K \backslash (\Omega_1 \cup \Omega_3 \cup \cdots \Omega_p) \subset \Omega_2$ and in the same way we built V_2 an open neighborhood of compact adhesion of S_2 such that $\overline{V_2} \subset \Omega_2$. Then, $K \subset \Omega_1 \cup V_2 \cup \Omega_3 \cup \cdots \cup \Omega_p$, from where by intersection, $K \subset V_1 \cup V_2 \cup \Omega_3 \cup \cdots \cup \Omega_p$.

Therefore, we construct by recurrence a sequence of open subsets $V_1, \cdots V_p$ with compact adherence such that for all j, $K \subset V_1 \cup \cdots \cup V_j \cup \Omega_{j+1} \cup \cdots \cup \Omega_p$. For any $j \in \{1, ..., p\}$, let $\tilde{\chi_j} \in \mathcal{D}(\Omega)$ such that supp $(\tilde{\chi_j}) \subset \Omega_j$, $\tilde{\chi_j} = 1$ on $\overline{V_j}$ and $0 \leq \tilde{\chi_j} \leq 1$. Finally, we set $\chi_1 = \tilde{\chi_1}$, $\chi_2 = \tilde{\chi_1}(1 - \tilde{\chi_2})$ and $\chi_j = (1 - \tilde{\chi_2}) \cdots (1 - \tilde{\chi_j})$.

The functions χ_j prove useful because we have, in addition,

$$1 - \sum_{j=1}^{p} \chi_j = \prod_{j=1}^{p} (1 - \tilde{\chi_j}) = 0 \text{ on } K. \qquad \square$$

5.1.2. *Distribution localization and recollement*

DEFINITION 5.2.– *Let $w \subset \Omega$ be an open subset and let $T \in \mathcal{D}'(\Omega)$.*

For $\varphi \in \mathcal{D}(w)$, we define $\tilde{\varphi} \in \mathcal{D}(\Omega)$ which is equal to φ on w and at 0 on $\Omega \setminus w$. Then, the linear form $T|_w$ defined on $\mathcal{D}(w)$ by

$$\forall \varphi \in \mathcal{D}(w), \langle T|_w, \varphi \rangle = \langle T, \tilde{\varphi} \rangle$$

is a distribution on w called the restriction of T to w.

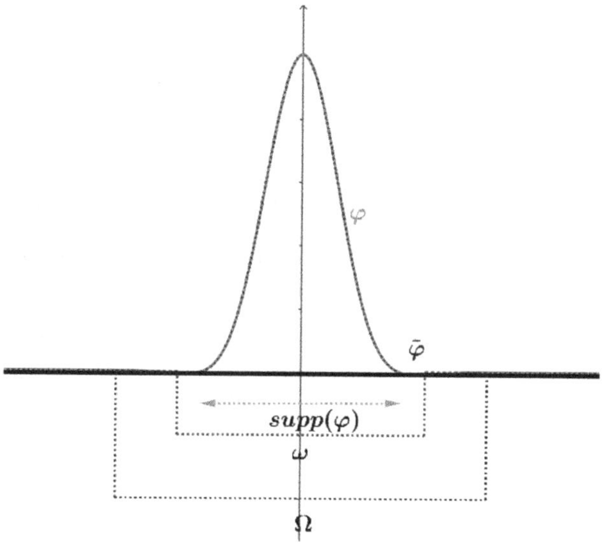

Figure 5.1. *Extension of the function $\varphi \in \mathcal{D}(w)$. For a color version of this figure, see www.iste.co.uk/aitbenhassi/distribution.zip*

It is clear by coupling, since $\operatorname{supp} \varphi \subset w \subset \Omega$ is compact, that $\tilde{\varphi} \in \mathcal{D}(\Omega)$. Therefore, the definition is well established.

DEFINITION 5.3.– *Let $T \in \mathcal{D}'(\Omega)$, and let $w \subseteq \Omega$ be an open subset. It is said that T is null in w if $T|_w = 0$.*

REMARK 5.1.– *We can also write*

$$T|_w = 0 \iff \langle T, \varphi \rangle = 0, \forall \varphi \in \mathcal{D}(\Omega) \text{ such that supp } \varphi \subset w.$$

EXAMPLE 5.1.–

1) Let $T = \delta \in \mathcal{D}'(\mathbb{R})$ and $w =]1, 2[\subset \mathbb{R}$. Let $\varphi \in \mathcal{D}(\mathbb{R})$ such that $\operatorname{supp} \varphi \subset w$. We have $\langle \delta, \varphi \rangle = \varphi(0) = 0$. Therefore, $T|_w = 0$.

2) Let $T = H \in \mathcal{D}'(\mathbb{R})$ is the Heaviside distribution. We have $T|_{\mathbb{R}^*_-} = 0$.

3) Let $f \in \mathcal{D}(\mathbb{R})$ such that $\operatorname{supp} f \subset [a, b]$. We have $T_f|_{]-\infty,a[} = T_f|_{]b,+\infty[} = 0$.

We then have a result from shifting from local to global for this notion of local nullity for a distribution.

LEMMA 5.3.– *Let $(w_i)_{i\in I}$ be a family of open sets of Ω and $w := \bigcup_{i\in I} w_i$ their union. Let $T \in \mathcal{D}'(\Omega)$ such that for all $i \in I$, $T|_{w_i} = 0$. Then $T|_w = 0$.*

Proof. We must show that for any $\varphi \in \mathcal{D}(w)$, $\langle T, \varphi \rangle = 0$.

Let thus $\varphi \in \mathcal{D}(w)$ and $K = \operatorname{supp} \varphi$. Since $K \subset \bigcup_{i\in I} w_i$, we can extract from this open cover of K a finite sub-cover indexed by finite $J \subset I$ (Borel–Lebesgue property). Let then $(\chi_i)_{i\in J}$ be a partition of the unit relative to the cover $(w_i)_{i\in J}$ of K. Then for all $i \in J$, $\chi_i \in \mathcal{D}(w_i)$ and $\sum_{i\in J} \chi_i(x) = 1$ for $x \in \bigcup_{i\in J} w_i$. Since φ has support in $K \subset \bigcup_{i\in J} w_i$, we have $\varphi = \sum_{j\in J} \chi_j \varphi$ and thus,

$$\langle T, \varphi \rangle = \sum_{j\in J} \langle T, \chi_i \varphi \rangle .$$

Since, for all $i \in J$, $\chi_i \varphi \in \mathcal{D}(w_i)$ and $T|_{w_i} = 0$, then $\langle T, \chi_i \varphi \rangle = 0$ and therefore $\langle T, \varphi \rangle = 0$. □

A consequence of this lemma is the following proposition.

PROPOSITION 5.1.– For any distribution T$\in \mathcal{D}'(\Omega)$, there exists a larger open subset $U \subset \Omega$ such that $T|_U = 0$.

5.2. Distribution support

5.2.1. *Definition*

By proposition 5.1, for any distribution T$\in \mathcal{D}'(\Omega)$, there exists a larger subset where T is null. This result allows us to define the support of a distribution as follows.

DEFINITION 5.4.– *Let $T \in \mathcal{D}'(\Omega)$ be a distribution. We call support of T the complementary of the largest open set where T is null. It is denoted by supp T.*

REMARK 5.2.– *As the complementary of an open set, supp T is always a closed set.*

By translating the definition, the following properties that characterize the support of a distribution can be obtained.

PROPOSITION 5.2.– Let $T \in \mathcal{D}'(\Omega)$. We have

1) $x_0 \notin \operatorname{supp} T \iff \exists\, V_{x_0}$, an open neighborhood of x_0 such that

$$\forall \varphi \in \mathcal{D}\big(V_{x_0}\big), \langle T, \varphi \rangle = 0;$$

2) $\operatorname{supp} T = \big\{x \in \Omega / \ T \text{ null near } x\big\}^c$;

3) $x_0 \in \operatorname{supp} T \iff \forall\, V_{x_0}$ open neighborhood of x_0, $\exists \varphi \in \mathcal{D}\big(V_{x_0}\big), \langle T, \varphi \rangle \neq 0$;

4) $\operatorname{supp} T \subset F \iff T = 0$ in F^c.

5.2.2. *Examples*

In this section, some examples of distribution supports are given. Others will be presented in more detail in exercises with solutions.

EXAMPLE 5.2.– Let f be a continuous function on Ω. Then, supp $T_f =$ supp f where supp f is the support in the classical sense of the continuous function f.

Actually, if $x_0 \notin$ supp f, there exists V_{x_0} an open neighborhood of x_0 such that, for all $x \in V_{x_0}, f(x) = 0$. So, for all $\varphi \in \mathcal{D}\big(V_{x_0}\big)$, we have $(\varphi f)(x) = 0$ in Ω. It follows that $\langle T_f, \varphi \rangle = \displaystyle\int_\Omega f(x)\varphi(x)\mathrm{d}x = 0$. Therefore, $x_0 \notin \operatorname{supp} T_f$. Hence the first inclusion: $\operatorname{supp} T_f \subset \operatorname{supp} f$.

Conversely, if $x_0 \notin \operatorname{supp} T_f$, there exists V_{x_0} an open neighborhood of x_0 such that

$$\forall \varphi \in \mathcal{D}\big(V_{x_0}\big), \ \int_\Omega f(x)\varphi(x)\mathrm{d}x = 0.$$

By the Dubois–Reymond lemma, f is zero almost everywhere (so everywhere since f is continuous) in V_{x_0}. Therefore, $x_0 \notin \operatorname{supp} f$, hence the second inclusion. Consequently, supp $T_f =$ supp f.

EXAMPLE 5.3.– For any $a \in \Omega$, we have $\operatorname{supp} \delta_a = \{a\}$.

Indeed, let $\varphi \in \mathcal{D}(\Omega\{a\})$, we have $\langle \delta_a, \varphi \rangle = \varphi(a) = 0$, thus $\operatorname{supp} \delta_a \subset \{a\}$.

Conversely, let V_a be an open neighborhood of a, and let ρ be a peak function in the neighborhood of a, such that $\rho(a) = 1$. Then, $\langle \delta_a, \rho \rangle = \rho(a) \neq 0$. Therefore, $a \in \operatorname{supp} \delta_a$, hence the converse inclusion. Consequently, $\operatorname{supp} \delta_a = \{a\}$.

EXAMPLE 5.4.– For any $a \in \Omega$ and for any $n \in \mathbb{N}$, we have $\operatorname{supp} \delta_a^{(n)} = \{a\}$.

Actually, let $\varphi \in \mathcal{D}(\Omega/\{a\})$, we have $\left\langle \delta_a^{(n)}, \varphi \right\rangle = (-1)^n \varphi^{(n)}(a) = 0$, thus $\operatorname{supp} \delta_a^{(n)} \subset \{a\}$.

Conversely, let V_a be an open neighborhood of a, and let $\chi \in \mathcal{D}(V_a)$ a plateau function in the neighborhood of a, such that $\chi(a) = 1$ in the neighborhood a. We consider the function φ defined by $\varphi(x) = \frac{(x-a)^n}{n!}\chi(x)$. By application of Leibniz's rule, it is clear that $\left\langle \delta_a^{(n)}, \varphi \right\rangle = 1 \neq 0$. Therefore, $a \in \operatorname{supp} \delta_a^{(n)}$, hence the converse inclusion. Consequently, $\operatorname{supp} \delta_a^{(n)} = \{a\}$.

EXAMPLE 5.5.– We have $\operatorname{supp} \mathrm{Vp}(\frac{1}{x}) = \mathbb{R}$.

Actually, let $x_0 \neq 0$. Let us assume that $x_0 > 0$, and the demonstration is the same for $x_0 < 0$. Let ρ_{x_0} be a peak function centered in x_0 and supported on $\left[\dfrac{x_0}{2}, \dfrac{3x_0}{2}\right]$. For any $\varepsilon > 0$ such that $\varepsilon < \dfrac{x_0}{2}$, we have

$$\int_{|x|\geq\varepsilon} \frac{\rho_{x_0}(x)}{x} \mathrm{d}x = \int_{x\geq\varepsilon} \frac{\rho_{x_0}(x)}{x} \mathrm{d}x = \int_{\frac{x_0}{2}}^{\frac{3x_0}{2}} \frac{\rho_{x_0}(x)}{x} \mathrm{d}x = C > 0.$$

Hence, $\left\langle \mathrm{Vp}(\frac{1}{x}), \rho_{x_0} \right\rangle \neq 0$ and $x_0 \in \operatorname{supp} \mathrm{Vp}(\frac{1}{x})$. Therefore, $\mathbb{R}^* \subset \operatorname{supp} \mathrm{Vp}(\frac{1}{x}) \subset \mathbb{R}$. Since the support of a distribution is closed, we necessarily have $\operatorname{supp} \mathrm{Vp}(\frac{1}{x}) = \mathbb{R}$.

5.2.3. *Properties of the support*

A useful practical result is given in the following manner.

PROPOSITION 5.3.– For any distribution $T \in \mathcal{D}'(\Omega)$, we have

$$\operatorname{supp} T = \emptyset \iff T = 0.$$

Proof. If $T = 0$, then clearly, by definition $\text{supp}\, T = \emptyset$.

Conversely, if $\text{supp}\, T = \emptyset$, then for all $x \in \Omega$, there exists an open set $w_x \in \Omega$ containing x such that $T|_{w_x} = 0$. By lemma 5.3, $T = 0$ because $\bigcup_{x \in \Omega} w_x = \Omega$. □

The corollary of this proposition is the following localization principle.

COROLLARY 5.1.– *Let $T \in \mathcal{D}'(\Omega)$. It is assumed that T is locally a $\mathcal{C}^k$-class function for $0 \leq k \leq \infty$ (i.e. $\forall x \in \Omega, \exists w_x$ open neighborhood of x, $\exists f_x \in \mathcal{C}^k(w_x)$, $T|_{w_x} = T_{f_x}$). Then there exists $f \in \mathcal{C}^k(\Omega)$ such that $T = T_f$.*

Proof. Since $\Omega = \bigcup_{x \in \Omega} w_x$, we can choose for all $x \in \Omega$, $f_x \in \mathcal{C}^k(w_x)$ such that $T|_{w_x} = T_{f_x}$. On $w_x \cap w_y$, we have $f_x = f_y$ because $T_{f_x}|_{w_x \cap w_y} = T|_{w_x \cap w_y} = T_{f_y}|_{w_x \cap w_y}$, then we use the continuity of f_x and f_y to deduce that $f_x = f_y$ everywhere on $w_x \cap w_y$ (and not only almost everywhere). Then, we can legitimately set $f : \Omega \to \mathbb{C}$ defined by $f(z) = f_x(z)$ if $z \in w_x$. The function f is of class $\mathcal{C}^k$ on Ω because it is of class $\mathcal{C}^k$ in the neighborhood of any $x \in \Omega$ and we have $\forall x \in \Omega, (T - T_f)|_{w_x} = 0$. By definition of the support, $\text{supp}\,(T - T_f) = \emptyset$, therefore by proposition 5.3, $T = T_f$. □

In the following, some results are given that reflect the result of the operations on the supports of the distributions. We start with the effect of multiplication.

PROPOSITION 5.4.– Let $T \in \mathcal{D}'(\Omega)$ and $a \in \mathcal{C}^\infty(\Omega)$. We have

$$\text{supp}\,(aT) \subset \text{supp}\, a \cap \text{supp}\, T.$$

Proof. Let $x_0 \in (\text{supp}\, a)^c \cup (\text{supp}\, T)^c$.

If $x_0 \in (\text{supp}\, a)^c$, there exists V_{x_0} a neighborhood of x_0 such that $a(x) = 0$ for all $x \in V_{x_0}$. Then if $\varphi \in \mathcal{D}(V_{x_0})$, we have for all $x \in \Omega$, $a(x)\varphi(x) = 0$.

Hence $\langle aT, \varphi \rangle = \langle T, a\varphi \rangle = 0$ and therefore, $x_0 \notin \text{supp}\,(aT)$.

If $x_0 \in (\text{supp}\, T)^c$, there exists V_{x_0} such that $\langle T, \psi \rangle = 0$ for all $\psi \in \mathcal{D}(V_{x_0})$.

Let $\varphi \in \mathcal{D}(V_{x_0})$. We have $a\varphi \in \mathcal{D}(V_{x_0})$, thus $\langle aT, \varphi \rangle = \langle T, a\varphi \rangle = 0$.

Then $x_0 \notin \text{supp}\,(aT)$. This completes the proof. □

For differentiation, the following result is obvious.

PROPOSITION 5.5.– Let $T \in \mathcal{D}'(\Omega)$. For all $\alpha \in \mathbb{N}^d$, $\operatorname{supp} \partial^\alpha T \subset \operatorname{supp} T$.

Proof. Let $\mathcal{O}$ the largest open set where $T = 0$ and W the largest open set where $\partial^\alpha T = 0$. By definition, we have supp $T = \mathcal{O}^c$ and supp $\partial^\alpha T = W^c$.

Since $T = 0$ on $\mathcal{O}$, then $\partial^\alpha T = 0$ on $\mathcal{O}$. As such, $W \supseteq \mathcal{O}$, consequently, $W^c \subseteq \mathcal{O}^c$. Hence, the desired result. □

REMARK 5.3.–

– *The inclusion in proposition 5.5 can be strict. As a matter of fact, let us consider the non-null constant distribution $T = c$. We have supp $T = \mathbb{R}$, while supp $T^{(n)} = \varnothing$, for all $n \in \mathbb{N}^*$.*

– *By applying this property, it can be found that supp $\delta_a^{(n)} = \{a\}$, for all $n \in \mathbb{N}^*$.*

PROPOSITION 5.6.– Let $T \in \mathcal{D}'(\Omega)$ and let $\varphi \in \mathcal{D}(\Omega)$. We have

$$\operatorname{supp} T \cap \operatorname{supp} \varphi = \emptyset \Longrightarrow \langle T, \varphi \rangle = 0.$$

Proof. The proof is mainly based on the Borel–Lebesgue property applied to an open cover of the compact supp φ. □

NOTE 5.1.– *Let $T \in \mathcal{D}'(\Omega)$ and $\varphi \in \mathcal{D}(\Omega)$.*

It is possible that $\varphi = 0$ on supp T but $\langle T, \varphi \rangle \neq 0$.

Counter example: We consider $T = \delta'$. Let $\varphi \in \mathcal{D}(\mathbb{R})$. We have supp $T = \{0\}$. On the other hand,

$$\langle T, \varphi \rangle = \langle \delta', \varphi \rangle = -\langle \delta, \varphi' \rangle = -\varphi'(0).$$

Is there $\varphi \in \mathcal{D}(\mathbb{R})$ such that $\varphi'(0) \neq 0$ and $\varphi(0) = 0$?

Let $\theta \in \mathcal{D}(\mathbb{R})$ be a plateau function that takes the value 1 on $[-1, 1]$ and null outside of $[-2, 2]$. We now consider $\varphi = x\theta$. We have $\varphi \in \mathcal{D}(\mathbb{R})$, $\varphi'(0) = 1 \neq 0$ and $\varphi(0) = 0$. The function φ thus constructed answers the question.

A result giving sufficient conditions for $\langle T, \varphi \rangle = 0$ is the following.

PROPOSITION 5.7.– Let $T \in \mathcal{D}'(\Omega)$ be a mth-order distribution and $\varphi \in \mathcal{D}(\Omega)$ null on the support of T as well as its partial derivatives of order less than or equal to m. Then $\langle T, \varphi \rangle = 0$.

Proof. Let K' be a compact neighborhood of the support of φ and $K := K' \cap \operatorname{supp} T$. Let $(\psi_n)_{n \in \mathbb{N}}$ a sequence of functions of $\mathcal{D}(\Omega)$ with values in $[0,1]$, null if $d(x,K) > \frac{3}{n}$ such that

– $\psi_n(x) = 1$ if $d(x,K) \leq \frac{3}{n}$;

– $\forall \alpha \in \mathbb{N}^d, \quad |\partial^\alpha \psi_n(x)| \leq c\, n^{|\alpha|}$.

Such a sequence does exist; in practice, we simply have to choose $\rho \in \mathcal{D}(\mathbb{R}^d)$ positive, with support in $B(0,1)$ and of integral 1 and we set $\psi_n := \mathbb{1}_{K_n} * \rho_n$ where $\rho_n(x) = n^d \rho(nx)$ and $K_n := \left\{ x \in \Omega, d(x,K) \leq \frac{2}{n} \right\}$.

The function $\varphi(1 - \psi_n)$ has its support in K' and is disjoint from K, thus disjoint from $\operatorname{supp} T$. By definition of the support of T, we thus have $\langle T, \varphi(1 - \psi_n) \rangle$. It follows that

$$\langle T, \varphi \rangle = \langle T, \varphi \psi_n \rangle .$$

Now, T is of order m, so there exists $M > 0$ such that

$$|\langle T, \varphi \psi_n \rangle| \leq M \sup_{|\alpha| \leq m} \sup_{x \in \Omega} |\partial^\alpha (\varphi \psi_n)(x)| .$$

Let us show that for $|\alpha| \leq m$, we have

$$\sup_{x \in \Omega} |\partial^\alpha (\varphi \psi_n)(x)| \leq C n^{|\alpha| - m - 1} \qquad [5.1]$$

for n large enough and with C a constant independent of n, which will show the desired result.

It should be noted that if $x \in \operatorname{supp}(\varphi \psi_n)$, then $x \in K$ in which case $\partial^\beta(\varphi)(x) = 0$ for all $|\beta| \leq m$, or $x \in \operatorname{supp}(\varphi \psi_n) \backslash K$ and then the Taylor formula (applied at a point $a \in K$ such that $d(x,K) = |x - a|$) implies that for all $|\beta| \leq m$, we have

$$\left| \partial^\beta \varphi(x) \right| \leq C \left| x - a \right|^{m+1-|\beta|} \leq C n^{|\beta| - m - 1}.$$

Then, by the Leibniz rule, we have

$$\partial^\alpha(\varphi\psi_n)(x) = \sum_{\beta\le\alpha}\binom{\alpha}{\beta}\partial^{\alpha-\beta}\psi_n(x)\partial^\beta\varphi(x).$$

Therefore, if $|\alpha| \le m$, we have

$$|\partial^\alpha(\varphi\psi_n)(x)| \le c\sum_{\beta\le\alpha} n^{|\alpha|-|\beta|} \sup_{d(x,K)\le\frac{3}{n}} \left|\partial^\beta\varphi(x)\right|.$$

So for n large enough, we have

$$|\partial^\alpha(\varphi\psi_n)(x)| \le c'\varepsilon\sum_{\beta\le\alpha} n^{|\alpha|-|\beta|}n^{|\alpha|-m-1} \le Cn^{|\alpha|-m-1}.$$

We have thus demonstrated inequality [5.1], which ends the demonstration. □

PROPOSITION 5.8.– Let $T \in \mathcal{D}'(\Omega)$ and $\varphi \in \mathcal{D}(\Omega)$ be null on the support of T as well as its partial derivatives of any order. Then $\langle T, \varphi\rangle = 0$.

Proof. Let $(K_n)_{n\in\mathbb{N}}$ be an exhaustive sequence of compacts Ω and for all $n \in \mathbb{N}$, let $\psi_n \in \mathcal{D}(\Omega)$ is equal to 1 in the neighborhood of K_n. Then $\psi_n T$ is a distribution with compact support. However, then $\psi_n T$ is of finite order m_n. Moreover, the support of $\psi_n T$ is included in the support of T. Hence, it can be deduced that φ and all its partial derivatives up to order m_n are null on the support of $\psi_n T$ and thus by proposition 5.8, $\forall n \in \mathbb{N}$, $\langle\psi_n T, \varphi\rangle = 0$. But there exists n_0 such that the support of φ is included in K_{n_0}, thus $\psi_{n_0}\varphi = \varphi$ and then $\langle T, \varphi\rangle = \langle T, \psi_{n_0}\varphi\rangle = \langle\psi_{n_0} T, \varphi\rangle = 0$. Hence, the result is looked for. □

PROPOSITION 5.9.– Let $T, S \in \mathcal{D}'(\Omega)$ denote two distributions. We have

$$\operatorname{supp}(T+S) \subset \operatorname{supp} T \cup \operatorname{supp} S.$$

5.3. Compact support distributions

Distributions with compact support have an important place in distribution theory. A distribution with compact support is actually a finite sum of (multiple) derivatives in the sense of distributions of continuous functions. This somehow confirms that the notion of distribution is the minimal extension of the notion of continuous function, which can be differentiable as many times as desired.

5.3.1. *Definition and properties*

DEFINITION 5.5.– *Let $T \in \mathcal{D}'(\Omega)$ be a distribution. It is said that T is compactly supported when supp T is compact. The set of distributions with compact support is denoted by $\mathcal{E}^{'}(\Omega)$.*

Obviously, $\mathcal{E}^{'}(\Omega)$ is a vector subspace of $\mathcal{D}'(\Omega)$ on $\mathbb{R}$ (or $\mathbb{C}$). *A priori*, a distribution with compact support is a continuous linear form on the space of functions of class $\mathcal{C}^\infty$ with compact supports. However, since the support of the distribution is itself compact, the test functions outside the support of the distribution can be ignored. This makes it possible to extend the duality and to consider distributions with compact support as continuous linear forms on the space of functions $\mathcal{C}^\infty(\Omega)$ endowed with the topology of uniform convergence on all compact derivatives of all orders; this is exactly the subject of section 5.3.1.1.

5.3.1.1. *Topology of* $\mathcal{C}^\infty(\Omega)$

Let $\varphi \in \mathcal{C}^\infty(\Omega)$. Based on the decomposition lemma, lemma 2.2, a family of semi-norms $(P_n)_n$ can be defined on $\mathcal{C}^\infty(\Omega)$ by

$$P_n(\varphi) = \max_{|\alpha|\leq n, x\in K_n} |\partial^\alpha \varphi(x)|.$$

THEOREM 5.1.– *The space $\mathcal{C}^\infty(\Omega)$ endowed with the P–topology associated with the family of semi-norms $(P_n)_n$ is a Fréchet space.*

DEFINITION 5.6.– *Let $T : \mathcal{C}^\infty(\Omega) \longrightarrow \mathbb{K} = (\mathbb{R}$ or $\mathbb{C})$ a linear form.*

$$T \text{ is continuous} \iff \exists c > 0, \exists n_0 \in \mathbb{N}, \forall \varphi \in \mathcal{C}^\infty(\Omega), |\langle T, \varphi\rangle| \leq c P_{n_0}(\varphi).$$

An equivalent sequential definition is given by definition 5.7.

DEFINITION 5.7.– *It is said that a sequence (φ_n) of $\mathcal{C}^\infty(\Omega)$ converges to φ in $\mathcal{C}^\infty(\Omega)$ if for any multi-index $\alpha \in \mathbb{N}^d$, $\partial^\alpha \varphi_n$ converges to $\partial^\alpha \varphi$ uniformly in any compact of Ω.*

REMARK 5.4.–

1) Unlike convergence in $\mathcal{D}(\Omega)$, no assumption is done here on the supports.

2) Continuity and linearity for a functional on $\mathcal{C}^\infty(\Omega)$ has been defined in the same way as in $\mathcal{D}(\Omega)$; the dual of $\mathcal{C}^\infty(\Omega)$ is denoted $\left(\mathcal{C}^\infty(\Omega)\right)'$, which is the space of continuous linear functionals defined on $\mathcal{C}^\infty(\Omega)$.

5.3.1.2. *Extension of a distribution with compact support*

The following density result is recalled.

PROPOSITION 5.10.– The space $\mathcal{D}(\Omega)$ is dense in $\mathcal{C}^\infty(\Omega)$.

The following result is intended to extend a compactly supported distribution.

PROPOSITION 5.11.– A distribution $\mathcal{E}'(\Omega)$ can be extended to $\mathcal{C}^\infty(\Omega)$.

Proof. Let K be the compact support of $T \in \mathcal{E}'(\Omega)$. Let K' be compact containing K and K'' a compact containing K' such that there exists χ a plateau function, which is equal to 1 on K' and that is supported in K''.

We verify that, for $\varphi \in \mathcal{D}(\Omega)$, we have

$$\langle T, \varphi \rangle = \langle T, \chi\varphi \rangle + \langle T, (1-\chi)\varphi \rangle ,$$

since $(1-\chi)\varphi$ is supported in a compact contained in the complementary of the support of T, $\langle T, (1-\chi)\varphi \rangle = 0$. Hence,

$$\forall \varphi \in \mathcal{D}(\Omega), \langle T, \varphi \rangle = \langle T, \chi\varphi \rangle . \qquad [5.2]$$

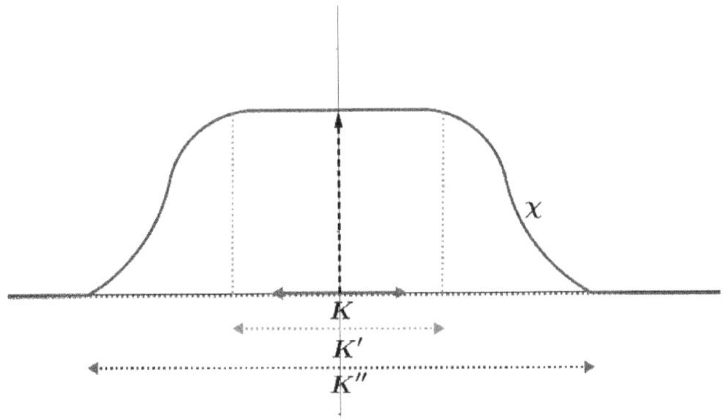

Figure 5.2. χ *function. For a color version of this figure, see www.iste.co.uk/aitbenhassi/distribution.zip*

Given the density of $\mathcal{D}(\Omega)$ in $\mathcal{C}^\infty(\Omega)$, we can define on $\mathcal{C}^\infty(\Omega)$ a functional $\tilde{T}$ extending T by

$$\forall\varphi \in \mathcal{C}^\infty(\Omega), \left\langle \tilde{T}, \varphi \right\rangle = \langle T, \chi\varphi \rangle . \qquad [5.3]$$

Since $\chi\varphi \in \mathcal{D}(\Omega)$, $\langle T, \chi\varphi \rangle$ is well defined, in addition,

$$\exists c_1 > 0, \exists n \in \mathbb{N}, \left|\left\langle \tilde{T}, \varphi \right\rangle\right| = |\langle T, \chi\varphi\rangle| \leq c_1 \max_{|\alpha|\leq n} \max_{x\in \text{supp}\,(\chi\varphi)} |\partial^\alpha(\chi\varphi)(x)| .$$

On the other hand, we have $|\partial^\alpha(\chi\varphi)| \leq c_2 \max\limits_{|\beta|\leq|\alpha|} \max\limits_{x\in K''} \left|\partial^\beta\varphi(x)\right|$. Consequently,

$$\exists c > 0, \exists n \in \mathbb{N}, \left|\left\langle \tilde{T}, \varphi \right\rangle\right| \leq c \max_{|\beta|\leq n} \max_{x\in K''} \left|\partial^\beta\varphi(x)\right| .$$

Finally,

$$\exists c > 0, \exists n \in \mathbb{N}, \forall\varphi \in \mathcal{C}^\infty(\Omega), \left|\left\langle \tilde{T}, \varphi \right\rangle\right| \leq cP_n(\varphi).$$

It is deduced that $\tilde{T} \in \left(\mathcal{C}^\infty(\Omega)\right)'$.

Consequence: We have just shown that for $T \in \mathcal{E}'(\Omega)$ there exists a linear mapping $\tilde{T}$ from $\mathcal{C}^\infty(\Omega)$ to $\mathbb{C}$ verifying:

i) $\left\langle \tilde{T}, \varphi \right\rangle = \langle T, \varphi \rangle , \forall\varphi \in \mathcal{D}(\Omega)$;

ii) $\left\langle \tilde{T}, \varphi \right\rangle = 0, \forall\varphi \in \mathcal{C}^\infty(\Omega), \text{supp}\,\varphi \cap \text{supp}\,T = \varnothing$;

iii) there exists $n \in \mathbb{N}$, a compact K (neighborhood of $K_0 = \text{supp}\,T$) and $c > 0$ such that

$$\forall\varphi \in \mathcal{C}^\infty(\Omega), \left|\left\langle \tilde{T}, \varphi \right\rangle\right| \leq c \max_{|\beta|\leq n} \max_{x\in K} \left|\partial^\beta\varphi(x)\right| .$$

□

REMARK 5.5.–

1) The expression [5.3] *defining* $\tilde{T}$ *is independent of the function* χ. *Actually, let* χ_1 *and* $\chi_2 \in \mathcal{D}(\Omega)$ *such that* $\chi_1|_{V_1} = 1$ *and* $\chi_2|_{V_2} = 1$ *where* V_1 *and* V_2 *are two open neighborhoods of* K *in* Ω. *Then*

$$\chi_1 - \chi_2|_{V_1\cap V_2} = 0.$$

Since $V_1 \cap V_2$ is an open neighborhood of supp (T) in Ω, for any function $\varphi \in \mathcal{D}(\Omega)$, the function $(\chi_1 - \chi_2)\varphi$, which is of class $\mathcal{C}^\infty$ on Ω, verifies supp $\left((\chi_1 - \chi_2)\varphi\right) \cap$ supp $T = \varnothing$. It follows that $\langle T, (\chi_1 - \chi_2)\varphi\rangle = 0$. Consequently, $\langle T, \chi_1\varphi\rangle = \langle T, \chi_2\varphi\rangle$.

2) From to this extension, $\mathcal{E}'(\Omega)$ can be identified with $(\mathcal{C}^\infty(\Omega))'$ just as we already saw for $\mathcal{D}'(\Omega) = \left(\mathcal{C}_c^\infty(\Omega)\right)'$. In fact, we have the following result.

PROPOSITION 5.12.– The mapping Φ from $\mathcal{E}'(\Omega)$ to $\left(\mathcal{C}^\infty(\Omega)\right)'$, which associates with $T \in \mathcal{E}'(\Omega)$ the continuous linear form $\tilde{T} \in \left(\mathcal{C}^\infty(\Omega)\right)'$, is a bijective linear mapping. It makes it possible to identify $\mathcal{E}'(\Omega)$ with $\left(\mathcal{C}^\infty(\Omega)\right)'$.

An important property of compact support distributions is the following.

PROPOSITION 5.13.– Any distribution with compact support is of finite order.

Proof. Let K be the compact support of $T \in \mathcal{E}'(\Omega)$. Consider K' a compact contained K and K'' a compact containing K', so that there exists χ a plateau function, which is equal to 1 on K' and which is supported in K''. We recall that

$$\forall \varphi \in \mathcal{D}(\Omega), \langle T, \varphi\rangle = \langle T, \chi\varphi\rangle + \langle T, (1-\chi)\varphi\rangle = \langle T, \chi\varphi\rangle .$$

Considering the compact K'' and since $T \in \mathcal{D}'(\Omega)$, then there exists an integer m_0 and a constant C_0, such that

$$\forall \phi \in D_{K''}(\Omega), |\langle T, \phi\rangle| \leq C_0 \max_{|\alpha|\leq m_0} \max_{x \in K''} |\partial^\alpha \phi(x)| .$$

On the other hand, since $\varphi\chi \in D_{K''}(\Omega)$, (i.e. supported in K''), then

$$|\langle T, \varphi\rangle| \leq C_0 \max_{|\alpha|\leq m_0} \max_{x \in K''} |\partial^\alpha (\chi\varphi)(x)| .$$

Applying Leibniz's rule yields $\partial^\alpha(\chi\varphi) = \sum\limits_{\gamma+\beta=\alpha} \binom{\alpha}{\beta} \partial^\gamma\chi\partial^\beta\varphi$ and the uniform upper bounds, for $|\gamma| \leq m_0$ of $|\partial^\gamma\chi|$ by $\|\chi\|_{m_0} = \max\limits_{|\gamma|\leq m_0} \max\limits_{x\in K''} |\partial^\gamma\chi(x)|$ and of $\sum\limits_{\beta+\gamma=\alpha} \binom{\alpha}{\beta} = 2^{|\alpha|}$ results in obtaining a constant C such that

$$C_0 \max_{|\alpha|\leq m_0} \max_{x \in K''} |\partial^\alpha(\chi\varphi)| \leq C \max_{|\alpha|\leq m_0} \max_{x \in \text{supp}\,\varphi} |\partial^\alpha\varphi(x)|,$$

and

$$|\langle T,\varphi\rangle| \le C \max_{|\alpha|\le m_0} \max_{x\in \operatorname{supp}\varphi} |\partial^\alpha \varphi(x)|.$$

This proves that T is of order m_0 at most, therefore it is of finite order. □

NOTE 5.2.– *The converse of this property is not always true. Simply consider the distribution* $Vp\left(\frac{1}{x}\right)$. *This distribution is a distribution of finite order, but* supp $Vp\left(\frac{1}{x}\right) = \mathbb{R}$ *is not compact.*

5.3.2. *Distributions with point support*

We have already seen in the examples above that the support of the Dirac derivatives is $\operatorname{supp}\delta_a^{(n)} = \{a\}$. Conversely, distributions whose support is a singleton admit a very simple characterization.

THEOREM 5.2.– *Let* $T \in \mathcal{D}'(\Omega)$ *and let* $x_0 \in \Omega$. *Let us assume that* $\operatorname{supp} T = \{x_0\}$. *Then there exists an integer* m *and some complex number* $(a_\alpha)_{|\alpha|\le m}$ *such that*

$$\forall \varphi \in \mathcal{D}(\Omega), \quad \langle T,\varphi\rangle = \sum_{|\alpha|\le m} a_\alpha \partial^\alpha \varphi(x_0),$$

which can also be written as $T = \sum_{|\alpha|\le m} \tilde{a}_\alpha \partial^\alpha \delta_{x_0}$ *where* $\tilde{a}_\alpha = (-1)^{|\alpha|} a_\alpha$.

For the proof of this theorem, see exercise 5.12.

5.4. Exercises with solutions

EXERCISE 5.1.– Consider the linear form T defined on T defined on $\mathcal{D}(\mathbb{R})$ by

$$T : \varphi \longmapsto \sum_{k=1}^{+\infty} \frac{1}{\sqrt{k}} \big(\varphi(\frac{1}{k}) - \varphi(-\frac{1}{k})\big).$$

1) Show that T is a distribution of order 1 at most.

2) Show that T is of order 1.

3) Show that the support of T is $S = \{\frac{1}{k} \ / \ k \in \mathbb{Z}\} \cap \{0\}$.

SOLUTION 5.1.–

1) Let $A > 0$ and $\varphi \in \mathcal{D}(\mathbb{R})$ such that $\operatorname{supp}\varphi \Subset [-A, A]$. By Taylor's formula, there exists $\psi \in \mathcal{C}^\infty(\mathbb{R})$ such that

$$\varphi(x) = \varphi(0) + x\psi(x).$$

Then, for all $N \in \mathbb{N}^*$, we have

$$\left|\sum_{k=1}^{N} \frac{1}{\sqrt{k}}(\varphi(\frac{1}{k}) - \varphi(-\frac{1}{k}))\right| \leq 2\left\|\varphi'\right\|_\infty \sum_{k=1}^{N} \frac{1}{k^{\frac{3}{2}}}.$$

The series $\sum \frac{1}{k^{\frac{3}{2}}}$ is convergent as a Riemann series. Evaluating the limit when $N \longrightarrow +\infty$, we obtain

$$|\langle T, \varphi\rangle| \leq \left(2\sum_{k=1}^{+\infty} \frac{1}{k^{\frac{3}{2}}}\right)\|\varphi'\|_\infty \leq \left(2\sum_{k=1}^{+\infty} \frac{1}{k^{\frac{3}{2}}}\right) P_1(\varphi).$$

Therefore, T is at most a first-order distribution.

2) Let $(\varphi_n)_{n\geq 1}$ be a sequence of positive functions of $\mathcal{D}(\mathbb{R})$ such that

$$\varphi_n(x) = 1 \text{ if } x \in \left[\tfrac{1}{n}, 1\right] \text{ and } \varphi_n(x) = 0 \text{ if } x \in \left]-\infty, \tfrac{1}{n+1}\right] \cup [2, +\infty[$$

For any $n \geq 1$, $\operatorname{supp}\varphi_n \subset K := [0, 2]$ and $\|\varphi_n\|_\infty = 1$. If we assume that T is of order 0, then there will exist a constant $C > 0$ such that

$$\forall n \in \mathbb{N}^*, \quad |\langle T, \varphi_n\rangle| \leq C.$$

However, $\langle T, \varphi_n\rangle = \sum_{k=1}^{n} \frac{1}{\sqrt{k}}$, thus $\forall n \in \mathbb{N}^*$, $\sum_{k=1}^{n} \frac{1}{\sqrt{k}} \leq C$. This is impossible because $\sum_{k=1}^{n} \frac{1}{\sqrt{k}} \underset{n\longrightarrow+\infty}{\longrightarrow} +\infty$. Therefore, T is a first-order distribution.

3) Let $k \in \mathbb{N}^*$. For any open set V containing the compact $\left\{\frac{1}{k}\right\}$, there exists a "peak" function, ψ_k such that $\operatorname{supp}\psi_k \subset \left]\frac{1}{k+1}, \frac{1}{k-1}\right[\cap V, \psi_k\left(\frac{1}{k}\right) = 1$ and $0 \leq \psi_k \leq 1$.

For any $k \in \mathbb{N}^*$, we have $\langle T, \psi_k\rangle = \frac{1}{\sqrt{k}} \neq 0$. Then $\frac{1}{k} \in \operatorname{supp} T, \forall k \in \mathbb{N}^*$. Similarly, we also show that $\forall k \in \mathbb{Z}_-^*$, $\frac{1}{k} \in \operatorname{supp} T$.

It follows that $\left\{\frac{1}{k}, k \in \mathbb{Z}^*\right\} \subset \operatorname{supp} T$. Since $\operatorname{supp} T$ is a closed set, then

$$S = \{0\} \cup \left\{\frac{1}{k}, k \in \mathbb{Z}^*\right\} \subset \operatorname{supp} T.$$

Conversely, if $x_0 \notin S$ then, there exists an open neighborhood V of x_0 such that $V \cap S = \varnothing$ and such that for any $\varphi \in \mathcal{D}(V)$, we have $\langle T, \varphi\rangle = 0$. Therefore, $x_0 \notin \operatorname{supp} T$. It is deduced that $\operatorname{supp} T \subset S$. Consequently, $\operatorname{supp} T = \left\{\frac{1}{k}, k \in \mathbb{Z}^*\right\} \cup \{0\}$.

EXERCISE 5.2.– For all $n \geq 0$, we set $T_n = \sum_{k=0}^{n} \delta_k$, where δ_k is the Dirac distribution at point $x = k$.

1) Show that T_n is a distribution with compact support.

2) What is its support?

SOLUTION 5.2.–

1) The distribution $T_n = \sum_{k=0}^{n} \delta_k$ is a sum of a finite number of distributions with compact support. Since the space $\mathcal{E}^{'}(\mathbb{R})$ of compactly supported distributions is a vector space, then T_n is itself with compact support.

2) Let $\varphi \in \mathcal{D}(\mathbb{R})$ with support in the complementary of $\{0, 1, ...n\}$. We have

$$\langle T_n, \varphi\rangle = \sum_{k=0}^{n}\langle \delta_k, \varphi\rangle = \sum_{k=0}^{n} \varphi(k) = 0.$$

Therefore, the support of T_n is contained in $\{0, 1, \ldots, n\}$.

It should be noted that proposition 5.9 allows us to easily obtain the inclusion of the support of T_n in $\{0, 1, \ldots, n\}$.

Let $k \in \{0, 1, \ldots, n\}$. So, V_k be an open neighborhood of k. It is considered that $\varphi \in \mathcal{D}(\mathbb{R})$ the "peak" function with support in $]k - \frac{1}{2}, k + \frac{1}{2}[\cap V_k$ and which equals 1 in k.

We have

$$\begin{aligned}\langle T_n, \varphi\rangle &= \langle \sum_{j=0}^{n} \delta_j, \varphi\rangle \\ &= \langle \delta_k, \varphi\rangle. \\ &= \varphi(k) \\ &= 1 \neq 0.\end{aligned}$$

Then $k \in \operatorname{supp} T_n$. As a result, we get the other inclusion. Finally,

$$\operatorname{supp} T_n = \{0, 1, \dots, n\}.$$

EXERCISE 5.3.– Consider the linear form T defined on $\mathcal{D}(\mathbb{R})$ by

$$T : \varphi \longmapsto \lim_{n \longrightarrow +\infty} \left(n\varphi(0) - \sum_{k=1}^{n} \varphi(\frac{1}{k^2}) \right).$$

1) Show that T is a distribution.

2) Determine the support of T.

SOLUTION 5.3.–

1) For all $n \in \mathbb{N}*$, we set

$$\langle T_n, \varphi\rangle = n\varphi(0) - \sum_{k=1}^{n} \varphi(\frac{1}{k^2}) = \sum_{k=1}^{n} u_k, \text{ where } u_k = \varphi(0) - \varphi(\frac{1}{k^2}).$$

The increment theorem implies that there exists $c_k \in \left]0, \frac{1}{k^2}\right[$ such that $u_k = -\varphi'(c_k)\frac{1}{k^2}$. In addition, $|\varphi'(c_k)| \leq \sup_{0 \leq x \leq 1} |\varphi'(x)|$. The series with general term u_k is thus upper bounded by a convergent Riemann sum (since of index 2). It is therefore convergent with limit $\langle T, \varphi\rangle$. Given that each T_n is linear in φ, the same can be said of the limit T defined by $\langle T, \varphi\rangle = \sum_{k=1}^{+\infty} u_k$, which is thus a linear form on

$\mathcal{D}(\mathbb{R})$. Moreover, for any compact $K \subset \mathbb{R}$ and any test function $\varphi \in \mathcal{D}(\mathbb{R})$ whose support is contained in K, we have

$$|\langle T, \varphi\rangle| \leq C \sup_{x\in K} |\varphi'(x)|\,, \text{ where } C := \sum_{k=1}^{+\infty} \frac{1}{k^2}.$$

It is then deduced that T is a distribution of order at most equal to 1.

2) Let $\varphi_k \in \mathcal{D}(\mathbb{R})$ verifying (for $k \geq 2$):

$$\varphi_k(\frac{1}{k^2}) = 1 \text{ and } \operatorname{supp}\varphi_k \subset \left]\frac{1}{(k+1)^2}, \frac{1}{(k-1)^2}\right[.$$

We obtain $\langle T, \varphi_k\rangle = \varphi_k(\dfrac{1}{k^2}) = 1$. Therefore, $\frac{1}{k^2} \in \operatorname{supp} T$. Since the support of T is closed, this implies that $H := \left\{\dfrac{1}{k^2}, k \in \mathbb{N}^*\right\} \cup \{0\} \subset \operatorname{supp} T$.

Then let $\varphi \in \mathcal{D}(\mathbb{R})$ verifying $\operatorname{supp}\varphi \cap H = \varnothing$. Since $\langle T_n, \varphi\rangle = 0$, we recover at the limit $\langle T, \varphi\rangle = 0$, which yields $\operatorname{supp} T = H$.

EXERCISE 5.4.– Let f be a function indefinitely differentiable on $\mathbb{R}$ and T a distribution and show that $\{x \in \mathbb{R} : f(x) \neq 0\} \cap \operatorname{supp} T \subset \operatorname{supp}(fT) \subset \operatorname{supp} f \cap \operatorname{supp} T$.

SOLUTION 5.4.– We set: $A = \{x \in \mathbb{R} : f(x) \neq 0\}$, $\operatorname{supp} f = \overline{A} =: B$, $\operatorname{supp} T = F$ and $\operatorname{supp}(fT) = E$. We know that, for any $\varphi \in \mathcal{D}(\mathbb{R})$,

$$\text{if } \operatorname{supp}\varphi \subset F^c, \text{ then } \langle T, \varphi\rangle = 0,$$

and

$$\text{if } \operatorname{supp}\varphi \subset E^c, \text{ then } \langle fT, \varphi\rangle = 0.$$

– Let us show that $A \cap F \subset E$. To this end, we have to show that if $u \in E^c$ and $u \in A$, then $u \in F^c$. Since E^c is an open set and $u \in E^c$, there exists a neighborhood of u, denoted $V(u)$, such that $V(u) \subset E^c$, $f(u) \neq 0$, $\langle fT, \varphi\rangle = 0$, for any function $\varphi \in \mathcal{D}(\mathbb{R})$, having its support in $V(u)$. We know that $f\varphi \in \mathcal{D}(\mathbb{R})$ and henceforth,

$$0 = \langle fT, \varphi\rangle = \langle T, f\varphi\rangle\,.$$

Consequently, $V(u) \subset F^c$ and thus $u \in F^c$.

– Let us show that $E \subset B \cap F$. To this end, we merely have to show that if $u \in B^c$ or $u \in F^c$, then $u \in E^c$. Let $u \in B^c$, then $f(u) = 0$ and $f(u)\varphi(u) = 0$, for all $\varphi \in \mathcal{D}(\mathbb{R})$. Henceforth, $0 = \langle T, f\varphi\rangle = \langle fT, \varphi\rangle$, and thus $u \notin E$. Recall that $u \in F$ if and only if T is not zero on any neighborhood $V(u)$ of u. Therefore, if $u \in F^c$, then $\langle T, \varphi\rangle = 0$, for any function $\varphi \in \mathcal{D}(\mathbb{R})$, having its support in $V(u)$. Hence,$\langle T, f\varphi\rangle = \langle fT, \varphi\rangle = 0$, and thus $u \in E^c$.

Conclusion: We have just shown that $A \cap F \subset E$ and that $E \subset B \cap F$.

Consequently, $A \cap F \subset E \subset B \cap F$, that is,

$$\{x \in \mathbb{R} : f(x) \neq 0\} \cap \operatorname{supp} T \subset \operatorname{supp}(fT) \subset \operatorname{supp} f \cap \operatorname{supp} T.$$

EXERCISE 5.5.– For $a \in \mathbb{R}^*$, we consider $T_a = \cos(ax)\mathrm{Vp}(\frac{1}{x}), \quad \forall x \in \mathbb{R}^*$.

1) Show that T_a defines a distribution on $\mathbb{R}$.

2) Determine supp T_a.

SOLUTION 5.5.–

1) It is known that Vp $\left(\frac{1}{x}\right)$ is a distribution on $\mathbb{R}$.

Since $\cos(ax) \in \mathcal{C}^\infty(\mathbb{R})$, then $T_a = \cos(ax)\mathrm{Vp}\left(\frac{1}{x}\right)$ is indeed a distribution on $\mathbb{R}$.

2) Let $A = \{x \in \mathbb{R} : \cos(ax) \neq 0\} = \left\{x \in \mathbb{R} : x \neq \dfrac{\pi}{2a} + \dfrac{k\pi}{a}, k \in \mathbb{Z}\right\}$. From exercise 5.4, we have

$$A \cap \operatorname{supp} \mathrm{Vp}\left(\tfrac{1}{x}\right) \subset \operatorname{supp} T_a \subset \operatorname{supp}\left(\cos(ax)\right) \cap \operatorname{supp} \mathrm{Vp}\left(\tfrac{1}{x}\right).$$

Now supp Vp $\left(\frac{1}{x}\right) = \mathbb{R}$ and supp $(\cos(ax)) = \overline{A} = \mathbb{R}$, thus $A \subset \operatorname{supp} T_a \subset \mathbb{R}$. Therefore, $\overline{A} \subset \overline{\operatorname{supp} T_a} \subset \mathbb{R}$, then $\overline{\operatorname{supp} T_a} = \mathbb{R}$. Consequently, supp $T_a = \mathbb{R}$.

EXERCISE 5.6.– For all $\varphi \in \mathcal{D}(\mathbb{R})$, we set $\langle T, \varphi\rangle = \displaystyle\int_{\mathbb{R}^2} e^{-x^2-y^2}\varphi(\sin(xy))dxdy$.

1) Show that T is a distribution.

2) What is the support of T?

SOLUTION 5.6.–

1) The integral $= \displaystyle\int_{\mathbb{R}^2} e^{-x^2-y^2}\varphi(\sin(xy))dxdy$ exists for all $\varphi \in \mathcal{D}(\mathbb{R})$ since φ is bounded. Therefore, T is indeed a functional, and its linearity is obvious. On the other hand, for $\varphi \in \mathcal{D}(\mathbb{R})$ such that $\text{supp}\varphi \Subset [-a, a]$, we have

$$
\begin{aligned}
|\langle T, \varphi\rangle| &= \int_{\mathbb{R}^2} e^{-x^2-y^2}\, |\varphi(\sin(xy))|\, dxdy \\
&\leq \|\varphi\|_\infty \int_{\mathbb{R}^2} e^{-x^2-y^2} dxdy \\
&\leq \pi\|\varphi\|_\infty.
\end{aligned}
$$

Then T is a zeroth-order distribution.

2) Consider the open set $U =]-\infty, -1[\cup]1, +\infty[$. If a function φ has a support included in U, then we have $\varphi(t) = 0$ for all $t \in [-1, 1]$. For such a function φ, we have $\langle T, \varphi\rangle = 0$. The distribution T is null in U. The support of T is therefore included in $[-1, 1]$. Let us show that supp $T = [-1, 1]$.

We denote by Ω the complementary of the support of T. We assume that there exists $u \in \Omega$ such that $-1 < u < 1$. Let us choose $\rho > 0$ such that

$$]u-\rho, u+\rho[\, \subset \Omega \cap]-1, 1[.$$

We consider a positive function $\varphi \in \mathcal{D}(\mathbb{R})$, with support included in $]u-\rho, u+\rho[$ and equal to 1 in $\left[u - \frac{\rho}{2}, u + \frac{\rho}{2}\right]$. The subset W of $\mathbb{R}^2$ defined by

$$W = \left\{(x, y) : |\sin(xy) - u| < \frac{\rho}{2}\right\}$$

is open (because the function which associates with (x,y) the real sin(xy) is a continuous function on $\mathbb{R}^2$). If $(x, y) \in W$, we have $\varphi(\sin(xy)) = 1$. Therefore,

$$\langle T, \varphi\rangle \geq \int_W e^{-x^2-y^2} dxdy > 0.$$

However, the support of φ is included in Ω and $\langle T, \varphi\rangle = 0$. This leads to a contradiction because such a real u does not exist. Since the support of T is closed, we correctly have $\operatorname{supp} T = [-1, 1]$.

EXERCISE 5.7.– Let T be the mapping from $\mathcal{D}(\mathbb{R}^2)$ into $\mathbb{R}$ defined by $T(\phi) = \displaystyle\int_{\mathbb{R}} \phi(x, -x)\mathrm{d}x$.

1) Show that T is a zeroth-order distribution on $\mathbb{R}^2$.

2) Determine the support of T.

SOLUTION 5.7.–

1) Let $\varphi \in \mathcal{D}(\mathbb{R}^2)$ with support in the compact $K \subset \mathbb{R}^2$. We set $K_1 = \big\{x \in \mathbb{R} : (x, -x) \in K\big\}$. Then K_1 is compact and we have

$$|\langle T, \varphi\rangle| = \left|\int_{\mathbb{R}} \varphi(x, -x)\mathrm{d}x\right| \leq \sup_{(x,y)\in K} |\varphi(x, y)| \int_{K_1} 1\mathrm{d}x \leq C.\|\varphi\|_\infty.$$

Therefore, T is a distribution and, in addition, of zeroth order.

2) Let us show that the support of T is the line $D := \big\{(x, -x), x \in \mathbb{R}\big\}$.

By definition, it is clear that $\operatorname{supp} T \subset D$ (because $T \equiv 0$ on $\mathbb{R}^2 \setminus D$). Moreover, for all $P_0 \in D$ and $\epsilon > 0$, there exists $\varphi \in \mathcal{D}(\mathbb{R}^2)$ such that $\operatorname{supp}(\varphi) \subset B(P_0, \epsilon)$ and $\langle T, \varphi\rangle = 0$ (choose φ positive, for example). The statement is deduced thereof.

EXERCISE 5.8.– We are given the function $g \in L^1(\mathbb{R})$ and $u \in C^1(\mathbb{R})$. We consider the mapping T defined on $\mathcal{D}(\mathbb{R}^2)$ by $\langle T, \varphi\rangle = \displaystyle\int_{\mathbb{R}} g(x)\varphi(x, u(x))\mathrm{d}x, \forall\varphi \in \mathcal{D}(\mathbb{R}^2)$.

1) Show that T is a distribution.

2) Show that the support of T is contained in the support of a $\mathbb{R}^2$ curve to be determined.

SOLUTION 5.8.–

1) Clearly, the mapping T is a linear form. For any compact $K \subset \mathbb{R}^2$ and for all $\varphi \in \mathcal{D}(\mathbb{R}^2)$ with compact support contained in K, we have

$$\langle T, \varphi\rangle \leq \|g\|_{L^1(\mathbb{R})} \sup_{(x,y)\in K} |\varphi(x, y)|.$$

Therefore, T is a distribution of order 0.

2) Let Γ denote the $\mathbb{R}^2$ curve of class $\mathcal{C}^1$ determined by $\Gamma := \{(x, u(x)), x \in \mathbb{R}\}$. The set Γ is a closed set of $\mathbb{R}^2$. Its complementary $\Omega := \Gamma^c$ is an open set $\mathbb{R}^2$. Let $\varphi \in \mathcal{D}(\mathbb{R}^2)$ with support in Ω. Then, $\forall x \in \mathbb{R}, \varphi(x, u(x)) = 0$ such that $\langle T, \varphi \rangle = 0$. This proves that the support of T is contained in Γ.

EXERCISE 5.9.– Let T be the distribution defined by

$$\langle T, \varphi \rangle = \int_0^{\infty} \left(\varphi\left(\frac{1}{t^2}, \sin t\right) - \varphi\left(0, \sin t\right) \right) dt, \ \forall \varphi \in \mathcal{D}(\mathbb{R}^2).$$

Determine the support of T.

SOLUTION 5.9.– Let us show that if $S = \left\{(x, y) \in \mathbb{R}^2 / y = \sin \frac{1}{\sqrt{x}}, x > 0\right\}$, then $\operatorname{supp} T = \overline{S} = S \cap \left(\{0\} \times [-1, 1]\right)$.

Let $x_0 \notin \overline{S}$. Then, there exists $\epsilon > 0$ such that $\overline{B}(x_0, \epsilon) \subset \overline{S}^c$.

Let $\varphi \in \mathcal{D}(\mathbb{R}^2)$ be such that supp $\varphi \subset \overline{B}(x_0, \epsilon)$.

By definition of T, we have $\langle T, \varphi \rangle = 0$ because $\forall t > 0, \left(\frac{1}{t^2}, \sin t\right) \notin \overline{B}(x_0, \epsilon)$ and $\left(0, \sin t\right) \notin \overline{B}(x_0, \epsilon)$. Therefore, $x_0 \notin \operatorname{supp} T$ and as such, we obtain $\operatorname{supp} T \subset \overline{S}$.

Conversely, let $x_0 \in S$ and let $\epsilon > 0$ such that $B(x_0, \epsilon) \cap \left(\{0\} \times [-1, 1]\right) = \varnothing$. Let $\varphi \in \mathcal{D}(\mathbb{R}^2)$ such that supp$T \subset B(x_0, \epsilon), \varphi = 1$ on $\overline{B}\left(x_0, \frac{\epsilon}{2}\right)$ and $\varphi \geq 0$. Then,

$$\langle T, \varphi \rangle = \int_0^{\infty} \left(\frac{1}{t^2}, \sin t\right) dt \geq (t_2 - t_1) > 0,$$

where t_1 verifies the equality $x_0 = \left(\frac{1}{t_1^2}, \sin t_1\right)$ and $t_2 = \inf \left\{t > t_1, \left(\frac{1}{t^2}, \sin t\right) \notin \overline{B}(x_0, \frac{\epsilon}{2})\right\}$. Then, T is non-null in the neighborhood of x_0 and $S \subset \operatorname{supp} T$.

Since $\operatorname{supp} T$ is closed, $\overline{S} \subset \operatorname{supp} T$ and finally, $\operatorname{supp} T = \overline{S}$.

EXERCISE 5.10.– Let $T \in \mathcal{D}'(\mathbb{R})$ be a distribution such that $\operatorname{supp} T \subset [a, +\infty[$, and let $\varphi \in \mathcal{C}^{\infty}(\mathbb{R})$ such that supp $\varphi \subset]-\infty, b]$.

1) We consider the following expression:

$$\langle T, \varphi \rangle = \langle T, \theta\varphi \rangle, \qquad [5.4]$$

where $\theta \in \mathcal{D}(\mathbb{R})$ is a function equal to 1 on $[-M, M]$ the interval that contains a and b. Show that this expression indeed defines a distribution.

2) Show that the function $\psi(t) = \langle T_x, \varphi(x+t)\rangle$ is defined for all $t \in \mathbb{R}$ and that $\operatorname{supp}\psi \subset]-\infty, b-a]$.

SOLUTION 5.10.–

1) Since $\varphi \in \mathcal{C}^\infty(\mathbb{R})$ and $\theta \in \mathcal{D}(\mathbb{R})$, then $\theta\varphi \in \mathcal{D}(\mathbb{R})$. Hence, the expression $\langle T, \theta\varphi\rangle$ where $\theta \in \mathcal{D}(\mathbb{R})$ indeed makes sense. Let us show that it is independent of θ. We assume that there exists $\theta_1 \in \mathcal{D}(\mathbb{R})$, which equals 1 on $[-M, M]$ and that verifies expression [5.4]. We have that $(\theta - \theta_1)\varphi$ cancels out on $]-M, +\infty[$, thus $\operatorname{supp}(\theta-\theta_1)\varphi \subset]-\infty, -M]$. On the other hand, $\operatorname{supp} T \subset [a, +\infty[$, then $\operatorname{supp}(\theta-\theta_1)\varphi \cap \operatorname{supp} T = \varnothing$. Consequently, $\langle T, (\theta-\theta_1)\varphi\rangle = 0$.

2) For any $t \in \mathbb{R}$, the function $\tau_{-t}\varphi$ belongs to $\mathcal{C}^\infty(\mathbb{R})$. In addition, we have $\operatorname{supp}\tau_{-t}\varphi \subset]-\infty, b-t]$. According to the first question, the bracket $\langle T_x, \varphi(x+t)\rangle$ thus makes sense. The function ψ cancels out if $\operatorname{supp}\tau_{-t}\varphi \cap \operatorname{supp} T = \varnothing$ that is, if $b - t < a$. Consequently, $\operatorname{supp}\psi \subset]-\infty, b-a]$.

EXERCISE 5.11.– Let T be a distribution on $\mathbb{R}^d$, and let f be a function of class $\mathcal{C}^\infty$ on $\mathbb{R}^d$, with values in $\mathbb{R}$.

1) Show that, if $fT = 0$, then the support of T is included in the set $Z(f) = \{x \in \mathbb{R}^d, f(x) = 0\}$.

2) In addition, it is assumed that T is of order 0. Show that the converse is true, if the support of T is included in the set $Z(f) = \{x \in \mathbb{R}^d, f(x) = 0\}$ then $fT = 0$.

3) Taking $T = \delta'$, show that the converse does not hold in general if T is not of zeroth order.

4) Characterize the functions f of class $\mathcal{C}^\infty$ on $\mathbb{R}$ such that $f\delta' = 0$.

SOLUTION 5.11.–

1) Let us assume that $fT = 0$. Let $\varphi \in \mathcal{D}(\mathbb{R}^d)$ such that $\operatorname{supp}\varphi \subset (Z(f))^c$. Then, $\dfrac{\varphi}{f} \in \mathcal{D}(\mathbb{R}^d)$ and we have $\langle T, \varphi\rangle = \left\langle fT, \frac{\varphi}{f}\right\rangle = 0$. Then, T cancels out on $(Z(f))^c$ and therefore $\operatorname{supp} T \subset Z(f)$.

2) Let $T \in \mathcal{D}'(\mathbb{R}^d)$ such that T is of zeroth order and $\operatorname{supp} T \subset Z(f)$. Let K be a compact of $\mathbb{R}^d$. Since T is of zeroth order, then there exists a positive constant c such that for any $\varphi \in \mathcal{D}(\mathbb{R}^d)$ such that $\operatorname{supp}\varphi \subset K$, we have $|\langle T, \varphi\rangle| \le c\,\|\varphi\|_\infty$. Since for

any $\varphi \in \mathcal{D}(\mathbb{R}^d)$ such that $\operatorname{supp}\varphi \subset K$, then we have $f\varphi \in \mathcal{D}(\mathbb{R}^d)$ and $\operatorname{supp} f\varphi \subset K$, thus

$$|\langle fT, \varphi\rangle| = |\langle T, f\varphi\rangle| \leq c\,\|f\varphi\|_\infty \,.$$

Since $\operatorname{supp} T \subset Z(f)$, then $f\varphi$ cancels out on $\operatorname{supp} T$ and thus $\langle fT, \varphi\rangle = 0$. Consequently, $fT = 0$.

3) Let us assume that $T = \delta'$ which is not of zeroth order. We take f such that $f(x) = x$. Then $\operatorname{supp} T \subset Z(f)$, because $\operatorname{supp} T = \{0\}$. However, if we take $\varphi \in \mathcal{D}(\mathbb{R}^d)$ such that $\varphi(0) \neq 0$, then

$$\langle x\delta', \varphi\rangle = -\left\langle \delta, (x\varphi)'\right\rangle = -(x\varphi)'(0) = -\varphi(0) \neq 0.$$

Therefore, $x\delta'$ is not a null distribution.

4) Let $\varphi \in \mathcal{D}(\mathbb{R}^d)$. We have $\langle f\delta', \varphi\rangle = -f'(0)\varphi(0) - f(0)\varphi'(0)$. Then $\langle f\delta', \varphi\rangle$ is null for any φ if $f'(0) = f(0) = 0$. Therefore, $f = x^2 g$ with $g \in \mathcal{D}(\mathbb{R})$.

EXERCISE 5.12.– Let T be a distribution on $\mathbb{R}$ such that $\operatorname{supp}(T) = \{0\}$.

1) Justify that T is of finite order. We will denote by m its order further in the text.

2) Let $\varphi \in \mathcal{D}(\mathbb{R})$ such that $\varphi(x) = o(x^m)$ in the neighborhood of 0.

Let ρ be a plateau function equal to 1 in the neighborhood of 0 and 0 outside $]-1, 1[$. We denote, for $r > 0$ and $x \in \mathbb{R}$, $\rho_r(x) = \rho(x/r)$.

a) Show that, if $l \leq m$, then $\lim\limits_{r\to 0} \sup\limits_{|x|\leq r} |(\rho_r\varphi)^{(l)}(x)| = 0$.

b) Therefrom, deduce that $\langle T, \varphi\rangle = 0$.

3) Let $\varphi \in \mathcal{D}(\mathbb{R})$. Show that φ is written in the form

$$\forall x \in \mathbb{R}, \varphi(x) = \sum_{k=0}^{m} \frac{\varphi^{(k)}(0)}{k!} x^k + \psi(x),$$

where ψ is a function of class $\mathcal{C}^\infty$ verifying $\psi(x) = o(x^m)$ in the neighborhood of 0.

4) Deduce the existence of complex numbers $\alpha_0, \ldots, \alpha_m$ such that, for all $\varphi \in \mathcal{D}(\mathbb{R})$, $\langle T, \varphi\rangle = \sum\limits_{k=0}^{m} \alpha_k \varphi^{(k)}(0)$, or still, $T = \sum\limits_{k=0}^{m} (-1)^k \alpha_k \delta_0^{(k)}$.

SOLUTION 5.12.–

1) Since T has compact support, it is of finite order.

2)a) The Leibniz rule gives

$$\forall \ell \leq m, \quad (\rho_r \varphi)^{(\ell)} = \sum_{k=0}^{\ell} C_\ell^k \left(\rho_r\right)^{(\ell-k)} (\varphi)^k .$$

Let $r > 0$. Since $\varphi(x) = o(x^m)$ in the neighborhood of 0, then $\varphi^{(k)}(x) = o(x^{m-k})$, by uniqueness of the limited expansion. Furthermore,

$$\forall x \in \mathbb{R}, \quad \left(\rho_r\right)^{(\ell-k)}(x) = \frac{1}{r^{\ell-k}} \rho^{(\ell-k)}(\frac{x}{r}).$$

For r small enough and for $|x| < r$, we have

$$\begin{aligned}\left|\left(\rho_r\right)^{(\ell-k)}(x)\left(\varphi\right)^k(x)\right| &\leq \frac{r^{m-k}}{r^{\ell-k}}\left\|\rho^{(\ell-k)}\right\|_\infty \\ &\leq r^{m-\ell}\left\|\rho^{(\ell-k)}\right\|_\infty .\end{aligned}$$

By combination, we obtain $\lim_{r \longrightarrow 0} \sup_{|x| \leq r} \left|(\rho_r\varphi)^{(\ell)}\right| = 0.$

b) Since T is a distribution having $\{0\}$ for support and that $\rho_r = 1$ in the neighborhood of 0, then $\forall r > 0, \forall \varphi \in \mathcal{D}(\mathbb{R}), \langle T, \varphi\rangle = \langle T, \rho_r\varphi\rangle$. Since T is of mth order and $\rho_r\varphi$ is a function with support in $[-r, r]$ with $[-r, r] \subset [-1, 1]$ which is a fixed compact. From question (a), we have $\lim_{r \longrightarrow 0} |\langle T, \rho_r\varphi\rangle| = 0$. Consequently, $\langle T, \varphi\rangle = \lim_{r \longrightarrow 0} |\langle T, \rho_r\varphi\rangle| = 0$.

3) By the Taylor rule with integral remainder, we have

$$\forall x \in \mathbb{R}, \quad \varphi(x) = \sum_{k=0}^{m} \frac{\varphi^{(k)}(0)}{k!} x^k + \frac{x^{m+1}}{m!} \int_0^1 (1-u)^m \varphi^{(m+1)}(xu) \mathrm{d}u.$$

Setting $\psi(x) = \dfrac{x^{m+1}}{m!} \displaystyle\int_0^1 (1-u)^m \varphi^{(m+1)}(xu) \mathrm{d}u$ and using the derivation under the integral sign, we have that ψ is of class $\mathcal{C}^\infty$ and $\psi(x) = o(x^m)$ in the neighborhood of 0.

4) Let $\varphi \in \mathcal{D}(\mathbb{R})$. From the previous question, there exists ψ of class $\mathcal{C}^\infty$ such that $\psi(x) = o(x^m)$ in the neighborhood of 0 and

$$\forall x \in \mathbb{R}, \quad \varphi(x) = \sum_{k=0}^{m} \frac{\varphi^{(k)}(0)}{k!} x^k + \psi(x).$$

We choose a function ρ as in the second question, then $\rho\varphi$ has a compact support; in addition, φ and $\rho\varphi$ are equal in the neighborhood of 0. Since $\operatorname{supp} T = \{0\}$, then

$$\langle T, \varphi \rangle = \langle T, \rho\varphi \rangle = \left\langle T, \sum_{k=0}^{m} \frac{\varphi^{(k)}(0)}{k!} \rho x^k \right\rangle + \langle T, \rho\psi(x) \rangle .$$

However, $\rho\psi$ is of class $\mathcal{C}^\infty$, has compact support and $\rho\psi(x) = o(x^m)$ in the neighborhood of 0, then, by Question 2, $\langle T, \rho\psi \rangle = 0$. Finally,

$$\langle T, \varphi \rangle = \sum_{k=0}^{m} \frac{\varphi^{(k)}(0)}{k!} \left\langle T, \rho x^k \right\rangle = \sum_{k=0}^{m} \frac{\left\langle T, \rho x^k \right\rangle}{k!} \varphi^{(k)}(0) = \sum_{k=0}^{m} \alpha_k \varphi^{(k)}(0),$$

with $\alpha_k = \dfrac{\left\langle T, \rho x^k \right\rangle}{k!}$. Finally,

$$T = \sum_{k=0}^{m} (-1)^k \alpha_k \delta^{(k)} = \sum_{k=0}^{m} a_k \delta^{(k)} \text{ where } a_k = (-1)^k \alpha_k.$$

6

Convolution of Distributions

The importance of the convolution product lies essentially in its regularization role. In this chapter, we introduce the convolution product in the space of distributions. To this end, we consider the convolution product of two integrable functions and then generalize to any two distributions by means of regular distributions.

6.1. Definition and examples

In this section, we define the convolution product of two distributions. To illustrate this definition, various examples are given.

6.1.1. *Convolution of two regular distributions*

Recall that according to proposition 2.4, the convolution product of two integrable functions on $\mathbb{R}^d$ can be defined almost everywhere. Actually, for $f, g \in L^1(\mathbb{R}^d)$, $f * g$ is defined almost everywhere; in addition, $f * g \in L^1(\mathbb{R}^d)$ and in particular we obtain $f * g \in \mathcal{D}'(\mathbb{R}^d)$.

In order to introduce the convolution product of two distributions, it is most natural to define it for regular distributions, namely locally integrable functions. For this purpose, we add an additional condition, which is sufficient to give a meaning to the convolution product within this framework. More precisely, we consider f and g two locally integrable functions on $\mathbb{R}^d$ such that at least one of the functions has a **compact support**. The fact that one of the functions has a compact support K that makes it possible to restrict the computation of the integral to the compact K and thus to address it as for integrable functions. The convolution product of f and g is well defined and given by

$$(f * g)(x) = \int_{\mathbb{R}^d} f(x-y)g(y)\mathrm{d}y = \int_K f(x-y)g(y)\mathrm{d}y.$$

In addition, we have

$$(f * g)(x) = \int_{\mathbb{R}^d} f(x-y)g(y)\mathrm{d}y = \int_{\mathbb{R}^d} f(y)g(x-y)\mathrm{d}y = (g * f)(x).$$

Hence, $f * g = g * f$.

Let $\varphi \in \mathcal{D}(\mathbb{R}^d)$. We have

$$\begin{aligned}
\langle T_{f*g}, \varphi\rangle &= \int_{\mathbb{R}^d} (f * g)(x)\varphi(x)\mathrm{d}x \\
&= \int_{\mathbb{R}^d} \left(\int_{\mathbb{R}^d} f(x-z)g(z)dz \right) \varphi(x)\mathrm{d}x \\
&= \int_{\mathbb{R}^d} \int_{\mathbb{R}^d} f(x-z)g(z)\varphi(x)dz\mathrm{d}x \\
&= \int_{\mathbb{R}^d} \int_{\mathbb{R}^d} f(y)g(z)\varphi(y+z)dz\mathrm{d}y \\
&= \int_{\mathbb{R}^d} \int_{\mathbb{R}^d} f(x)g(y)\varphi(x+y)\mathrm{d}x\mathrm{d}y \\
&= \int_{\mathbb{R}^d} f(x) \left(\int_{\mathbb{R}^d} g(y)\varphi(x+y)\mathrm{d}y \right) \mathrm{d}x \\
&= \langle T_f, \langle T_g, \varphi(x+y)\rangle\rangle .
\end{aligned}$$

Then

$$\forall \varphi \in \mathcal{D}(\mathbb{R}^d), \langle T_{f*g}, \varphi\rangle = \langle T_f, \langle T_g, \varphi(x+y)\rangle\rangle . \qquad [6.1]$$

Note that it can also be written that $\langle T_{f*g}, \varphi\rangle := \langle f \otimes g, \varphi(x+y)\rangle$ where $f \otimes g$ is the tensor product of f and g and verifies

$$\langle f \otimes g, \psi\rangle = \int_{\mathbb{R}^d} \int_{\mathbb{R}^d} f(x)g(y)\psi(x,y)\mathrm{d}x\mathrm{d}y, \quad \forall \psi \in \mathcal{D}(\mathbb{R}^2 \times \mathbb{R}^2).$$

Consequently,

$$\forall\varphi \in \mathcal{D}(\mathbb{R}^d), \langle T_{f*g}, \varphi\rangle = \langle T_f, \langle T_g, \varphi(x+y)\rangle\rangle = \langle f \otimes g, \varphi(x+y)\rangle . \qquad [6.2]$$

DEFINITION 6.1.– *Let f and g be two locally integrable functions on $\mathbb{R}^d$ such that at least one of the functions has compact support. The convolution product of T_f and T_g is the distribution denoted by $T_f * T_g$ and defined by*

$$\forall\varphi \in \mathcal{D}(\mathbb{R}^d), \langle T_f * T_g, \varphi\rangle = \langle T_f, \langle T_g, \varphi(x+y)\rangle\rangle .$$

Otherwise, $\forall\varphi \in \mathcal{D}(\mathbb{R}^d), \langle T_f * T_g, \varphi\rangle = \langle f \otimes g, \varphi(x+y)\rangle$.

6.1.2. *Convolution of a distribution and a function in* $\mathcal{D}(\mathbb{R}^d)$

The definition of the convolution product can be extended with no difficulty to any problem to the convolution product of a distribution by a function of class $\mathcal{C}^\infty$ with compact support.

DEFINITION 6.2.– *Let $T \in \mathcal{D}'(\mathbb{R}^d)$ be a distribution and let $\varphi \in \mathcal{D}(\mathbb{R}^d)$. The convolution product of T by φ is the function $T * \varphi$ defined by*

$$T * \varphi(x) = \langle T, \varphi(x - \cdot)\rangle , \forall x \in \mathbb{R}^d.$$

PROPOSITION 6.1 (Support of $T * \varphi$).– Let $T \in \mathcal{D}'(\mathbb{R}^d)$ be a distribution and let $\varphi \in \mathcal{D}(\mathbb{R}^d)$. We have $\text{supp}\,(T * \varphi) \subset \text{supp}\,T + \text{supp}\,\varphi$.

Proof. Let $x \in \mathbb{R}^d \setminus (\text{supp}\,T + \text{supp}\,\varphi)$. We have $\text{supp}\,\varphi(x - .) = \{x\} - \text{supp}\,\varphi$. Thus, $\text{supp}\,\varphi(x - .) \cap \text{supp}\,T = \varnothing$. This results in $T * \varphi(x) = \langle T, \varphi(x - \cdot)\rangle = 0$. Consequently, $T * \varphi = 0$ on the open set $\mathbb{R}^d \setminus (\text{supp}\,T + \text{supp}\,\varphi)$, which shows that this open set is included in the complementary support of $T * \varphi$. It is deduced that

$$\text{supp}\,(T * \varphi) \subset \text{supp}\,T + \text{supp}\,\varphi. \qquad \square$$

As for the convolution of a locally integrable function by a test function, the product of convolution of a distribution by a test function can be derived by obtaining the derivative of any of the two factors.

PROPOSITION 6.2 (Regularity of $T * \varphi$).– Let $T \in \mathcal{D}'(\mathbb{R}^d)$ be a distribution and let $\varphi \in \mathcal{D}(\mathbb{R}^d)$. The convolution product $T * \varphi$ of T by φ is a function of class $\mathcal{C}^\infty$ on $\mathbb{R}^d$; in addition, we have $\forall\alpha \in N^d, \partial^\alpha (T * \varphi) = (\partial^\alpha T) * \varphi = T * \partial^\alpha (\varphi)$.

Proof. Let $\alpha \in \mathbb{N}^d$. We have

$$\begin{aligned}(\partial^\alpha T) * \varphi(x) &= \langle \partial^\alpha T, \varphi(x - \cdot)\rangle \\ &= (-1)^{|\alpha|} \left\langle T, \partial_y^\alpha \left(\varphi(x - \cdot)\right)\right\rangle.\end{aligned}$$

Since $(-1)^{|\alpha|}\partial_y^\alpha \left(\varphi(x - \cdot)\right) = (\partial^\alpha \varphi)(x - \cdot)$, then

$$(\partial^\alpha T) * \varphi(x) = T * \partial^\alpha \varphi(x).$$

On the other hand, let $x_0 \in \mathbb{R}^d$ and let $\chi \in \mathcal{D}(\mathbb{R}^d)$ be the plateau function such that $\chi = 1$ on $B(x_0, 1)$. The function Φ defined by $\Phi : (x, y) \longmapsto \chi(x)\varphi(x - y)$ is a function of class $\mathcal{C}^\infty$ on $\mathbb{R}^d \times \mathbb{R}^d$ and supp $\Phi \subset$ supp $(\chi) \times$ (supp (χ) + supp (φ)).

According to proposition 4.11, addressing differentiation under the bracket, the function $x \longmapsto \langle T, \chi(x)\varphi(x - \cdot)\rangle = \langle T, \Phi(x, y)\rangle$ is of class $\mathcal{C}^\infty$ on $\mathbb{R}^d$ and we have

$$\partial^\alpha \langle T, \chi(x)\varphi(x - \cdot)\rangle = \langle T, \partial_x^\alpha \left(\chi(x)\varphi(x - \cdot)\right)\rangle.$$

The function $x \longmapsto \langle T, \Phi(x, y)\rangle$ coincide with $T * \varphi$ on $B(x_0, 1)$, then $T * \varphi$ is of class $\mathcal{C}^\infty$ on $B(x_0, 1)$ and we have $T * \partial^\alpha \varphi(x) = (\partial^\alpha T) * \varphi(x) = \partial^\alpha (T * \varphi)(x)$ for all $x \in B(x_0, 1)$. We conclude by observing that the result found is valid for all $x_0 \in \mathbb{R}^d$. □

REMARK 6.1.– *The convolution product of a distribution with compact support by a function of class $\mathcal{C}^\infty$ is defined by the same formula as that in definition 6.2.*

PROPOSITION 6.3 (A different notation).– Let $T \in \mathcal{D}'(\mathbb{R}^d)$ and $\varphi \in \mathcal{D}(\mathbb{R}^d)$. For all $\psi \in \mathcal{D}(\mathbb{R}^d)$, we have $\langle T * \varphi, \psi\rangle = \langle T, \check{\varphi} * \psi\rangle = \langle T, \langle\varphi, \psi(x + y)\rangle\rangle$.

Proof. Let $\psi \in \mathcal{D}(\mathbb{R}^d)$. Using the integration property, under the bracket, of proposition 4.12, we have

$$\begin{aligned}\langle T * \varphi, \psi\rangle &= \int_{\mathbb{R}^d} T * \varphi(x)\psi(x)\mathrm{d}x \\ &= \int_{\mathbb{R}^d} \langle T_y, \varphi(x - y)\rangle\, \psi(x)\mathrm{d}x \\ &= \left\langle T_y, \int_{\mathbb{R}^d} \varphi(x - y)\psi(x)\mathrm{d}x\right\rangle \\ &= \langle T_y, \check{\varphi} * \psi\rangle.\end{aligned}$$

Then, $\langle T * \varphi, \psi \rangle = \langle T, \check{\varphi} * \psi \rangle$. By establishing that $z = x - y$, we get

$$\int_{\mathbb{R}^d} \varphi(x-y)\psi(x)\mathrm{d}x = \int_{\mathbb{R}^d} \varphi(z)\psi(y+z)\mathrm{d}z = \langle \varphi, \psi(y+\cdot) \rangle_x .$$

It follows that $\langle T * \varphi, \psi \rangle = \langle T_y, \langle \varphi, \psi(y+\cdot) \rangle_x \rangle_y = \langle T, \langle \varphi, \psi(y+x) \rangle \rangle .$ □

6.1.3. *Density of* $\mathcal{D}(\Omega)$ *in* $\mathcal{D}'(\Omega)$

Using proposition 6.2, a distribution can be approximated by a sequence of functions of class $\mathcal{C}^\infty$. This is the goal of the following proposition.

PROPOSITION 6.4 (Density of $\mathcal{C}^\infty(\mathbb{R}^d)$ in $\mathcal{D}'(\mathbb{R}^d)$).– Let $T \in \mathcal{D}'(\mathbb{R}^d)$ be a distribution and ρ_ε be a regularizing sequence (i.e. $\rho_\varepsilon(x) = \varepsilon^{-d}\rho\left(\frac{x}{\varepsilon}\right)$ where $\rho \in \mathcal{C}^\infty(\mathbb{R}^d)$ with support in $B(0,1)$, $\rho \geq 0$ and $\displaystyle\int_{\mathbb{R}^d} \rho(x)\mathrm{d}x = 1$). We set $T_\varepsilon = T * \rho_\varepsilon$. Then $T_\varepsilon \in \mathcal{C}^\infty(\mathbb{R}^d)$ and $T_\varepsilon \underset{\varepsilon \longrightarrow 0}{\longrightarrow} T$ in $\mathcal{D}'(\mathbb{R}^d)$.

Proof. The fact that $T_\varepsilon \in \mathcal{C}^\infty(\mathbb{R}^d)$ directly follows from proposition 6.2.

In addition to proposition 6.2, the rest of the proof is based on the property of integration under the duality bracket seen in proposition 4.12. □

REMARK 6.2.– *Obviously, the above regularization property is easily localized in an open set of* $\mathbb{R}^d$.

By combining the truncation principle and the regularization principle, a distribution can be approximated by a sequence of functions of class $\mathcal{C}^\infty$ with compact support. This is the concern of the following proposition.

PROPOSITION 6.5 (Density of $\mathcal{D}(\Omega)$ in $\mathcal{D}'(\Omega)$).– Let $T \in \mathcal{D}'(\Omega)$ be a distribution. Then there exists a sequence $(T_n)_{n \geq 1}$ of functions of class $\mathcal{C}^\infty$ with compact supports such as $T_n \underset{n \longrightarrow +\infty}{\longrightarrow} T$ in $\mathcal{D}'(\Omega)$.

In proposition 6.6, we show the effect of convergence on a sequence of convoluted distributions.

PROPOSITION 6.6 (Convolution and convergence).– Let $\varphi \in \mathcal{D}(\mathbb{R}^d)$ denote a test function and $(T_n)_{n \geq 1}$ a sequence of distributions on $\mathbb{R}^d$ converging to T in the sense of distributions. Then, for any multi-index $\alpha \in \mathbb{N}^d$, $\partial^\alpha (T_n * \varphi) \underset{n \longrightarrow +\infty}{\longrightarrow} \partial^\alpha (T * \varphi)$ uniformly on any compact $\mathbb{R}^d$.

Proof. Without restricting generality, it can be assumed that $T = 0$. On the other hand, we simply have to show that $(T_n * \varphi) \underset{n \longrightarrow +\infty}{\longrightarrow} 0$ and to apply this statement by recurrence by changing T_n into $\partial^\alpha T_n$ since, we have $\partial^\alpha (T_n * \varphi) = \partial^\alpha (T_n) * \varphi$. By hypothesis, we have

$$\langle T_n, \varphi(x - \cdot)\rangle \underset{n \longrightarrow +\infty}{\longrightarrow} 0, \quad \forall x \in \mathbb{R}^d.$$

Let us set $K = \overline{B(0,1)} + \operatorname{supp}(\check{\varphi})$ where $\check{\varphi}(x) = \varphi(-x)$, which is compact in $\mathbb{R}^d$. According to the principle of uniform boundedness (proposition 3.5), there exists $C > 0$ and $p \in \mathbb{N}$ such that

$$\sup_{|x| \leq R} |T_n * \varphi(x)| = \sup_{|x| \leq R} |\langle T_n, \varphi(x - \cdot)\rangle| \leq C \sup_{|\alpha| \leq p} \sup_{x \in K} |\partial^\alpha \varphi(x)|.$$

Let us show that $\sup_{|x| \leq R} |T_n * \varphi(x)| \underset{n \longrightarrow +\infty}{\longrightarrow} 0$. Otherwise, there would exist $\eta > 0$ and a sequence $x_n \in \overline{B(0,1)}$ such that $|T_n * \varphi(x_n)| > \eta, \forall n \geq 1$. Since $\overline{B(0,1)}$ is compact, there exists $n_k \geq 1$ such that the sub-sequence $x_{n_k} \longrightarrow x^* \in \overline{B(0,1)}$,when $k \longrightarrow +\infty$.

Then

$$\begin{aligned}
|T_{n_k} * \varphi(x^*)| &\geq |T_{n_k} * \varphi(x_{n_k})| - |T_{n_k} * \varphi(x_{n_k}) - T_{n_k} * \varphi(x^*)| \\
&\geq \eta - N |x^* - x_{n_k}| \max_{1 \leq j \leq N} \sup_{|y| \leq R} |T_{n_k} * \partial_j \varphi(y)| \\
&\geq \eta - N |x^* - x_{n_k}| \max_{1 \leq j \leq N} (C \max_{|\alpha| \leq p} \sup_{z \in K} |\partial^\alpha \partial_j \varphi(z)|) \\
&\geq \eta - C' |x^* - x_{n_k}| \geq \frac{\eta}{2}.
\end{aligned}$$

with $C' = N \max_{1 \leq j \leq N} \left(C \max_{|\alpha| \leq p} \sup_{z \in K} |\partial^\alpha \partial_j \varphi(z)| \right)$ and where the second inequality above follows from the increment theorem, while the third is a consequence of the uniform estimation on $T_n * \varphi$ obtained above.

Therefore, $\lim_{k \longrightarrow +\infty} |T_{n_k} * \varphi(x^*)| \geq \frac{\eta}{2} > 0$, but this contradicts the fact that $T_n * \varphi(x) \longrightarrow 0$ for all $x \in \mathbb{R}^d$ when $n \longrightarrow +\infty$. It is deduced that the above assumption concerning the existence of the real η is false, that is,

$$\sup_{|x| \leq R} |T_n * \varphi(x)| \underset{n \longrightarrow +\infty}{\longrightarrow} 0.$$

Now, this means that $T_n * \varphi \underset{n \longrightarrow +\infty}{\longrightarrow} 0$ uniformly $\overline{B(0,R)}$ for all $R > 0$. □

6.1.4. *Convolution of two distributions*

In a similar way, the notion of the tensor product and the convolution product can be generalized to distributions that are not necessarily regular. The derivation property under the bracket already seen in proposition 4.11 supports and gives a meaning to the notation defining this tensor product.

THEOREM 6.1.– *Let $T \in \mathcal{D}'(\Omega)$ be a distribution on Ω and $S \in \mathcal{D}'(\Omega')$ a distribution on Ω'. Then, there exists a unique distribution on $\Omega \times \Omega'$ denoted $T \otimes S$, called tensor product such that*

$$\forall \varphi \in \mathcal{D}(\Omega), \forall \psi \in \mathcal{D}(\Omega'), \langle T \otimes S, \varphi \otimes \psi \rangle = \langle T, \varphi \rangle \langle S, \psi \rangle .$$

In addition, for any φ in $\mathcal{D}(\Omega \times \Omega')$ we have

$$\langle T \otimes S, \varphi \rangle = \left\langle T_x, \langle S_y, \varphi \rangle_y \right\rangle_x = \langle S_y, \langle T_x, \varphi \rangle_x \rangle_y .$$

Based on proposition 6.3, the convolution product can be generalized to two distributions, of which at least one has compact support.

DEFINITION 6.3.– *Let T and S be two distributions in $\mathcal{D}'(\mathbb{R}^d)$; it is assumed that at least one has compact support. The convolution $T * S$ is a distribution on $\mathbb{R}^d$ defined by*

$$\forall \varphi \in \mathcal{D}(\mathbb{R}^d), \langle T * S, \varphi \rangle = \langle T_x \otimes S_y, \varphi(x+y) \rangle = \langle T_x, \langle S_y, \varphi(x+y) \rangle \rangle .$$

We simply write

$$\forall \varphi \in \mathcal{D}(\mathbb{R}^d), \langle T * S, \varphi \rangle = \langle T, \langle S, \varphi(x+y) \rangle \rangle . \qquad [6.3]$$

REMARK 6.3.–

1) According to lemma 4.2, the function $\Phi : x \longmapsto \langle S, \varphi(x+y) \rangle$ is of class $\mathcal{C}^\infty$ on $\mathbb{R}^d$, in addition if $S \in \mathcal{D}'(\mathbb{R}^d)$ has support compact, then $\Phi \in \mathcal{D}(\mathbb{R}^d)$, which indeed gives a meaning to the convolution product defined by formula [6.3].

2) By generalizing the convolution product given in proposition 6.3 to the case of two distributions of which at least one is compactly supported, it can be written that $\forall \varphi \in \mathcal{D}(\mathbb{R}^d), \langle T * S, \varphi \rangle = \langle T, \check{S} * \varphi \rangle$.

REMARK 6.4.– *Why assume that at least one of the distributions has compact support? In order to simplify our development, only the case in which d $= 1$ is considered here.*

1) We defined a test function on $\mathbb{R}^2$ by $\psi : (x, y) \longmapsto \psi(x, y) = \varphi(x + y)$ where $\varphi \in \mathcal{D}(\mathbb{R})$. Let $(a, b) \in \mathbb{R}^2$ such that $\operatorname{supp} \varphi \subset [a, b]$, we have $\psi(x, y) = \varphi(x + y)$, is constant for $x + y = c$, with arbitrary $c \in [a, b]$. Therefore, $\operatorname{supp} \psi$ contains all lines of slope -1 that intersect the $y-$axis (as well as the $x-$axis) at a point $c \in \operatorname{supp} \varphi$, as represented here below. As such, ψ does not have compact support, so it is not a test function. However, the action of extended distributions to test functions $\varphi \notin \mathcal{D}(\mathbb{R})$ can be used.

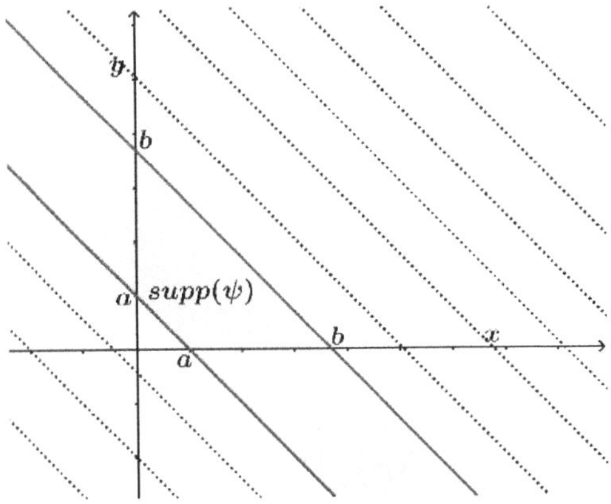

Figure 6.1. *supp ψ is unbounded. For a color version of this figure, see www.iste.co.uk/aitbenhassi/distribution.zip*

Let us assume that the distribution S has a compact support. Then let a and b be such that $\operatorname{supp} S \subset [a, b]$; in Figure 6.2, it can be seen that the test function ψ has compact support. In this figure, $\operatorname{supp} \varphi \subset [\alpha, \beta]$.

2) The definition of the convolution product $T * S$ of two distributions remains valid for two distributions T and S, "causal", having two left bound supports: that is, $\operatorname{supp} T \subset [a, +\infty[$ and $\operatorname{supp} S \subset [b, +\infty[$.

3) The result remains true if the two distributions have their supports limited to the right: $\operatorname{supp} T \subset]-\infty, a]$ and $\operatorname{supp} S \subset]-\infty, b]$.

6.1.5. *Some examples*

EXAMPLE 6.1.– Let T be a distribution in $\mathcal{D}'(\mathbb{R}^d)$.

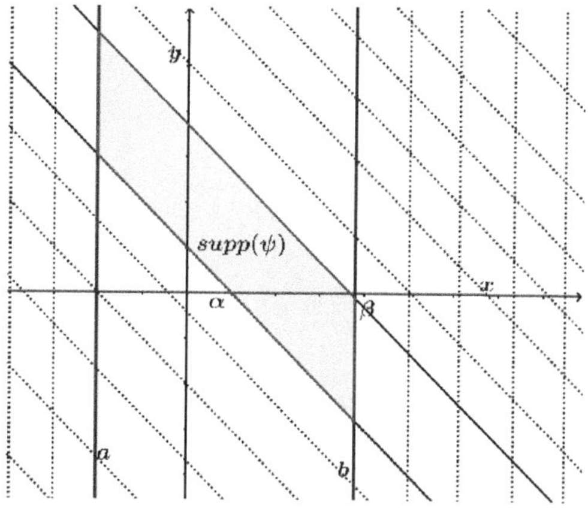

Figure 6.2. *First example in which suppψ is bounded. For a color version of this figure, see www.iste.co.uk/aitbenhassi/distribution.zip*

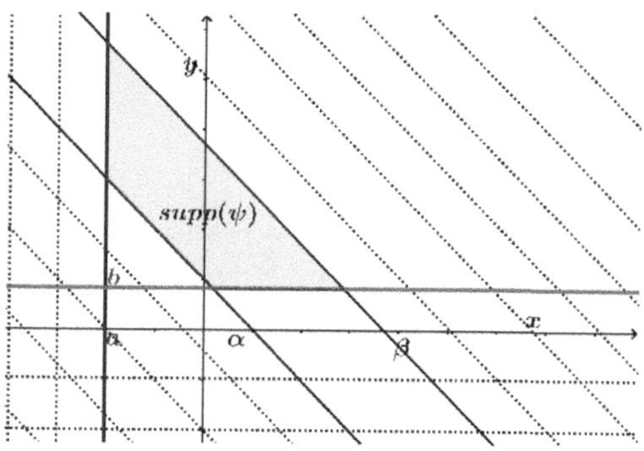

Figure 6.3. *Second example in which suppψ is bounded. For a color version of this figure, see www.iste.co.uk/aitbenhassi/distribution.zip*

1) We have $T * \delta_0 = T$.

Indeed, for any φ in $\mathcal{D}(\mathbb{R}^d)$, we have

$$
\begin{aligned}
\langle T * \delta, \varphi \rangle &= \left\langle T_x, \langle \delta^y, \varphi(x+y) \rangle_y \right\rangle_x \\
&= \langle T_x, \varphi(x+0) \rangle \\
&= \langle T_x, \varphi(x) \rangle \\
&= \langle T, \varphi \rangle .
\end{aligned}
$$

Therefore, $T * \delta = T$. Similarly, we have $\delta * T = T$.

2) We have $\delta_a * T = \tau_a T$.

For any φ in $\mathcal{D}(\mathbb{R}^d)$, we actually have

$$\begin{aligned}\langle \delta_a * T, \varphi \rangle &= \langle \delta_a \otimes T, \varphi(x+y) \rangle \\ &= \left\langle \delta_a^x, \langle T_y, \varphi(x+y) \rangle_y \right\rangle_x \\ &= \langle T_y, \varphi(a+y) \rangle \\ &= \langle T, \tau_{-a}\varphi \rangle \\ &= \langle \tau_a T, \varphi \rangle .\end{aligned}$$

3) We have $\delta_a * \delta_b = \delta_{a+b}$.

Furthermore, for any φ in $\mathcal{D}(\mathbb{R}^d)$, we have

$$\begin{aligned}\langle \delta_a * \delta_b, \varphi \rangle &= \langle \delta_a \otimes \delta_b, \varphi(x+y) \rangle \\ &= \left\langle \delta_a^x, \langle \delta_b^y, \varphi(x+y) \rangle_y \right\rangle_x \\ &= \langle \delta_b^y, \varphi(a+y) \rangle \\ &= \varphi(a+b) \\ &= \langle \delta_{a+b}, \varphi \rangle .\end{aligned}$$

REMARK 6.5.–

1) The convolution product «$*$ » is an internal composition law on $\mathcal{E}'(\mathbb{R}^d)$ of neutral element $\delta = \delta_0$.

2) The convolution product «$*$ » is not an internal law on $\mathcal{D}'(\mathbb{R}^d)$.

EXAMPLE 6.2.– Let T be a distribution in $\mathcal{D}'(\mathbb{R})$. We have $\delta' * T = T'$.

Indeed, for any φ in $\mathcal{D}(\mathbb{R})$, we have

$$\begin{aligned}\langle \delta' * T, \varphi \rangle &= \langle \delta' \otimes T, \varphi(x+y) \rangle \\ &= \left\langle \delta'_x, \langle T_y, \varphi(x+y) \rangle_y \right\rangle_x \\ &= -\frac{d}{dx} \left(\langle T_y, \varphi(x+y) \rangle \right) |_{x=0}\end{aligned}$$

$$= -\left\langle T_y, \frac{d}{\mathrm{d}x}(\varphi(x+y)|_{x=0}\right\rangle \text{ (we use differentiation under the bracket)}$$

$$= -\langle T_y, \varphi'(0+y)\rangle$$

$$= -\langle T, \varphi'\rangle$$

$$= \langle T', \varphi\rangle .$$

EXAMPLE 6.3.– We consider $\mathcal{D}'_+(\mathbb{R}) = \{T \in \mathcal{D}'(\mathbb{R}), \operatorname{supp} T \subset [a, +\infty[\,, a \in \mathbb{R}\}$ (respectively, $\mathcal{D}'_-(\mathbb{R}) = \{T \in \mathcal{D}'(\mathbb{R}), \operatorname{supp} T \subset \,]-\infty, b]\,, b \in \mathbb{R}\}$) the set of distributions with left-hand limited supports "also called causal" (respectively, right-hand limited).

The set $\mathcal{D}'_+(\mathbb{R})$ (respectively, $\mathcal{D}'_-(\mathbb{R})$) is a real vector space. In addition, the convolution product "$*$" is well defined on $\mathcal{D}'_+(\mathbb{R})$ (respectively, on $\mathcal{D}'_-(\mathbb{R})$).

6.2. Properties of convolution

In this section, we give some properties of the convolution of two distributions.

6.2.1. *Support of a convolution*

PROPOSITION 6.7.– Let S and T be two distributions on $\mathbb{R}^d$ of which at least one has compact support. We have $\operatorname{supp}(T * S) \subset \operatorname{supp} T + \operatorname{supp} S$.

Proof. We introduce the mapping

$$\begin{aligned} s: \operatorname{supp} T \times \operatorname{supp} S &\longrightarrow \mathbb{R}^d \\ (x, y) &\longmapsto x + y. \end{aligned}$$

Let $\varphi^{\triangle}$ denote the mapping $\varphi^{\triangle} : (x, y) \longmapsto \varphi(x + y)$.

Let $x \notin \operatorname{supp} T + \operatorname{supp} S$. Since $\operatorname{supp} T + \operatorname{supp} S$ is a closed set, there exists $\delta > 0$ such that $B(x, \delta) \cap (\operatorname{supp} T + \operatorname{supp} S) = \varnothing$. Then, for $\varphi \in \mathcal{D}\big(B(x, \delta)\big)$, we have

$$\operatorname{supp} \varphi^{\triangle} \cap (\operatorname{supp} T \times \operatorname{supp} S) = s^{-1}(\operatorname{supp} \varphi) \subset s^{-1}\big(B(x, \delta)\big) = \varnothing.$$

Therefore, $\langle T * S, \varphi\rangle = \langle T \otimes S, \varphi^{\triangle}\rangle = 0$ and $x \notin \operatorname{supp}(T * S)$.

Consequently, $\operatorname{supp}(T * S) \subset \operatorname{supp} T + \operatorname{supp} S$. □

REMARK 6.6.– *The reciprocal inclusion does not hold in general. Two locally integrable functions u and v should be taken with $u = 1$ and $\int_{\mathbb{R}^d} v(x)dx = 0$. We then have $u * v = 0$, hence supp $(T_u * T_v) = \varnothing$ while supp T_u + supp $T_v = \mathbb{R}^d$ since supp $T_u = \mathbb{R}^d$.*

However, we can have a form of reciprocal inclusion by considering distributions with compact support only and convex envelopes. If $A \subset \mathbb{R}^d$, we denote Conv(A) its convex envelope, that is, the smallest convex $\mathbb{R}^d$ set that contains A. We then have the following result.

THEOREM 6.2.– *Let $S, T \in \mathcal{E}'(\mathbb{R}^d)$ be two distributions on $\mathbb{R}^d$ with compact supports. We have Conv$\big($supp $(T * S)\big)$ = Conv$\big($supp $T\big)$ + Conv$\big($supp $S\big)$.*

6.2.2. Sequential continuity of the convolution product

THEOREM 6.3.– *Let $\left(T_n\right)_{n\geq 1}$ be a sequence of $\mathcal{D}'(\mathbb{R}^d)$ and $\left(S_n\right)_{n\geq 1}$ a sequence of $\mathcal{E}'(\mathbb{R}^d)$ such that $T_n \underset{n\to+\infty}{\longrightarrow} T$ and $S_n \underset{n\to+\infty}{\longrightarrow} S$ in $\mathcal{D}'(\mathbb{R}^d)$. Furthermore, let us assume that there exists a compact $K \subset \mathbb{R}^d$ such that supp$(S_n) \subset K$, for all $n \geq 1$. Then,*

$$T_n * S_n \underset{n\to+\infty}{\longrightarrow} T * S \text{ in } \mathcal{D}'\big(\mathbb{R}^d\big).$$

Proof. Let $\varphi \in \mathcal{D}(\mathbb{R}^d)$. We have $\langle T_n * S_n, \varphi\rangle = \big\langle T_n, \check{S}_n * \varphi\big\rangle$. By hypothesis for all $n \geq 1$, we have supp $\check{S}_n * \varphi \subset$ supp $\check{S}_n$ + supp $\varphi \subset$ supp $\varphi - K$. Therefore, supp $\check{S}_n * \varphi$ is compact as a closed set included in the sum of two compacts. By proposition 6.6, for any multi-index $\alpha \in \mathbb{N}^d$, we have

$$\partial^\alpha\left(\check{S}_n * \varphi\right) = \partial^\alpha\left(\check{S}_n\right) * \varphi \underset{n\longrightarrow+\infty}{\longrightarrow} \partial^\alpha\left(\check{S}\right) * \varphi = \partial^\alpha\left(\check{S} * \varphi\right)$$

uniformly on any compact of $\mathbb{R}^d$. Otherwise,

$$\check{S}_n * \varphi \underset{n\longrightarrow+\infty}{\longrightarrow} \check{S} * \varphi, \quad \text{in } \mathcal{D}(\mathbb{R}^d).$$

Corollary 3.2 gives

$$\langle T_n * S_n, \varphi\rangle = \big\langle T_n, \check{S}_n * \varphi\big\rangle \underset{n\longrightarrow+\infty}{\longrightarrow} \big\langle T, \check{S} * \varphi\big\rangle = \langle T * S, \varphi\rangle\,.$$

The test function $\varphi \in \mathcal{D}(\mathbb{R}^d)$ being arbitrary, it is deduced that

$$T_n * S_n \underset{n \longrightarrow +\infty}{\longrightarrow} T * S \text{ in } \mathcal{D}'(\mathbb{R}^d).$$ □

6.2.3. *Associativity and convolution*

The convolution product "$*$" is not associative in $\mathcal{D}^{'}(\mathbb{R}^n)$ (see the counterexample below). However, we have the following result.

PROPOSITION 6.8.– Let R, S and $T \in \mathcal{D}'(\mathbb{R}^d)$, verifying one of the following two conditions, namely,

i) at least two of them have compact support;

ii) or they are all causal.

Then we have $R * \big(S * T\big) = \big(R * S\big) * T$.

Proof. By virtue of commutativity, the problem can always be reduced to the case where S and T have compact supports. If we take $r > 0$ and large enough, we will assume that $\operatorname{supp} S, \operatorname{supp} T \subset B(0, r)$.

We consider the regularizing sequence ρ_n defined by $\rho_n(x) = n^d \rho(nx)$ where $\rho \in \mathcal{D}(\mathbb{R}^d)$ is a function such that $\operatorname{supp} \rho \subset B(0,1), \rho \geq 0, \displaystyle\int_{\mathbb{R}^d} \rho(x)\mathrm{d}x = 1$.

We set $T_n = T * \rho_n$ and $S_n = S * \rho_n$. Let us verify that $R * \big(S_n * T_n\big) = \big(R * S_n\big) * T_n$. We have

$$\begin{aligned} R * \big(S_n * T_n\big)(x) &= \langle R, (S_n * T_n)(x - \cdot)\rangle \\ &= \left\langle R, \int_{\mathbb{R}^d} S_n\left((x - \cdot) - z\right) T_n(z)\mathrm{d}z \right\rangle. \end{aligned}$$

The proposition of integration under the duality bracket makes it possible to have

$$\begin{aligned} R * \big(S_n * T_n\big)(x) &= \int_{\mathbb{R}^d} \langle R, S_n\left((x - \cdot) - z\right) T_n(z)\rangle \,\mathrm{d}z \\ &= \int_{\mathbb{R}^d} R * S_n\left(x - z\right) T_n(z)\mathrm{d}z \\ &= (R * S_n) * T_n(x). \end{aligned}$$

Furthermore, $R * (S_n * T_n) = (R * S_n) * T_n$. By construction, for all $n \geq 1$, we have

$$\operatorname{supp} S_n \subset B(0, r+1) \quad \text{and} \quad \operatorname{supp} T_n \subset B(0, r+1)$$

and

$$S_n \underset{n \to +\infty}{\longrightarrow} S, \quad T_n \underset{n \to +\infty}{\longrightarrow} T \quad \text{in } \mathcal{D}'(\mathbb{R}^d).$$

Theorem 6.3 of sequential continuity of the convolution product ensures, on the one hand, that $R{*}S_n \underset{n \to +\infty}{\longrightarrow} R{*}S$, then $(R{*}S_n){*}T_n \underset{n \to +\infty}{\longrightarrow} (R{*}S){*}T$ in $\mathcal{D}'(\mathbb{R}^d)$. On the other hand, it is similarly shown that

$$S_n * T_n \underset{n \to +\infty}{\longrightarrow} S * T \quad \text{and } \operatorname{supp}(S_n * T_n) \subset \operatorname{supp} S_n + \operatorname{supp} T_n \subset B(0, 2r+2).$$

By sequential continuity of the convolution product, it is further deduced that

$$R * (S_n * T_n) \underset{n \to +\infty}{\longrightarrow} R * (S * T) \text{ in } \mathcal{D}'(\mathbb{R}^d).$$

Consequently, $R * (S * T) = (R * S) * T$. □

REMARK 6.7.– *The convolution product of n distributions of which $(n-1)$ at least have compact support is associative and commutative.*

We should bear in mind that this result may not hold if only one of the three distributions has compact supports, although the right- and left-hand members of the previous equality make sense. Here follows an example of this phenomenon.

EXAMPLE 6.4 (Counterexample of non-associativity).– We consider the distributions H, δ' and 1.

The products $(H * \delta') * 1$ and $H * (\delta' * 1)$ do exist but are different. As a matter of fact,

$$(H * \delta') * 1 = (H' * \delta) * 1 = \delta * 1 = 1$$

$$H * (\delta' * 1) = H * (\delta * 1') = H * 1' = H * 0 = 0.$$

It should be observed that H, δ' and 1 do not satisfy the conditions ensuring associativity.

6.2.4. *Differentiation and convolution*

In this section, we present the action of differentiation on the convolution of two distributions.

PROPOSITION 6.9.– Let S and T in $\mathcal{D}'(\mathbb{R})$. If $S * T$ exists, then we have

$$S' * T = S * T' = \big(S * T\big)'.$$

In general, we have

$$\forall (p,q) \in \mathbb{N}^2, S^{(p)} * T^{(q)} = S^{(p+q)} * T = S * T^{(p+q)} = \big(S * T\big)^{(p+q)}.$$

Proof. Let S and T in $\mathcal{D}'(\mathbb{R})$ such that $S * T$ exists, that is, either they are both causal, or at least one of them has compact support, then δ', S and T satisfy the previous conditions (actually, in the first case, they are all causal, including δ', and in the second, two have indeed compact support, in particular δ'). We thus have $\big(\delta' * S\big) * T = S * \big(\delta' * T\big) = \delta' * \big(S * T\big)$. From example 6.2, we obtain $S' * T = S * T' = \big(S * T\big)'$. Reasoning by recurrence leads to easily generalizing the formula to any order, and we can write that

$$\forall (p,q) \in \mathbb{N}^2, S^{(p)} * T^{(q)} = S^{(p+q)} * T = S * T^{(p+q)} = \big(S * T\big)^{(p+q)}. \qquad \square$$

REMARK 6.8.– *To derive a convolution product, one of the factors has simply to be derived.*

The result is given for distributions on $\mathbb{R}$, but it remains valid for distributions defined on $\mathbb{R}^d$. A generalization in the space is given by the following proposition.

PROPOSITION 6.10.– Let S and T in $\mathcal{D}'(\mathbb{R}^d)$ and $\alpha \in \mathbb{N}^d$. If $S * T$ exists, we then have

1) $\partial^\alpha T = (\partial^\alpha \delta) * T$;

2) $\partial^\alpha (S * T) = \partial^\alpha S * T = S * \partial^\alpha T$.

APPLICATION EXERCISE.– Compute $x^m \delta_0^{(n)} * x^p \delta_0^{(p)}$.

EXAMPLE 6.5.– Let $T \in \mathcal{D}'(\mathbb{R})$ be a causal distribution, then a primitive P of T is simply written as $P = H * T$. In fact, $P' = H' * T = \delta * T = T$.

6.2.5. *Translation and convolution*

PROPOSITION 6.11.– Let $S, T \in \mathcal{D}'(\mathbb{R}^d)$ denote two distributions such that $S * T$ exists. For all $a \in \mathbb{R}^d$, we have $\tau_a S * T = S * \tau_a T = \tau_a(S * T)$.

Proof. The distributions δ_a, S and T verify the conditions for applying associativity. We have $(\delta_a * S) * T = S * (\delta_a * T) = \delta_a * (S * T)$. Using the fact that $\delta_a * T = \tau_a T$ already seen in example 6.1, we get

$$\tau_a S * T = S * \tau_a T = \tau_a(S * T).$$

□

6.2.6. *Algebraic study of* $\mathcal{D}'_+(\mathbb{R})$

PROPOSITION 6.12.– The convolution produit "$*$" is an internal law in $\mathcal{D}'_+(\mathbb{R})$.

Proof. For any $T, S \in \mathcal{D}'_+(\mathbb{R})$, such that $\operatorname{supp} T \subset [a, +\infty[$ and $\operatorname{supp} S \subset [b, +\infty[$, the convolution product $T * S$ is well defined. By proposition 6.7 on the support of the convoluted of two distributions, we have $\operatorname{supp}(T * S) \subset \operatorname{supp} T + \operatorname{supp} S$. Therefore, $\operatorname{supp}(T * S) \subset [a + b, +\infty[$. Consequently, $T * S \in \mathcal{D}'_+(\mathbb{R})$. □

PROPOSITION 6.13.– $\mathcal{D}'_+(\mathbb{R})$ is a commutative unitary algebra for the convolution product.

Proof. The mapping $(T, S) \longmapsto T * S$ is bilinear. The commutativity follows from the fact that $\varphi(x+y) = \varphi(y+x)$. The neutral element is δ. Associativity is the subject of proposition 6.8. □

REMARK 6.9.–

1) The proposition holds for $\mathcal{D}'_-(\mathbb{R})$ and consequently for $\mathcal{D}'_+(\mathbb{R}) \cap \mathcal{D}'_-(\mathbb{R}) = \mathcal{E}'(\mathbb{R})$.

2) An interesting special case: the spaces $\mathcal{D}'_{0,+}(\mathbb{R}) = \{T \in \mathcal{D}'(\mathbb{R}), \operatorname{supp} T \subset [0, +\infty[\}$, $\mathcal{D}'_{0,-}(\mathbb{R}) = \{T \in \mathcal{D}'(\mathbb{R}), \operatorname{supp} T \subset]-\infty, 0]\}$ and $\mathcal{D}'_0(\mathbb{R}) = \mathcal{D}'_{0,+}(\mathbb{R}) \cap \mathcal{D}'_{0,-}(\mathbb{R})$ are commutative unitary algebras for the convolution product.

Any element T of $\mathcal{D}'_0(\mathbb{R})$ has point support $\{0\}$. By theorem 5.2, the distribution T is uniquely written as a differential polynomial in δ. Then, the following result is deduced.

PROPOSITION 6.14.– $\mathcal{D}'_0(\mathbb{R})$ endowed with the law "$*$" is isomorphic to the algebra of polynomials.

Proof. We consider the mapping $\Theta : \sum_{0\leq k\leq m} a_k\delta^{(k)} \longmapsto \sum_{0\leq k\leq m} a_k X^k.$ □

REMARK 6.10.– *Since the algebra of polynomials is integral, $\mathcal{D}'_0(\mathbb{R})$ is also integral and admits a fraction field. The latter is identified with a subalgebra of $\mathcal{D}'_{0,+}(\mathbb{R})$.*

THEOREM 6.4.– *The subalgebra of $\mathcal{D}'_{0,+}(\mathbb{R})$ generated by $\mathcal{D}'_0(\mathbb{R})$ and distributions similar to $H(x)e^{\lambda x}$ with $\lambda \in \mathbb{C}$ is a commutative field for the law "$*$". It is the fraction field of $\mathcal{D}'_0(\mathbb{R})$. Moreover, it is the direct sum of $\mathcal{D}'_0(\mathbb{R})$ and the space of distributions similar to $H(x)\left(\sum_{j=0}^{N} P_j e^{\lambda_j x}\right)$ (the $\lambda_j \in \mathbb{C}$ and the P_j are polynomials).*

This field is isomorphic to the field $\mathbb{C}(X)$ of rational fractions.

6.3. Exercises with solutions

EXERCISE 6.1.– Let $a, b \in \mathbb{R}^d$. Compute the convolution product in each of the following cases:

1) $\delta_a * \delta_b$;

2) $\delta'_a * \delta'_b$;

3) $\delta'_a * u$ where $u \in \mathcal{D}'(\mathbb{R}^d)$.

SOLUTION 6.1.–

1) Let $\varphi \in \mathcal{D}(\mathbb{R}^d)$. By definition, we have

$$\begin{aligned}\langle \delta_a * \delta_b, \varphi\rangle &= \langle \delta_a, \langle \delta_b, \varphi(x+y)\rangle\rangle \\ &= \langle \delta_a, \varphi(x+b)\rangle \\ &= \varphi(a+b)\end{aligned}$$

which shows that $\delta_a * \delta_b = \delta_{a+b}$.

2) Let us recall that $\partial^\alpha(u * v) = (\partial^\alpha u) * v = u * (\partial^\alpha v)$. Thus,

$$\delta'_a * \delta'_b = (\delta_a * \delta_b)'' = \delta''_{a+b}.$$

3) We have $\delta'_a * u = \delta_a * u'$, and a calculation similar to that of the first question shows that $\delta_a * u' = \tau_a(u')$.

EXERCISE 6.2.– Calculate the convolution product $(H(x)\cos x) * (\delta' + H(x))$, where $H(x)$ is the Heaviside function.

SOLUTION 6.2.– We have

$$\begin{aligned}(H(x)\cos x) * (\delta' + H(x)) &= (H(x)\cos x) * \delta' + (H(x)\cos x) * H(x)\\ &= (H(x)\cos x)' * \delta + H(x) * (H(x)\cos x)\\ &= (H(x)\cos x)' + H(x) * (H(x)\cos x)\\ &= H'(x)\cos x - H(x)\sin x + H(x) * (H(x)\cos x)\\ &= \delta\cos x - H(x)\sin x + H(x) * (H(x)\cos x)\\ &= \delta - H(x)\sin x + H(x) * (H(x)\cos x).\end{aligned}$$

We compute $H(x) * (H(x)\cos x)$. For $\varphi \in \mathcal{D}(\mathbb{R})$, we have

$$\begin{aligned}\langle H(x) * (H(x)\cos x), \varphi\rangle &= \langle H(x), \langle (H(y)\cos y), \varphi(x+y)\rangle\rangle\\ &= \left\langle H(x), \int_0^{+\infty} \cos y\varphi(x+y)\mathrm{d}y\right\rangle\\ &= \int_0^{+\infty}\int_0^{+\infty} \cos y\varphi(x+y)\mathrm{d}y\mathrm{d}x\\ &= \int_0^{+\infty}\left(\int_0^{+\infty} \varphi(x+y)\mathrm{d}x\right)\cos y\mathrm{d}y.\end{aligned}$$

We set $\psi(y) = \displaystyle\int_0^{+\infty} \varphi(x+y)\mathrm{d}x$. Then

$$\begin{aligned}\langle H(x) * (H(x)\cos x), \varphi\rangle &= \int_0^{+\infty} \psi(y)\cos y\mathrm{d}y\\ &= [\psi(y)\cos y]_0^{+\infty} - \int_0^{+\infty} \psi'(y)\sin y\mathrm{d}y\\ &= -\int_0^{+\infty} \psi'(y)\sin y\mathrm{d}y.\end{aligned}$$

By setting $u = x + y$, we have $\psi(y) = \displaystyle\int_y^{+\infty} \varphi(u)\mathrm{d}u$. Then, $\psi'(y) = -\varphi(y)$. It follows that

$$\langle H(x) * (H(x)\cos x), \varphi\rangle = \int_0^{+\infty} \varphi(y)\sin y\mathrm{d}y = \int_{-\infty}^{+\infty} H(y)\sin y\varphi(y)\mathrm{d}y.$$

Consequently, $\langle H(x) * (H(x)\cos x), \varphi\rangle = \langle H(x)\sin x, \varphi\rangle$.

Therefore, $H(x) * (H(x)\cos y) = H(x)\sin x$. Finally,

$$(H(x)\cos x) * (\delta' + H(x)) = \delta.$$

EXERCISE 6.3.– Let $S, T \in \mathcal{D}'_+(\mathbb{R})$ from the space add two distributions and $a \in \mathbb{C}$ be a complex number. Show that $e^{ax}(S * T) = e^{ax}S * e^{ax}T$.

SOLUTION 6.3.– By definition, for any test function $\varphi \in \mathcal{D}(\mathbb{R})$, we have

$$\begin{aligned}\langle e^{ax}S * e^{ax}T, \varphi\rangle &= \langle e^{ax}S_x, \langle e^{ay}T_y, \varphi(x+y)\rangle\rangle \\ &= \langle S_x, e^{ax}\langle e^{ay}T_y, \varphi(x+y)\rangle\rangle \\ &= \langle S_x, \langle T_y, e^{a(x+y)}\varphi(x+y)\rangle\rangle \\ &= \langle S * T, e^{ax}\varphi\rangle \\ &= \langle e^{ax}(S * T), \varphi\rangle.\end{aligned}$$

Consequently, $e^{ax}(S * T) = e^{ax}S * e^{ax}T$.

EXERCISE 6.4.– Let H be a Heaviside distribution, and let H^{*n} denote the convolution product of H by itself n–times: $H^{*n} = H * H * \cdots * H$.

1) Show that for any natural number $n \geq 1$, we have $H^{*n} = \dfrac{x^{n-1}}{(n-1)!}H$.

2) Therefrom, deduce the derivative of order n of H^{*n}, for all $n \geq 1$.

SOLUTION 6.4.–

1) The relation is obvious for $n = 1$. By induction, assume that

$$H^{*(n-1)} = \frac{x^{n-2}}{(n-2)!}H.$$

For any test function $\varphi \in \mathcal{D}(\mathbb{R})$, we have

$$\langle H^{*n}, \varphi\rangle = \langle H_x, \langle H_y^{*(n-1)}, \varphi(x+y)\rangle\rangle.$$

By induction hypothesis, we have

$$\langle H_y^{*(n-1)}, \varphi(x+y)\rangle = \int_0^{+\infty} \frac{y^{n-2}}{(n-2)!}\varphi(x+y)\mathrm{d}y = \int_x^{+\infty} \frac{(s-x)^{n-2}}{(n-2)!}\varphi(s)\mathrm{d}s.$$

As a result, it follows that

$$\langle H^{*n}, \varphi\rangle = \int_0^{+\infty}\int_x^{+\infty} \frac{(s-x)^{n-2}}{(n-2)!}\varphi(s)\mathrm{d}s\mathrm{d}x.$$

By inverting the order of the integrals, we obtain

$$\begin{aligned}\langle H^{*n}, \varphi\rangle &= \int_0^{+\infty}\left(\int_0^s \frac{(s-x)^{n-2}}{(n-2)!}\mathrm{d}x\right)\varphi(s)\mathrm{d}s \\ &= \int_0^{+\infty} \frac{s^{n-1}}{(n-1)!}\varphi(s)\mathrm{d}s \\ &= \langle \frac{x^{n-1}}{(n-1)!}H, \varphi\rangle.\end{aligned}$$

The proof is thus complete.

2) A simple calculation gives $(H^{*n})^{(n)} = \delta$, for all $n \geq 1$.

EXERCISE 6.5.– Let $P \in \mathbb{C}[X]$ be a polynomial of degree d and let $u \in \mathcal{E}'(\mathbb{R})$.

Show that $u * P$ is also a polynomial of degree less than or equal to d.

SOLUTION 6.5.– Since $v \longmapsto u * v$ is a linear mapping, it can be assumed that: $P = X^k$, with $0 \leq k \leq d$. We have

$$\begin{aligned}(u * X^k)(x) &= \langle u, (x-y)^k\rangle \\ &= \langle u, \sum_{i=0}^{k}\binom{k}{i} x^i(-y)^{k-i}\rangle \\ &= \sum_{i=0}^{k}\binom{k}{i} x^i \langle u, (-y)^{k-i}\rangle.\end{aligned}$$

Since the $\langle u, (-y)^{k-i} \rangle$ are constants in $\mathbb{C}$, it is deduced that $u * P$ is a polynomial of degree $\leq d$.

EXERCISE 6.6.–

1) Let $T \in \mathcal{D}'(\mathbb{R})$. Show that there exists a distribution $E \in \mathcal{D}'(\mathbb{R})$, with compact support, such that $E * T = T^{(k)}$.

2) Let T and S in $\mathcal{D}'(\mathbb{R})$, and$S$ is assumed to have compact support. For $n \in \mathbb{N}$, we denote by X^n the function, from $\mathbb{R}$ in $\mathbb{R}$, $x \longmapsto x^n$. Demonstrate the following formula: $X^n(T * S) = \sum_{k=0}^{n} \binom{n}{k} (X^k T) * \left(X^{n-k} S\right)$.

SOLUTION 6.6.–

1) The distribution $E = \delta_0^{(k)}$ is suitable. It happens that if we derive k-times the identity $\delta_0 * T = T$, we will obtain $\delta_0^{(k)} * T = T^{(k)}$.

2) Let $\varphi \in \mathcal{D}(\mathbb{R})$. If we set $\varphi^{\triangle}(x, y) = \varphi(x + y)$, we have

$$\begin{aligned}
\langle X^n(T * S), \varphi \rangle &= \langle T * S, X^n \varphi \rangle \\
&= \left\langle T \otimes S, (x + y)^n \varphi^{\triangle}(x, y) \right\rangle \\
&= \sum_{k=0}^{n} \binom{n}{k} \left\langle T \otimes S, x^k y^{n-k} \varphi^{\triangle}(x, y) \right\rangle \\
&= \sum_{k=0}^{n} \binom{n}{k} \left\langle (x^k T) \otimes (y^{n-k} S), \varphi^{\triangle}(x, y) \right\rangle \\
&= \langle \sum_{k=0}^{n} \binom{n}{k} (X^k T) * (X^{n-k} S, \varphi \rangle.
\end{aligned}$$

Consequently, $X^n(T * S) = \sum_{k=0}^{n} \binom{n}{k} (X^k T) * \left(X^{n-k} S\right)$.

EXERCISE 6.7.– We write $\mathcal{D}'_{0,+}(\mathbb{R}) = \left\{ T \in \mathcal{D}'(\mathbb{R}), \operatorname{supp}(T) \subset \mathbb{R}^+ \right\}$.

Let $\chi \in \mathcal{C}^\infty(\mathbb{R})$ such that, $\chi = 1$ on $\left] -\frac{1}{2}, +\infty \right[$ and $\chi = 0$ on $\left] -\infty, -1 \right[$.

1)a) Let $\varphi \in \mathcal{D}(\mathbb{R})$. Show that the mapping $\varphi^{\triangle} : (x, y) \to \chi(x)\chi(y)\varphi(x + y)$ is in $\mathcal{D}(\mathbb{R}^2)$.

b) Let $T, S \in \mathcal{D}'_{0,+}(\mathbb{R})$. We define: $\langle T * S, \varphi \rangle = \left\langle T_x \otimes S_y, \varphi^{\triangle} \right\rangle$.

Show that $T * S$ is independent of the choice of χ. Then deduce that $T * S \in \mathcal{D}'_{0,+}(\mathbb{R})$.

2) We say that $T \in \mathcal{D}'_{0,+}(\mathbb{R})$ is invertible, if there exists $S \in \mathcal{D}'_{0,+}(\mathbb{R})$ such that $T * S = \delta_0$. We write $S = T^{-1}$.

a) Show that δ'_0 is invertible and compute its inverse.

b) Show that if $T \in \mathcal{D}(]0, +\infty[)$, then T is not invertible.

SOLUTION 6.7.–

1)a) Since φ has compact support, there exists $a > 0$ such that $\mathrm{supp}(\varphi) \subset [-a, a]$. It can be observed that if $\chi(x)\chi(y)\varphi(x+y) \neq 0$, then we have $x, y \geq -1$ and $x + y \leq a$. Hence, $x, y \in [-1, a+1]$. Therefore, we obtain $\mathrm{supp}(\varphi^{\triangle}) \subset [-1, a+1]^2$, and it follows that $\varphi^{\triangle} \in \mathcal{D}(\mathbb{R}^2)$.

b) According to the previous question, the distribution $T * S$ is well defined. Let then χ_1 denote another function verifying the same assumptions as χ. We have

$$\left\langle T_x \otimes S_y, \varphi^{\triangle}(x, y) \right\rangle - \langle T_x \otimes S_y, \chi_1(x)\chi_1(y)\varphi(x+y) \rangle =$$

$$\langle T_x \otimes S_y, (\chi(x) - \chi_1(x))\, \chi(y)\varphi(x+y) \rangle$$

$$+ \langle T_x \otimes S_y, \chi_1(x)\, (\chi(y) - \chi_1(y))\, \varphi(x+y) \rangle .$$

Since $\mathrm{supp}\,(\chi - \chi_1) \subset \left]-\infty, -\frac{1}{2}\right]$, then $\mathrm{supp}\,(T) \cap \mathrm{supp}\,(\chi - \chi_1) = \emptyset$, which implies that the first bracket of the second member of the equality is zero. In addition, since $\mathrm{supp}(S) \cap \mathrm{supp}\,(\chi - \chi_1) = \emptyset$, the second bracket is also zero. Therefore, $\left\langle T_x \otimes S_y, \varphi^{\triangle} \right\rangle = \left\langle T_x \otimes S_y, \varphi_1^{\triangle} \right\rangle$ (corresponding to χ_1) and $T * S$ does not depend on the choice of χ.

2) Let $\gamma \in \left]0, \frac{1}{2}\right[$ and let $\chi_\gamma \equiv 1$ on $]-\gamma, +\infty[$ and $\chi_\gamma \equiv 0$ on $]-\infty, -2\gamma[$.

From Question (a, b), we have

$$\forall \varphi \in C_c^\infty(\mathbb{R}), \langle T * S, \varphi \rangle = \langle T \otimes, \varphi(x+y)\chi_\gamma(x)\chi_\gamma(y) \rangle .$$

We choose φ with support in $]-\infty, 0[$, then there exists $a > 0$ such that: $\mathrm{supp}(\varphi) \subset]-\infty, -a[$. Therefore, $(x, y) \longrightarrow \varphi(x+y)\chi_\gamma(x)\chi_\gamma(y)$ is non-zero only if $x + y \leq -a$ and $x, y \geq -2\gamma$, or still $x, y \leq -a + 2\gamma$.

However, for γ small enough, we can have $-a + 2\gamma < 0$. Since $\text{supp}\,(T) \subset [0, +\infty[$ and $\text{supp}(S) \subset [0, +\infty[$, we then obtain: $\langle T * S, \varphi\rangle = 0$. As such, $\text{supp}\,(T * S) \subset [0, +\infty[$.

3) a) Let H denote a Heaviside distribution, that is the indicator function of $]0, +\infty[$. Then, $H \in \mathcal{D}'_{0,+}(\mathbb{R})$ and we have

$$H * \delta_0' = H' * \delta_0 = \delta_0 * \delta_0 = \delta_0.$$

Therefore, H is the inverse of δ_0' in $\mathcal{D}'_{0,+}(\mathbb{R})$.

b) We assume by contradiction that there exists $S \in \mathcal{D}'_{0,+}(\mathbb{R})$ such that $T * S = \delta_0$.

Since $T \in \mathcal{D}\,(]0, +\infty[)$, we should have that $T * S$ is the distribution associated with a function of $\mathcal{D}\,(]0, +\infty[)$. However, δ_0 is not. It is a contradiction! We have the following result: if $T \in \mathcal{D}\,(]0, +\infty[)$, then T is not invertible in $\mathcal{D}'_{0,+}(\mathbb{R})$.

EXERCISE 6.8.– We consider, in $\mathcal{D}'_+(\mathbb{R})$, the following equation:

$$(E): \quad (\delta'' + \delta) * X = \delta.$$

We set $X = H(t)Z(t)$.

1) Verify that in order to solve (E), we simply have to solve the differential equation (E') $\;Z'' + Z = 0$ with $Z(0) = 0$ and $Z'(0) = 1$.

2) Solve (E'), then deduce X.

SOLUTION 6.8.–

1) We have

$$\begin{aligned}
(\delta'' + \delta) * X = \delta &\iff \delta'' * X + \delta * X = \delta \\
&\iff \delta * X''(t) + \delta * X(t) = \delta \\
&\iff X''(t) + X(t) = \delta \\
&\iff H''(t)Z(t) + 2H'(t)Z'(t) + Z''(t)H(t) + H(t)Z(t) = \delta \\
&\iff \delta' Z(t) + 2\delta Z'(t) + Z''(t)H(t) + H(t)Z(t) = \delta \\
&\iff (\delta Z(t))' + \delta Z'(t) + H(t)Z''(t) + H(t)Z(t) = \delta \\
&\iff Z(0)\delta' + Z'(0)\delta + (Z''(t) + Z(t))H(t) = \delta \\
&\iff Z''(t) + Z(t) = 0, Z(0) = 0, Z'(0) = 1.
\end{aligned}$$

We point out here that shifting from the penultimate equivalence to the last one is due to the fact that the three distributions H, δ and δ' form a free family.

Then, to solve (E) the differential equation (E'): $Z'' + Z = 0$ with $Z(0) = 0$ and $Z'(0) = 1$ should be solved.

2) The solution of the equation (E') is $Z(t) = \alpha e^{-it} + \beta e^{it}$. Since $Z(0) = 0$ and $Z'(0) = 1$, then $\alpha + \beta = 0$ and $(-\alpha + \beta)i = 1$. Hence, $\alpha = -\frac{1}{2i}$ and $\beta = -\frac{1}{2i}$. It follows that $Z(t) = \sin t$.

We have $Z(t) = \sin t$, thus $X(t) = H(t) \sin t$ is a solution to the equation (E).

Conclusion: From equation (E), it is deduced that $(\delta'' + \delta)$ is invertible in $\mathcal{D}'_+(\mathbb{R})$ and $(\delta'' + \delta)^{-1} = H(t) \sin t$.

EXERCISE 6.9.– Compute the inverses, in $\mathcal{D}'_+(\mathbb{R})$, of the following distributions:

1) δ';

2) $\delta' + H$.

SOLUTION 6.9.–

1) By definition, the inverse of δ' in $\mathcal{D}'_+(\mathbb{R})$ is obtained by solving the equation $\delta' * X = \delta$.

We have $(\delta * X)' = \delta$, d'où $X' = \delta$ and consequently, $X = H$.

Finally, in $\mathcal{D}'_+$, we have $(\delta')^{-1} = H$.

2) As previously, we look for X in $\mathcal{D}'_+(\mathbb{R})$ such that $(\delta' + H) * X = \delta$.

We have $((\delta' + H) * X)' = \delta'$, hence $(\delta' + H)' * X = \delta'$. Therefore, $(\delta'' + \delta) * X = \delta'$, thus $X = (\delta'' + \delta)^{-1} * \delta'$.

From exercise 6.8, we have $(\delta'' + \delta)^{-1} = H(t) \sin t$, thus

$$\begin{aligned} X &= H \sin t * \delta' \\ &= (H \sin t * \delta)' \\ &= (H \sin t)' \\ &= H' \sin t + H \cos t \\ &= \delta \sin t + H \cos t. \end{aligned}$$

Consequently, $X = H(t) \cos t$. Finally, in $\mathcal{D}'_+(\mathbb{R})$, we have

$$(\delta' + H)^{-1} = H(t) \cos t.$$

EXERCISE 6.10.– In the following, we study the solutions to the equation

$$T'' - T' - 2T = F.$$

I) The homogeneous equation $F = 0$.

1) Give the set of solutions to the homogeneous equation

$$y'' - y' - 2y = 0. \quad [6.4]$$

2) Give the unique solution to the homogeneous equation [6.4] such that $y(0) = 0, y'(0) = 1$. This solution will be denoted by y_0.

3) Specify the set of distributions solutions to the equation

$$T'' - T' - 2T = 0.$$

II) Search for a particular solution.

1) Let us assume that F is a function-type distribution, for example $F = [x \to e^{3x}]$. Find, by application of the variation of the constant, a particular solution to the equation: $y'' - y' - 2y = e^{3x}$. Deduce a distribution which is solution of: $T'' - T' - 2T = [x \to e^{3x}]$.

2) Let us examine the more general case of a distribution F with compact support.

a) Recall the definition of the support of a distribution.

b) Give the support of each of the following distributions: δ_1 and $[\mathbb{1}_{[-1,1]}]$.

c) Consider $E = [y_0 H]$. After calculating E' and E'', verify that E is a solution to: $T'' - T' - 2T = \delta_0$.

d) What is $F * \delta_0$ equal to?

e) It should be noted that since F has compact support, $E*F$ is well defined. Verify that $E * F$ is a particular solution to the equation

$$T'' - T' - 2T = F.$$

f) Give a particular solution for $F = [\mathbb{1}_{[0,1]}]$.

III) The conclusion. Let S be a solution distribution of $T'' - T' - 2T = F$. Show that $S - F * E$ is a solution to the homogeneous equation: $T'' - T' - 2T = 0$. Then conclude the solution set.

IV) Example. Solve: $T'' - 2T = \delta_0 + \mathbb{1}_{[0,1]}$.

SOLUTION 6.10.–

I) The homogeneous equation $F = 0$.

1) The characteristic equation is: $r^2 - r - 2 = 0$. It has two possible roots $r_1 = -1, r_2 = 2$. The real solutions of the homogeneous equation are given by $S_H = \left\{t \longrightarrow c_1 e^{-t} + c_2 e^{2t} | c_1, c_2 \in \mathbb{R}\right\}$.

2) Let us look for the solution y verifying the conditions

$$(*) : \begin{cases} \text{y(0)=0} \\ \text{y}'(0) = 1. \end{cases}$$

We have $(*) \iff \begin{cases} C_1 + C_2 = 0 \\ -C_1 + 2C_2 = 1 \end{cases}$ and therefore $C_1 = -C_2 = -\dfrac{1}{3}$.

Hence,

$$y_0 : t \longrightarrow \frac{1}{3}\left(e^{2t} - e^{-t}\right).$$

3) The induced distributions $[y]$ with $y \in S_H$ simply have to be considered.

II) Search for a particular solution.

1) Let us look directly for a solution in the form $y_p : t \longrightarrow ce^{3t}$ (since 3 is not root). Thus:

$$y_p''(t) - y_p'(t) - 2y_p(t) = (9c - 3c - 2c)e^{3t} = e^{3t} \Longrightarrow c = \frac{1}{4}.$$

Finally, $[y_p]$ is a solution distribution.

2) a) The support of a distribution is the smallest closed set F such that for any φ with support in $\mathbb{R} \setminus F$, we have $\langle T, \varphi \rangle = 0$.

b) Here we have $\text{supp}(\delta_1) = \{1\}$ and $\text{supp}(\mathbb{1}_{[-1,1]}) = [-1, 1]$.

c) Using the fact that for a function such that $f(0) = 0$, we have $f\delta_0 = f(0)\delta_0 = 0$, the derivatives are

$$E' = [y_0 H]' = y_0'[H] + y_0[H]' = y_0'[H] + y_0(0)\delta_0 = y_0'[H]$$

$$E'' = [y_0 H]'' = [y_0'[H]]' = y_0''[H] + y_0'(0)\delta_0 = y_0''[H] + \delta_0.$$

By summation and using the fact that $y_0'' - y_0' - 2y_0 = 0$, we have

$$E'' - E' + 2E = (y_0'' - y_0' - 2y_0)[H] + \delta_0 = \delta_0.$$

E is therefore indeed a solution.

d) Consider two distributions T and S, then $(T * S)' = T' * S = T * S'$. By applying this property of the convolution operator,

$$E'' - E' + 2E = \delta_0 \Longrightarrow F * (E'' - E' + 2E) = F * \delta_0.$$

We have $F * \delta_0 = F$ and $F * (E'' - E' + 2E) = (F * E)'' - (F * E)' + 2(F * E)$. Since $F * E = E * F$, it is deduced that this distribution, $E * F$, is indeed a solution to our problem.

e) A particular solution is given by the distribution:

$$[y_0 H] * [\mathbb{1}_{[0,1]}] = [y_0 H \mathbb{1}_{[0,1]}],$$

that is, the distribution induced by the function

$$x \longrightarrow \frac{1}{3} \int_{\min(x-1,0)}^{\min(x+1,0)} e^{2t} - e^{-t} \mathrm{d}t.$$

f) We simply have to achieve the difference of the two following equations:

$$S'' - S' + 2S = F, (F * E)'' - (F * E)' + 2(F * E) = F.$$

Then $(S - F * E)'' - (S - F * E)' + 2(S - F * E) = F - F = 0$

III) Conclusion.

From the above, if S is a solution distribution, then $S - F * E$ is a solution to the homogeneous equation. Now, we have already described the space of solutions of the homogeneous equation. We know that $S - F * E$ is of $[y_H]$ type with $y_H \in S_H$. The set of solutions is therefore

$$\mathbb{S} = \left\{F * E + [t \to c_1 e^{-t} + c_2 e^{2t}] | (c_1, c_2) \in \mathbb{R}^2\right\}.$$

The space is said to possess an affine space structure.

IV) Example.

The characteristic equation is $r^2 - 2 = 0$. The set of solutions of the homogeneous equation is $\mathbb{S} = \{F * E + [t \to c_1 e^{\sqrt{2}t} + c_2 e^{-\sqrt{2}t}] | (c_1, c_2) \in \mathbb{R}^2\}$. The fundamental solution is $y_0(t) = \dfrac{1}{2\sqrt{2}}(e^{\sqrt{2}t} - e^{-\sqrt{2}t})$. A particular solution is given by $[y_0 H] * (\delta_1 + [\mathbb{1}_{[0,1]}])$. Now, for any function $f \in L^1_{loc}$,

$$[f] * \delta_1 = [t \to f(t-1)] \Rightarrow [\delta_0 H] * \delta_1 = [t \to (t-1)e^{\sqrt{2}(t-1)} \mathbb{1}_{[1,+\infty[}(t)]$$

$$[\delta_0 H] * [\mathbb{1}_{[-1,1]}] = [[\delta_0 H] * \mathbb{1}_{[-1,1]}] = \left[x \to \frac{1}{2\sqrt{2}} \int_{\min(x,0)}^{\min(x+1,0)} e^{\sqrt{2}t} - e^{-\sqrt{2}t} \mathrm{d}t\right].$$

Finally, the solutions to the equation are given by the particular solution (found with the fundamental solution) and the solutions to the homogeneous equation

$$\mathbb{S} = \left\{[y_0 H] * (\delta_1 + [\mathbb{1}_{[0,1]}]) + \left[t \to c_1 e^{\sqrt{2}t} + c_2 e^{-\sqrt{2}t}\right] \mid (c_1, c_2) \in \mathbb{R}^2\right\}.$$

7

Schwartz Spaces and Tempered Distributions

So far, we have defined the distributions on the space of test functions $\mathcal{D}(\mathbb{R}^d)$. There are, however, particular distributions that are defined as linear continuous functionals on spaces larger than $\mathcal{D}(\mathbb{R}^d)$. Among these spaces, the Schwartz spaces are introduced here, which will be denoted by $\mathcal{S}(\mathbb{R}^d)$. These spaces will play an important role in defining a restricted class of distributions called tempered distributions, which are very useful for solving physical problems.

7.1. $\mathcal{S}(\mathbb{R}^d)$ Schwartz spaces

Several definitions can be found in the literature on Schwartz spaces and they are all equivalent. This section starts by presenting some of these definitions.

7.1.1. *Definitions and examples*

DEFINITION 7.1.– *Let φ be a $\mathbb{R}^d$ definite function with complex values.*

The function φ is said to be **rapidly decreasing** *if for any $\alpha \in \mathbb{N}^d$, there exists a constant $c_\alpha > 0$ such that: $\forall\, x \in \mathbb{R}^d$, $|x^\alpha \varphi(t)| \leq c_\alpha$.*

REMARK 7.1.– *It can be seen that a function is rapidly decreasing if and only if its product by any polynomial tends to 0 at infinity.*

EXAMPLE 7.1.–

1) The function $\varphi_1 : x \longmapsto \exp(-x^2)$ is a rapidly decreasing function.

2) The function $\varphi_2 : x \longmapsto \exp(-|x|)$ is a rapidly decreasing function but not indefinitely differentiable.

3) A non-zero polynomial does not define a rapidly decreasing function.

4) The function $\frac{1}{1+x^2}$ is of class $\mathcal{C}^\infty$ but not rapidly decreasing.

REMARK 7.2.– *Contrary to what its name might suggest, definition 7.1 implies no monotonicity for f, even in a neighborhood of infinity as in $f(x) = e^{-|x|}\sin x$.*

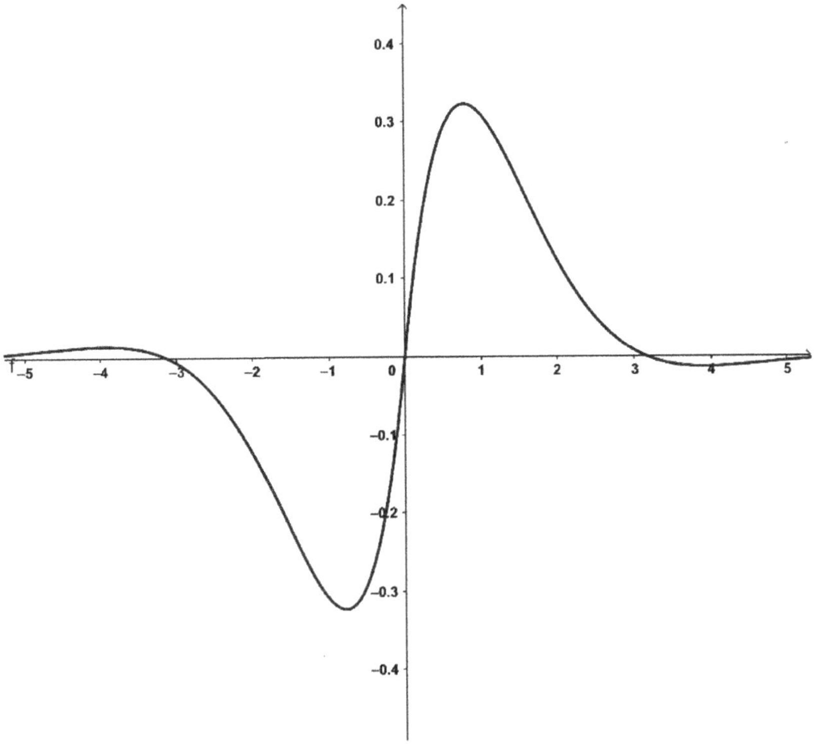

Figure 7.1. *Function* $f(x) = e^{-|x|}\sin x$

PROPOSITION 7.1.– Let φ be a $\mathbb{R}^d$ definite function with complex values. The function φ is rapidly decreasing if only if, for all $m \in \mathbb{N}$, there exists a constant $c_m > 0$ such that: $|\varphi(x)| \leq c_m \frac{1}{(1+|x|)^m}, \forall x \in \mathbb{R}^d$.

$|\cdot|$ denotes the Euclidean norm of $\mathbb{R}^d$.

Proof. Let us assume that φ is rapidly decreasing. Let $m \in \mathbb{N}$ be fixed. For all $x \in \mathbb{R}^d$, we have $|x| \leq \sqrt{d} \max_{|\alpha|=1} |x^\alpha|$. On the other hand, by a convexity argument of the function $x \mapsto x^m$, there exists $a_m > 0$ such that $\forall t \geq 0, (1+t)^m \leq a_m(1+t^m)$. Therefore, $(1+|x|)^m \leq a_m(1 + d^{\frac{m}{2}} \max_{|\alpha|=m} |x^\alpha|)$. Hence,

$$(1+|x|)^m |\varphi(x)| \leq a_m(1 + d^{\frac{m}{2}} \max_{|\alpha|=m} |x^\alpha|) |\varphi(x)| .$$

Given that φ is rapidly decreasing, then there exists a constant $b_m > 0$ such that

$$\forall x \in \mathbb{R}^d, (1 + d^{\frac{m}{2}} \max_{|\alpha|=m} |x^\alpha|) |\varphi(x)| \leq b_m.$$

Consequently, $\forall x \in \mathbb{R}^d, (1+|x|)^m |\varphi(x)| \leq c_m = a_m b_m$.

For the reciprocal, for all $\alpha \in \mathbb{N}^d$, we have $\forall x \in \mathbb{R}^d, |x^\alpha| \leq |x|^{|\alpha|} \leq (1+|x|)^{|\alpha|}$. Furthermore, $\forall x \in \mathbb{R}^d, |x^\alpha| |\varphi(x)| \leq |x|^{|\alpha|} \leq (1+|x|)^{|\alpha|} |\varphi(x)| < \infty$. This completes the proof. □

REMARK 7.3.– *It has just been shown that*

$$\forall \alpha \in \mathbb{N}^d, \sup_{x \in \mathbb{R}^d} |x^\alpha| |\varphi(x)| < \infty \Longleftrightarrow \forall m \in \mathbb{N}, \sup_{x \in \mathbb{R}^d} (1+|x|)^m |\varphi(x)| < \infty \qquad [7.1]$$

DEFINITION 7.2.– *Let φ be a $\mathbb{R}^d$ definite function with complex values. The function φ is said to be* ***tempered or at most polynomially growing****, if there exists $N \in \mathbb{N}$, and a constant $c_N > 0$ such that $|\varphi(x)| \leq c_N (1+|x|)^N$, $\forall x \in \mathbb{R}^d$.*

EXAMPLE 7.2.–

1) Any polynomial function is at most a polynomially growing function.

2) The two functions $f_1 : x \longmapsto 1 + t^2 \cos(t)$ and $f_2 : x \longmapsto \sin(\exp(t))$ are tempered.

DEFINITION 7.3.–

– A complex-valued function φ defined on $\mathbb{R}^d$ is said to be a Schwartz function if it is indefinitely differentiable and if it is, as well as all its derivatives rapidly decreasing.

– The Schwartz space $\mathcal{S}(\mathbb{R}^d)$ is the set of functions of $C^\infty(\mathbb{R}^d)$ which are as well as all their derivatives rapidly decreasing. We write

$$\mathcal{S}(\mathbb{R}^d) = \left\{ \varphi \in C^\infty(\mathbb{R}^d) \mid \ \forall \alpha, \beta \in \mathbb{N}^d, \ N_{\alpha,\beta}(\varphi) = \sup_{x \in \mathbb{R}^d} \left| x^\alpha \partial^\beta \varphi(x) \right| < \infty \right\}.$$

An equivalent definition of the Schwartz space is given in the following manner.

PROPOSITION 7.2 (Equivalent definitions).– We have

$$\mathcal{S}(\mathbb{R}^d) = \left\{ \varphi \in \mathcal{C}^\infty(\mathbb{R}^d) \mid \forall p \in \mathbb{N},\, \mathbf{N}_p(\varphi) = \sum_{|\alpha|,|\beta| \le p} \sup_{x \in \mathbb{R}^d} \left| x^\beta \partial^\alpha \varphi(x) \right| < \infty \right\}$$

$$= \left\{ \varphi \in \mathcal{C}^\infty(\mathbb{R}^d) \mid \forall p \in \mathbb{N},\, \mathbf{Q}_p(\varphi) = \sup_{|\beta| \le p} \sup_{x \in \mathbb{R}^d} (1+|x|)^p \left| \partial^\beta \varphi(x) \right| < \infty \right\}.$$

Proof.

– For the first equality, it should be observed that $\mathbf{N}_p(\varphi)$ is a finite sum of $N_{\alpha,\beta}(\varphi)$. More precisely, we have $\mathbf{N}_p(\varphi) = \sum_{|\alpha|,|\beta| \le p} N_{\alpha,\beta}(\varphi)$.

– For the second equality, we use the equivalence [7.1] for the functions $\partial^\beta \varphi$.

□

REMARK 7.4.–

– *The two expressions of* $\mathbf{N}_p$ *and* $\mathbf{Q}_p$, *respectively, define two equivalent semi-norms.*

– *In the expression of* $\mathbf{N}_p$, $\sum_{|\alpha|,|\beta| \le p}$ *can be replaced by* $\sup_{|\alpha|,|\beta| \le p}$.

Conversely, in the expression of $\mathbf{Q}_p$, $\sup_{|\beta| \le p}$ *can be replaced by* $\sum_{|\beta| \le p}$.

– *Let* $\varphi \in \mathcal{C}^\infty(\mathbb{R})$. *We have*

$$\varphi \in \mathcal{S}(\mathbb{R}) \iff \forall n, m \in \mathbb{N},\ \left| x^n \varphi^{(m)}(x) \right| \underset{|x| \longrightarrow +\infty}{\longrightarrow} 0 .$$

– *Any function* $\varphi \in \mathcal{S}(\mathbb{R}^d)$ *is obviously bounded and summable.*

EXAMPLE 7.3.–

1) $\mathcal{D}(\mathbb{R}^d) = \mathcal{C}_c^\infty(\mathbb{R}^d) \subset \mathcal{S}(\mathbb{R}^d)$.

2) $\varphi(x) = e^{-|x|^2} \in \mathcal{S}(\mathbb{R}^d)$.

3) $\varphi(x) = P(x)e^{-\alpha|x|^2} \in \mathcal{S}(\mathbb{R}^d)$ *with* P *a polynomial and* $\alpha > 0$.

4) No rational function (other than the null function) belongs to $\mathcal{S}(\mathbb{R}^d)$.

7.1.2. *Topology and convergence in* $\mathcal{S}(\mathbb{R}^d)$

7.1.2.1. $\mathcal{S}(\mathbb{R}^d)$ *topology*

On the space $\mathcal{S}(\mathbb{R}^d)$, several families of semi-norms have been defined; the three types that are often used are given.

First type	**See (Zuily 1988)**
$N_{\alpha,\beta}(\varphi) = \sup\limits_{x\in\mathbb{R}^d} \lvert x^\beta \partial^\alpha \varphi(x)\rvert,$	$\forall \alpha, \beta \in \mathbb{N}^d.$
$N'_{n,p}(\varphi) = \sum\limits_{\lvert\alpha\rvert\le n, \lvert\beta\rvert\le p} N_{\alpha,\beta}(\varphi),$	$\forall n, p \in \mathbb{N}.$
$N''_{n,p}(\varphi) = \sup\limits_{\lvert\alpha\rvert\le n, \lvert\beta\rvert\le p} N_{\alpha,\beta}(\varphi),$	$\forall n, p \in \mathbb{N}.$
$\mathbf{N}_p(\varphi) = \sum\limits_{\lvert\alpha\rvert, \lvert\beta\rvert\le p} \sup\limits_{x\in\mathbb{R}^d} \lvert x^\beta \partial^\alpha \varphi(x)\rvert,$	$\forall p \in \mathbb{N}.$

Second type	**See (Khoan 1972)**
$m_{n,p}(\varphi) = \sup\limits_{x\in\mathbb{R}^d, \lvert\beta\rvert\le p} \left(1+\lvert x\rvert^2\right)^n \partial^\beta \varphi(x)\rvert,$	$\forall n, p \in \mathbb{N}.$
$M_{n,p}(\varphi) = \sum\limits_{\lvert\beta\rvert\le p} \sup\limits_{x\in\mathbb{R}^d} \left(1+\lvert x\rvert^2\right)^n \lvert\partial^\beta \varphi(x)\rvert,$	$\forall n, p \in \mathbb{N}.$
$\mathbf{M}_p(\varphi) = \sup\limits_{x\in\mathbb{R}^d, \lvert\alpha\rvert\leqslant n} \left(1+\lvert x\rvert^2\right)^p \lvert\partial^\alpha \varphi(x)\rvert,$	$\forall p \in \mathbb{N}.$

Third type	**See (Guillopé 2008)**
$q_{n,p}(\varphi) = \sup\limits_{x\in\mathbb{R}^d, \lvert\beta\rvert\le p} (1+\lvert x\rvert)^n \partial^\beta \varphi(x)\rvert,$	$\forall n, p \in \mathbb{N}.$
$Q_{n,p}(\varphi) = \sum\limits_{\lvert\beta\rvert\le p} \sup\limits_{x\in\mathbb{R}^d} (1+\lvert x\rvert)^n \lvert\partial^\beta \varphi(x)\rvert,$	$\forall n, p \in \mathbb{N}.$
$\mathbf{Q}_p(\varphi) = \sup\limits_{x\in\mathbb{R}^d, \lvert\alpha\rvert\leqslant p} (1+\lvert x\rvert)^p \lvert\partial^\alpha \varphi(x)\rvert,,$	$\forall p \in \mathbb{N}.$

Table 7.1. *Several types of semi-norms*

The topology of the $\mathcal{S}(\mathbb{R}^d)$ space is defined by the countable family of norms, defined by $\mathbf{N}_p(\varphi) = \sum\limits_{\lvert\alpha\rvert, \lvert\beta\rvert\le p} \sup\limits_{x\in\mathbb{R}^d} \left\lvert x^\beta \partial^\alpha \varphi(x)\right\rvert.$

REMARK 7.5.–

1) The topology of the $\mathcal{S}(\mathbb{R}^d)$ *space can also be defined by the countable family of norms*

$$\mathbf{Q}_p(\varphi) = \sup_{x\in\mathbb{R}^d, \lvert\alpha\rvert\leqslant p} (1+\lvert x\rvert)^p \lvert\partial^\alpha \varphi(x)\rvert$$

or

$$\mathbf{M}_p(\varphi) = \sup_{x \in \mathbb{R}^d, |\alpha| \leqslant p} \left(1 + |x|^2\right)^p |\partial^\alpha \varphi(x)|.$$

2) For any $p \in \mathbb{N}$, we have $\mathbf{Q}_p(\varphi) \leq \mathbf{N}_p(\varphi)$ and $\mathbf{M}_p(\varphi) \leq \mathbf{N}_{2p}(\varphi)$.

3) Endowed with the family of semi-norms $(\mathbf{N}_p)_{p \in \mathbb{N}}$ (or $(\mathbf{M}_p)_{p \in \mathbb{N}}$, $(\mathbf{Q}_p)_{p \in \mathbb{N}}$), the Schwartz space is a Fréchet space.

4) For the introduction of Sobolev spaces, some authors use the family $(\mathbf{M}_p)_{p \in \mathbb{N}}$, as basis family (see (Bony 2001), for instance).

With the family of norms $(\mathbf{N}_p)$, a distance can be defined on $\mathcal{S}(\mathbb{R}^d)$ by

$$d(f, g) = \sum_{p=0}^{+\infty} 2^{-p} \frac{\mathbf{N}_p(f - g)}{1 + \mathbf{N}_p(f - g)}.$$

$(\mathcal{S}(\mathbb{R}^d), d)$ is a complete metric space whose distance d is translation invariant. It is said that $\mathcal{S}(\mathbb{R}^d)$ is a metrizable space.

7.1.2.2. *Convergence in $\mathcal{S}(\mathbb{R}^d)$*

With the family of norms $(\mathbf{N}_p)$, the convergence of a sequence in $\mathcal{S}(\mathbb{R}^d)$ can be defined as follows.

DEFINITION 7.4.– *Let (φ_n) be a sequence of elements of $\mathcal{S}(\mathbb{R}^d)$. It is said that the sequence (φ_n) converges to $\varphi \in \mathcal{S}(\mathbb{R}^d)$ in the space $\mathcal{S}(\mathbb{R}^d)$ if*

$$\mathbf{N}_p(\varphi_n - \varphi) \longrightarrow 0, \quad \forall p \in \mathbb{N}.$$

NOTE 7.1.– *There is no norm $\mathcal{N}$ on $\mathcal{S}(\mathbb{R}^d)$ which is adequate to express this convergence itself, that is, such that $\varphi_n \longrightarrow \varphi$ in $\mathcal{S}(\mathbb{R}^d)$ if and only if, $\mathcal{N}(\varphi_n - \varphi) \longrightarrow 0$. This shows why the topology of $\mathcal{S}(\mathbb{R}^d)$ is defined by **the family of norms** and not by a single norm.*

7.1.3. *First properties of $\mathcal{S}(\mathbb{R}^d)$*

PROPOSITION 7.3.– For all $q \in [1, +\infty]$, we have: $\mathcal{S}(\mathbb{R}^d) \subset L^q(\mathbb{R}^d)$. More precisely, for any $\varphi \in \mathcal{S}(\mathbb{R}^d)$, we have

$$\left\| x^\alpha \partial^\beta \varphi \right\|_{L^q} \leq C_q (\mathbf{N}_p(\varphi))^{1 - \frac{1}{q}} (\mathbf{N}_{p+d+1}(\varphi))^{\frac{1}{q}}, \forall\, |\alpha|, |\beta| \leq p. \qquad [7.2]$$

Proof. It should be noted that for $\alpha = \beta = 0$, the inequality [7.2] makes it possible to deduce that $\mathcal{S}(\mathbb{R}^d) \subset L^q(\mathbb{R}^d)$. Let $\alpha, \beta \in \mathbb{N}^d$ such that $|\alpha|, |\beta| \leq p$. We set $g(x) = x^\alpha \partial^\beta \varphi(x)$. Then,

$$\begin{aligned}
\left\|x^\alpha \partial^\beta \varphi\right\|_{L^q}^q &\leq \int_{\mathbb{R}^d} |g(x)|^q \mathrm{d}x \\
&\leq \sup_{x\in\mathbb{R}^d} |g(x)|^{q-1} \int_{\mathbb{R}^d} |g(x)| \mathrm{d}x \\
&\leq \sup_{x\in\mathbb{R}^d} |g(x)|^{q-1} \sup_{x\in\mathbb{R}^d} (1+|x|)^{d+1} |g(x)| \int_{\mathbb{R}^d} \frac{1}{(1+|x|)^{d+1}} \mathrm{d}x \\
&\leq C_d \mathbf{N}_0(g)^{q-1} \mathbf{N}_{d+1}(g) \\
&\leq C_d (\mathbf{N}_p(\varphi))^{q-1} \mathbf{N}_{p+d+1}(\varphi).
\end{aligned}$$

Consequently, $\left\|x^\alpha \partial^\beta \varphi\right\|_{L^q} \leq C_q (\mathbf{N}_p(\varphi))^{1-\frac{1}{q}} (\mathbf{N}_{p+d+1}(\varphi))^{\frac{1}{q}}, \forall\, |\alpha|, |\beta| \leq p$.

We have $\mathbf{N}_p(\varphi) \leq \mathbf{N}_{p+d+1}(\varphi)$, then

$$\left\|x^\alpha \partial^\beta \varphi\right\|_{L^q} \leq C_q \mathbf{N}_{p+d+1}(\varphi), \forall\, |\alpha|, |\beta| \leq p.$$

Special case: If $q = \infty$, then $\left\|x^\alpha \partial^\beta \varphi\right\|_{L^\infty} \leq C \mathbf{N}_p(\varphi)$. □

Using the density results given in section 2.2.5 and the fact that

$$\mathcal{D}(\mathbb{R}^d) \subset \mathcal{S}(\mathbb{R}^d) \subset L^p(\mathbb{R}^d),$$

the following density results can be shown.

PROPOSITION 7.4.– We have the following properties:

1) $\overline{\mathcal{D}}(\mathbb{R}^d) = \mathcal{S}(\mathbb{R}^d)$;

2) $\overline{\mathcal{S}(\mathbb{R}^d)} = L^p(\mathbb{R}^d)$ for all $p \in [1, +\infty[$;

3) $\overline{\mathcal{S}(\mathbb{R}^d)} = \mathcal{C}^0(\mathbb{R}^d)$.

Proof.

1) Let $\chi \in \mathcal{D}(\mathbb{R}^d)$, with support in $B(0,2)$, with values in $[0,1]$, and equal to 1 on $\overline{B(0,1)}$. For all $n \geq 1$, we set $\chi_n = \chi\left(\frac{x}{n}\right)$. Let $\varphi \in \mathcal{S}(\mathbb{R}^d)$ and we set $\varphi_n = \chi_n \varphi$.

Then $\varphi_n \in \mathcal{D}(\mathbb{R}^d)$ has support in $B(0,2n)$ and coincides with φ sur $\overline{B(0,1)}$. Let p denote a fixed integer and $\alpha, \beta \in \mathbb{N}^d$ such that $|\alpha| \leq p$ and $|\beta| \leq p$. Based on the Leibniz rule, it can be easily shown that

$$\partial^\beta (\varphi - \varphi_n)(x) = (1 - \chi_n(x))\, \partial^\alpha \varphi(x) + \sum_{\gamma \leq \beta, |\gamma| > 1} \binom{\beta}{\gamma} \frac{1}{n^{|\gamma|}} \partial^\gamma \chi\left(\frac{x}{n}\right) \partial^{\beta-\gamma} \varphi(x).$$

Then,

$$\begin{aligned}\left|x^\alpha \partial^\beta (\varphi - \varphi_n)(x)\right| &\leq \left|x^\alpha (1 - \chi_n(x))\, \partial^\beta \varphi(x)\right| \\ &+ \frac{1}{n} \sum_{\gamma \leq \beta, |\gamma| \geq 1} \binom{\beta}{\gamma} \sup_{z \in \mathbb{R}^d} |\partial^\gamma \chi(z)| \sup_{x \in \mathbb{R}^d} \left|x^\alpha \partial^{\beta-\gamma} \varphi(x)\right| \\ &\leq \sup_{|x|>n} \left|x^\alpha \partial^\beta \varphi(x)\right| + \frac{C}{n} \sum_{\gamma \leq \beta} \sup_{x \in \mathbb{R}^d} \left|x^\alpha \partial^{\beta-\gamma} \varphi(x)\right| \\ &\leq \sup_{|x|>n} \left|x^\alpha \partial^\beta \varphi(x)\right| + \frac{C}{n} \mathbf{N}_p(\varphi).\end{aligned}$$

with $C = 2^p \sum_{|\gamma| \leq p} \sup_{\mathbb{R}^d} \left|\partial^\beta \chi(x)\right|$.

On the other hand,

$$\sup_{|x|>n} \left|x^\alpha \partial^\beta \varphi(x)\right| \leq \frac{1}{n^2} \sup_{|x|>n} \left||x|^2 x^\alpha \partial^\beta \varphi(x)\right| \leq \frac{1}{n^2} \mathbf{N}_{p+2}(\varphi).$$

It can be deduced that

$$\max_{|\alpha|,\, |\beta| \leq p} \sup_{x \in \mathbb{R}^d} \left|x^\alpha \partial^\beta (\varphi - \varphi_n)(x)\right| \leq \frac{1}{n^2} \mathbf{N}_{p+2}(\varphi) + \frac{C}{n} \mathbf{N}_p(\varphi).$$

Then, $\mathbf{N}_p(\varphi - \varphi_n) \leq \frac{1}{n^2} \mathbf{N}_{p+2}(\varphi) + \frac{C}{n} |\mathbf{N}_p(\varphi)| \underset{n \longrightarrow +\infty}{\longrightarrow} 0$. This result holds for $p \in \mathbb{N}$; it is deduced that $\varphi_n \underset{n \longrightarrow +\infty}{\longrightarrow} \varphi$ in $\mathcal{S}(\mathbb{R}^d)$, which expresses the result sought for, namely $\overline{\mathcal{D}}(\mathbb{R}^d) = \mathcal{S}(\mathbb{R}^d)$.

2) This result is derived from the inclusion $\mathcal{D}(\mathbb{R}^d) \subset \mathcal{S}(\mathbb{R}^d) \subset L^p(\mathbb{R}^d)$ and from theorem 2.2.

3) We apply proposition 2.11. □

7.1.4. *Operators in* $\mathcal{S}(\mathbb{R}^d)$

7.1.4.1. *Differentiation and multiplication operators*

PROPOSITION 7.5 (Derivative and multiplication operators).– *Let* $\varphi \in \mathcal{S}(\mathbb{R}^d)$*, and we have:*

1) for any $\beta \in \mathbb{N}^d$, $\partial^\beta \varphi \in \mathcal{S}(\mathbb{R}^d)$*;*

2) for any function $\rho \in \mathcal{C}^\infty(\mathbb{R}^d)$ *whose derivatives are all polynomially increasing (tempered),* $\rho\varphi \in \mathcal{S}(\mathbb{R}^d)$.

Proof.

1) For any $p, q \in \mathbb{N}$, and for any $\alpha, \beta \in \mathbb{N}^d$ such that $|\alpha|, |\beta| \leq q$, we have

$$\begin{aligned}\mathbf{N}_p(x^\alpha \partial^\beta \varphi) &= \sum_{|\lambda|,|\mu|\leq p} \sup_{x\in\mathbb{R}^d} \left| x^\lambda \partial^\mu \left(x^\alpha \partial^\beta \varphi(x) \right) \right| \\ &\leq \mathbf{N}_{p+q}(\varphi).\end{aligned}$$

For all $\beta \in \mathbb{N}^d$, there exists $q \in \mathbb{N}$ such that $|\beta| \leq q$. By applying the above estimate for $\alpha = 0$, we obtain

$$\forall p \in \mathbb{N}, \mathbf{N}_p(\partial^\beta \varphi) \leq \mathbf{N}_{p+q}(\varphi) < \infty.$$

because $\varphi \in \mathcal{S}(\mathbb{R}^d)$. Hence, $\partial^\beta \varphi \in \mathcal{S}(\mathbb{R}^d)$.

2) This point follows from the Leibniz rule and the fact that $\rho \in \mathcal{C}^\infty(\mathbb{R}^d)$ is a function whose derivatives are all polynomially increasing. □

The Leibniz rule is used to obtain the stability of the $\mathcal{S}(\mathbb{R}^d)$ space by multiplication.

PROPOSITION 7.6.– If $f, g \in \mathcal{S}(\mathbb{R}^d)$, then $fg \in \mathcal{S}(\mathbb{R}^d)$.

NOTE 7.2.– $\rho \in \mathcal{C}^\infty(\mathbb{R}^d)$ *and* $f \in \mathcal{S}(\mathbb{R}^d)$ *does not always imply that* $\rho f \in \mathcal{S}(\mathbb{R}^d)$. *In reality, we simply have to take* $f(x) = e^{-x^2} \in \mathcal{S}(\mathbb{R})$ *and* $\rho(x) = e^{2x^2} \in \mathcal{C}^\infty(\mathbb{R})$.

7.1.4.2. *Convolution in* $\mathcal{S}(\mathbb{R}^d)$

DEFINITION 7.5.– *Let* $\varphi_1, \varphi_2 \in \mathcal{S}(\mathbb{R}^d)$*. Their convolution product is the function* $\varphi_1 * \varphi_2$ *defined by* $\forall x \in \mathbb{R}^d$, $\varphi_1 * \varphi_2(x) = \int_{\mathbb{R}^d} \varphi_1(x-t)\varphi_2(t)dt$.

EXAMPLE 7.4.– For the Gaussian γ_1 such that $\gamma_1(t) = e^{-t^2}, \forall t \in \mathbb{R}$, we have

$$\gamma_1 * \gamma_1(t) = \sqrt{\tfrac{\pi}{2}} e^{-\frac{t^2}{2}}.$$

PROPOSITION 7.7.– Let $\varphi_1, \varphi_2, \varphi_3 \in \mathcal{S}(\mathbb{R}^d)$. We have

i) $\varphi_1 * \varphi_2$ is a Schwartz function; that is $\varphi_1 * \varphi_2 \in \mathcal{S}(\mathbb{R}^d)$;

ii) $\varphi_1 * \varphi_2 = \varphi_2 * \varphi_1$; that is the law "$*$" is commutative in $\mathcal{S}(\mathbb{R}^d)$;

iii) $(\varphi_1 * \varphi_2) * \varphi_3 = \varphi_1 * (\varphi_2 * \varphi_3)$; that is the law "$*$" is associative in $\mathcal{S}(\mathbb{R}^d)$.

Proof.

1) The convolution product $\varphi_1 * \varphi_2$ is well defined because $\varphi_1, \varphi_2 \in L^1(\mathbb{R}^d)$. Let $p \in N$ and $\alpha, \beta \in \mathbb{N}^d$ with $|\alpha|, |\beta| \leq p$. For all $x \in \mathbb{R}^d$, we have

$$x^\alpha \partial^\beta (\varphi_1 * \varphi_2)(x) = x^\alpha \partial_x^\beta \int_{\mathbb{R}^d} \varphi_1(x-y)\varphi_2(y)\mathrm{d}y.$$

Since the function $y \longmapsto \partial_x^\beta \varphi_1(x-y)\varphi_2(y)$ is integrable for all x and is dominated by $\sup |\partial_x^\beta \varphi_1||\varphi_2| \in L^1(\mathbb{R}^d)$, we can derive under the integral sign to obtain

$$\begin{aligned}
x^\alpha \partial^\beta (\varphi_1 * \varphi_2)(x) &= x^\alpha \partial_x^\beta \int_{\mathbb{R}^d} \varphi_1(x-y)\varphi_2(y)\mathrm{d}y \\
&= \int_{\mathbb{R}^d} x^\alpha \partial_x^\beta \varphi_1(x-y)\varphi_2(y)\mathrm{d}y \\
&= \int_{\mathbb{R}^d} (x-y+y)^\alpha \partial_x^\beta \varphi_1(x-y)\varphi_2(y)\mathrm{d}y \\
&= \sum_{\gamma \leq \alpha} \binom{\alpha}{\gamma} \int_{\mathbb{R}^d} (x-y)^\gamma (y)^{\alpha-\gamma} \partial_x^\beta \varphi_1(x-y)\varphi_2(y)\mathrm{d}y \\
&= \sum_{\gamma \leq \alpha} \binom{\alpha}{\gamma} \int_{\mathbb{R}^d} (x-y)^\gamma \partial_x^\beta \varphi_1(x-y) y^{\alpha-\gamma}\varphi_2(y)\mathrm{d}y \\
&= \sum_{\gamma \leq \alpha} \binom{\alpha}{\gamma} y^\gamma \partial^\beta \varphi_1 * y^{\alpha-\gamma}\varphi_2(x).
\end{aligned}$$

According to Young's inequality, we have

$$\left\| x^\alpha \partial^\beta (\varphi_1 * \varphi_2) \right\|_{L^\infty} \leq \sum_{\gamma \leq \alpha} \binom{\alpha}{\gamma} \left\| y^\gamma \partial^\beta \varphi_1 \right\|_{L^1} \|y^{\alpha-\gamma}\varphi_2\|_{L^\infty}.$$

It follows that $\mathbf{N}_p(\varphi_1 * \varphi_2) \leq C\mathbf{N}_{p+d+1}(\varphi_1)\mathbf{N}_p(\varphi_2)$.

Consequently, $\varphi_1 * \varphi_2 \in \mathcal{S}(\mathbb{R}^d)$. □

PROPOSITION 7.8.– Let $T \in \mathcal{E}'(\mathbb{R}^d)$. If $\varphi \in \mathcal{S}(\mathbb{R}^d)$, then $\varphi * T \in \mathcal{S}(\mathbb{R}^d)$. Moreover, there exists $m \in \mathbb{N}$ such that $\forall p \in \mathbb{N}, \exists C_p > 0, \mathbf{N}_p(\varphi * T) \leq C_p \mathbf{N}_{p+m}(\varphi)$.

Proof. Since $\varphi \in \mathcal{C}^\infty(\mathrm{R}^\mathrm{d})$, then $\varphi * T$ is an indefinitely differentiable function and we have

$$x^\alpha \partial^\beta (\varphi * T)(x) = \langle T_y, x^\alpha \partial^\beta \varphi(x-y)\rangle.$$

On the other hand, since T is compactly supported, then T is of finite order m, and there exists $C > 0$ such that, if supp $T \subset B(0, R)$, we have

$$\begin{aligned}
\left\|x^\alpha \partial^\beta (\varphi * T)(x)\right\| &\leq C \sum_{|\mu|\leq m} \sup_{|y|\leq R} \left\|\partial_y^\mu x^\alpha \partial_x^\beta \varphi(x-y)\right\| \\
&\leq C \sum_{|\mu|\leq m} \sup_{|y|\leq R} \left\|x^\alpha \partial^{\beta+\mu} \varphi(x-y)\right\| \\
&\leq C \sum_{|\mu|\leq m} \sup_{|y|\leq R} \left\|(x-y+y)^\alpha \, \partial^{\beta+\mu} \varphi(x-y)\right\| \\
&\leq C' \, \mathbf{N}_{p+m}(\varphi) \text{ (binomial formula).}
\end{aligned}$$

Hence, for all $\alpha, \beta \in \mathbb{N}^d$ such that $|\alpha|, |\beta| \leq p$, we have

$$\left\|x^\alpha \partial^\beta (\varphi * T)(x)\right\| \leq C' \mathbf{N}_{p+m}(\varphi).$$

Therefore, there exists $m \in \mathbb{N}$ such that

$$\forall p \in \mathbb{N}, \exists C_p > 0, \mathbf{N}_p(\varphi * T) \leq C_p \mathbf{N}_{p+m}(\varphi).$$ □

7.2. Tempered distributions

This section begins with the definition of a tempered distribution followed by examples illustrating this definition.

7.2.1. *Definition and examples*

DEFINITION 7.6.– *A tempered distribution is a definite linear and continuous form on $\mathcal{S}(\mathbb{R}^d)$, that is, a linear form T on $\mathcal{S}(\mathbb{R}^d)$ such that there exists $p \geq 0$ and there exists $c > 0$ such that*

$$\forall \varphi \in \mathcal{S}(\mathbb{R}^d), \quad |\langle T, \varphi \rangle| \leq c\mathbf{N}_p(\varphi). \qquad [7.3]$$

The set of tempered distributions is a vector space that we denote by $\mathcal{S}'(\mathbb{R}^d)$.

REMARK 7.6.–

1) Given the density of $\mathcal{D}(\mathbb{R}^d)$ in $\mathcal{S}(\mathbb{R}^d)$, it suffices that inequality [7.3] *is verified for $\varphi \in \mathcal{D}(\mathbb{R}^d)$.*

2) In the definition, the family of norms $(\mathbf{N}_p)$ can be replaced by the previously defined family $(\mathbf{Q}_p)$.

3) A linear form T on $\mathcal{S}(\mathbb{R}^d)$ is a tempered distribution if and only if, for any sequence (φ_n) of elements of $\mathcal{S}(\mathbb{R}^d)$,

$$\varphi_n \overset{\mathcal{S}(\mathbb{R}^d)}{\longrightarrow} 0 \textit{ leads to } \langle T, \varphi_n \rangle \longrightarrow 0 \textit{ in } \mathbb{C}.$$

PROPOSITION 7.9.– For a distribution T to belong to $\mathcal{S}'(\mathbb{R}^d)$, it is necessary and sufficient that T be a continuous linear form on $\mathcal{D}(\mathbb{R}^d)$ for the convergence induced by that on $\mathcal{S}(\mathbb{R}^d)$.

Proof. The necessary condition is immediate.

The converse is true because since T is continuous linear on $\mathcal{D}(\mathbb{R}^d)$ which is dense in $\mathcal{S}(\mathbb{R}^d)$, it uniquely extends into a continuous linear form on $\mathcal{S}(\mathbb{R}^d)$. □

It should be noted that here we are talking about a density extension of a continuous linear form on $\mathcal{D}(\mathbb{R}^d)$ that is dense in $\mathcal{S}(\mathbb{R}^d)$, which is not the case for Banach spaces. For more details, see Bony (2001).

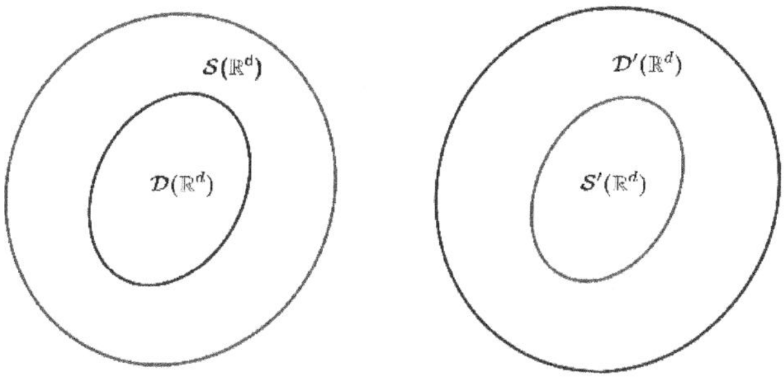

Figure 7.2. *Spaces and their duals. For a color version of this figure, see www.iste.co.uk/aitbenhassi/distribution.zip*

EXAMPLE 7.5.– We have $\delta_a \in \mathcal{S}'(\mathbb{R}^d)$. Actually,

$$\forall \varphi \in \mathcal{S}(\mathbb{R}^d), |\langle \delta_a, \varphi \rangle| = |\varphi(a)| \leqslant \sup_{x \in \mathbb{R}^d} |\varphi(x)| \leq \mathbf{N}_0(\varphi).$$

EXAMPLE 7.6.– Any polynomial function defines a tempered distribution (that is to say $\mathbb{R}[X] \subset \mathcal{S}'(\mathbb{R}^d)$.

We simply have to show the result for a function such as $f(x) = x^\alpha$ where $\alpha \in \mathbb{N}^d$. A linear combination will make it possible to conclude. Let $\alpha \in \mathbb{N}^d$ such that $|\alpha| \leq N$ and let $\varphi \in \mathcal{S}(\mathbb{R}^d)$. We have

$$\begin{aligned}
|\langle f, \varphi \rangle| &\leq \int_{\mathbb{R}^d} |f(x)\varphi(x)| \, \mathrm{d}x \\
&\leq \int_{\mathbb{R}^d} |x^\alpha \varphi(x)| \, \mathrm{d}x \\
&\leq \int_{\mathbb{R}^d} \frac{1}{(1+|x|)^{d+1}} (1+|x|)^{d+1} |x^\alpha \varphi(x)| \, \mathrm{d}x \\
&\leq \sup_{x \in \mathbb{R}^d} (1+|x|)^{d+1} |x^\alpha \varphi(x)| \int_{\mathbb{R}^d} \frac{1}{(1+|x|)^{d+1}} \mathrm{d}x \\
&\leq \left(\int_{\mathbb{R}^d} \frac{1}{(1+|x|)^{d+1}} \mathrm{d}x \right) \mathbf{N}_{N+d+1}(\varphi) \\
&\leq C \, \mathbf{N}_{N+d+1}(\varphi).
\end{aligned}$$

Hence, f defines a tempered distribution.

EXAMPLE 7.7.–

1) If f is a measurable function and if there exists a polynomial function P such that $|f(x)| \leqslant |P(x)|$, then $f \in \mathcal{S}'(\mathbb{R}^d)$.

2) The principal value $\mathrm{Vp}\frac{1}{x}$ is a tempered distribution (see exercise 7.2).

3) We have $\mathcal{E}'(\mathbb{R}^d) \subset \mathcal{S}'(\mathbb{R}^d)$.

4) The distribution defined by $T = \sum_{n\in\mathbb{Z}} a_n \delta_{na}, a > 0$ is a tempered distribution if the sequence $(a_n)_{n\in\mathbb{Z}}$ is at most polynomially growing. (i.e. $\exists p > 0$, $a_k \leq (1+|k|)^p), \forall k \in \mathbb{Z}$) (see exercise 7.5).

PROPOSITION 7.10.– Any function in $L^p(\mathbb{R}^d)$ defines a tempered distribution (i.e. $L^p(\mathbb{R}^d) \subset \mathcal{S}'(\mathbb{R}^d), \forall p \in [1, \infty]$).

Proof. Let $f \in L^p(\mathbb{R}^d)$. For any $\varphi \in \mathcal{S}(\mathbb{R}^d)$, we have

$$\langle f, \varphi \rangle = \int_{\mathbb{R}^d} f(x)\varphi(x)\mathrm{d}x = \int_{\mathbb{R}^d} \frac{f(x)}{(1+|x|)^m}(1+|x|)^m \varphi(x)\mathrm{d}x.$$

– For $1 < p < +\infty$: let m be large enough so that $\int_{\mathbb{R}^d} \frac{\mathrm{d}x}{(1+|x|)^{qm}} < +\infty$ with q such that $\left(\frac{1}{p}+\frac{1}{q}=1\right)$. We have

$$\begin{aligned}
|\langle f, \varphi\rangle| &\leqslant \left(\int_{\mathbb{R}^d} \frac{|f(x)|}{(1+|x|)^m}\mathrm{d}x\right) \sup_{x\in\mathbb{R}^d} (1+|x|)^m |\varphi(x)| \\
&\leqslant \|f\|_{L^p} \left\|\tfrac{1}{(1+|x|)^m}\right\|_{L^q} \sup_{x\in\mathbb{R}^d} (1+|x|)^m |\varphi(x)| \\
&\leq \|f\|_{L^p} \left\|\tfrac{1}{(1+|x|)^m}\right\|_{L^q} \mathbf{Q}_m(\varphi) \\
&\leq C\mathbf{N}_m(\varphi).
\end{aligned}$$

– For $p = 1$: (the result is obvious). □

– For $\mathrm{p} = \infty$: the result is derived from example 7.6.

PROPOSITION 7.11.– Any continuous function f at most polynomially growing defines a tempered distribution on $\mathbb{R}^d$.

Proof. Since f is continuous and at most polynomially growing, then there exists $N \in \mathbb{N}$ and it exists $C > 0$ such that $|f(x)| \leq C\,(1 + |x|)^N$, $\forall x \in \mathbb{R}^d$.

Let $\varphi \in \mathcal{S}(\mathbb{R}^d)$. We have

$$\begin{aligned}
|\langle f, \varphi\rangle| &\leq \int_{\mathbb{R}^d} |f(x)\varphi(x)|\,\mathrm{d}x \\
&\leq \int_{\mathbb{R}^d} C\,(1 + |x|)^N\,|\varphi(x)|\,\mathrm{d}x \\
&\leq C \sup_{x \in \mathbb{R}^d} (1 + |x|)^{N+d+1}\,|\varphi(x)| \int_{\mathbb{R}^d} \frac{1}{(1 + |x|)^{(d+1)}}\mathrm{d}x \\
&\leq C \left(\int_{\mathbb{R}^d} \frac{1}{(1 + |x|)^{d+1}}\mathrm{d}x \right) \mathbf{N}_{N+d+1}(\varphi) \\
&\leq C_N\,\mathbf{N}_{N+d+1}(\varphi).
\end{aligned}$$

Consequently, f defines a tempered distribution. □

EXAMPLE 7.8 (Counterexamples).–

1) The function $f(x) = e^{x^2}$ defines a distribution but it is not a tempered distribution.

As a matter of fact, for $\varphi(x) = e^{-\frac{x^2}{2}} \in \mathcal{S}(\mathbb{R})$, we have

$$\begin{aligned}
\langle f, \varphi\rangle &= \int_{\mathbb{R}} f(x)\varphi(x)\mathrm{d}x \\
&= \int_{\mathbb{R}} e^{x^2} e^{-\frac{x^2}{2}}\,\mathrm{d}x \\
&= \int_{\mathbb{R}} e^{\frac{x^2}{2}}\,\mathrm{d}x \\
&= +\infty.
\end{aligned}$$

2) The distributions T_{e^x}, T_{shx} and T_{chx} are not tempered distributions (see exercise 7.3).

7.2.2. *Convergence in* $\mathcal{S}'(\mathbb{R}^d)$

The set of tempered distributions $\mathcal{S}'(\mathbb{R}^d)$ is naturally endowed with the weak dual topology, defined by the weak convergence $*$ of sequences.

DEFINITION 7.7.– *We say that a sequence* $(T_n)_{n\geq 1}$ *of tempered distributions converges to* $T \in \mathcal{S}'(\mathbb{R}^d)$ *if,* $\langle T_n, \varphi\rangle \underset{n\to+\infty}{\longmapsto} \langle T, \varphi\rangle, \forall \varphi \in \mathcal{S}(\mathbb{R}^d)$.

PROPOSITION 7.12.– Let $(T_n)_{n\geq 1}$ be a sequence of tempered distributions. If for any $\varphi \in \mathcal{S}(\mathbb{R}^d)$, the sequence $(\langle T_n, \varphi\rangle)_{n\geq 1}$ converges, then there exists $T \in \mathcal{S}'(\mathbb{R}^d)$ such that $(T_n)_{n\geq 1}$ converges to T in $\mathcal{S}'(\mathbb{R}^d)$.

REMARK 7.7.–

1) If $f_n \longrightarrow f$ *in* $L^p(\mathbb{R}^d)$*, then* $f_n \longrightarrow f$ *in* $\mathcal{S}'(\mathbb{R}^d)$.

2) If $T_n \longrightarrow T$ *in* $\mathcal{S}'(\mathbb{R}^d)$*, then* $T_n \longrightarrow T$ *in* $\mathcal{D}'(\mathbb{R}^d)$*. The converse does not hold. Furthermore, for any real sequence* (a_n)*, the sequence* $(a_n\delta_n) \longrightarrow 0$ *in* $\mathcal{D}'(\mathbb{R})$*, whereas it converges in* $\mathcal{S}'(\mathbb{R})$*, only if there exists* $c > 0$ *and* $p \in \mathbb{N}$ *such that* $|a_n| \leq c(1+n)^p$.

7.2.3. *First properties of* $\mathcal{S}'(\mathbb{R}^d)$

A first property of tempered distributions is the following.

PROPOSITION 7.13.– The space $\mathcal{S}'(\mathbb{R}^d)$ injects into $\mathcal{D}'(\mathbb{R}^d)$.

Proof. Let $T \in \mathcal{S}'(\mathbb{R}^d)$ be a tempered distribution. By definition, there exists $m \in \mathbb{N}$ and a constant $C > 0$ such that: $|\langle T, \varphi\rangle| \leq C\mathbf{N}_m(\varphi), \quad \forall \varphi \in \mathcal{S}(\mathbb{R}^d)$.

In particular, for any compact $K \subset \mathbb{R}^d$ and for all $\varphi \in \mathcal{D}(\mathbb{R}^d)$ with support in K, we have $\sup\limits_{x\in\mathbb{R}^d} |x^\alpha \partial^\beta \varphi| \leq C_{K,\alpha} \sup\limits_{x\in K} |\partial^\beta \varphi|$. Therefore,

$$\begin{aligned}
|\langle T, \varphi\rangle| &\leq C\big(\sum_{|\alpha|\leq m} C_{K,\alpha}\big) \sum_{|\beta|\leq m} \sup_{x\in K} |\partial^\beta \varphi| \\
&\leq C' P_m(\varphi),
\end{aligned}$$

which shows that $T_{|\mathcal{D}(\mathbb{R}^d)} \in \mathcal{D}'(\mathbb{R}^d)$. Therefore, the mapping $\Psi : T \longmapsto T_{|\mathcal{D}(\mathbb{R}^d)}$ is well defined. Moreover, by density of $\mathcal{D}(\mathbb{R}^d)$ in $\mathcal{S}(\mathbb{R}^d)$, if $T_{|\mathcal{D}(\mathbb{R}^d)} = 0$, then $T = 0$ on $\mathcal{S}(\mathbb{R}^d)$. Hence the injection of the mapping Ψ. □

COROLLARY 7.1.– *Any tempered distribution is of finite order.*

Proof. Let $T \in \mathcal{S}'(\mathbb{R}^d)$ be a tempered distribution. In the proof of proposition 7.13, it has been shown that there exists $m \in \mathbb{N}$ such that for any compact $K \subset \mathbb{R}^d$, there exists a constant $C' > 0$ such that: $|\langle T, \varphi\rangle| \leq C' P_m(\varphi), \forall \varphi \in \mathcal{D}(\mathbb{R}^d), \operatorname{supp}\varphi \subset K$. Thus, T is at most an mth-order distribution. □

Application: In exercise 3.8, it has been shown that the distribution $\sum_{n=0}^{+\infty} \delta_n^{(n)}$ is of infinite order; it is then deduced that this distribution can never be tempered.

7.2.4. *Operators in* $\mathcal{S}'(\mathbb{R}^d)$

All the operations already seen in the space $\mathcal{S}(\mathbb{R}^d)$ are transmitted by duality to the space $\mathcal{S}'(\mathbb{R}^d)$.

7.2.4.1. *Differentiation and multiplication operators*

The properties of the derivation and multiplication operators in $\mathcal{S}(\mathbb{R}^d)$ are transposed to its dual space $\mathcal{S}'(\mathbb{R}^d)$.

PROPOSITION 7.14.– Let $T \in \mathcal{S}'(\mathbb{R}^d)$, and we have

1) for all $\alpha \in \mathbb{N}^d$, $\partial^\alpha T \in \mathcal{S}'(\mathbb{R}^d)$;

2) for any function ρ at most polynomially growing and all its derivatives, we have $\rho T \in \mathcal{S}'(\mathbb{R}^d)$.

COROLLARY 7.2.– *Let $T \in \mathcal{S}'(\mathbb{R}^d)$. For all $\alpha, \beta \in \mathbb{N}^d$, we have $x^\alpha \partial^\beta T \in \mathcal{S}'(\mathbb{R}^d)$.*

EXAMPLE 7.9 (Infinite growth and tempered character).– To see if a distribution defined by a function, for example, is tempered or not, studying the infinity growth of the modulus of this function does not suffice. Consider, for example, the function $x \longmapsto ie^x e^{ie^x}$. We have $|ie^x e^{ie^x}| = e^x$, which has the same infinite growth as the above counterexamples. However,

$$ie^x e^{ie^x} = \frac{d}{dx}(e^{ie^x}),$$

and since the function $x \longmapsto e^{ie^x}$ is a function of class C^1 bounded on $\mathbb{R}$, it defines a tempered distribution on $\mathbb{R}$.

On the other hand, its derivative in the sense of distributions coincides with the distribution defined by its derivative function in the usual sense, so that, according to proposition 7.14 this, derivative function $x \longmapsto ie^x e^{ie^x}$ properly defines an element of $\mathcal{S}'(\mathbb{R}^d)$.

Intuitively, the fast oscillations of the function $x \longmapsto e^{ie^x}$ in the neighborhood of $+\infty$ are the underlying cause that annihilate the growth of $x \longmapsto e^{ie^x}$ for x in the neighborhood of $+\infty$.

REMARK 7.8.– *In practice, to decide if a distribution $T \in \mathcal{D}'(\mathbb{R}^d)$ is tempered, it will be obviously necessary to find out if it belongs to the classes of examples above, namely distributions with compact support, functions of L^p, functions with polynomial growth, etc. If this is not the case, we will then have to find out whether the distribution under consideration is a derivative (of any order) of a distribution, which is already known to be tempered (this is obviously the case in the previous example). If none of these approaches allows us to conclude, we must then return to the definition of temperate distributions, and verify the continuity property. In fact, we have the following characterization of temperate distributions.*

THEOREM 7.1 (Characterization of $\mathcal{S}'(\mathbb{R}^d)$ distributions).– *Any distribution $T \in \mathcal{S}'(\mathbb{R}^d)$ is of the form $T = \partial_x^\alpha(1+|x|^2)^n f$ in the sense of distributions where $\alpha \in \mathbb{N}^d$ is a natural number, n is a natural number, and f is a continuous function on $\mathbb{R}^d$.*

7.2.4.2. *Convolution in $\mathcal{S}'(\mathbb{R}^d)$*

PROPOSITION 7.15 (Convolution).– Let $T \in \mathcal{S}'(\mathbb{R}^d)$ be a tempered distribution and $S \in \mathcal{E}'(\mathbb{R}^d)$ a distribution with compact support. We have $T * S \in \mathcal{S}'(\mathbb{R}^d)$.

Proof. Let $\varphi \in \mathcal{D}(\mathbb{R}^d)$. There exists a sequence $(\psi_n) \in \mathcal{D}(\mathbb{R}^d)$ such that $\psi_n \longrightarrow S$ in $\mathcal{D}'(\mathbb{R}^d)$. Therefore, $\langle T * \psi_n, \varphi\rangle = \langle T, \check{\psi}_n * \varphi\rangle$. Since $T * \psi_n \longrightarrow T * S$ in $\mathcal{D}'(\mathbb{R}^d)$ and $\check{\psi}_n * \varphi \longrightarrow \check{S} * \varphi$ in $\mathcal{D}(\mathbb{R}^d)$, then

$$\langle T * S, \varphi\rangle = \lim_{n \mapsto +\infty} \langle T, \check{\psi}_n * \varphi\rangle = \lim_{n \mapsto +\infty} \langle T * \psi_n, \varphi\rangle = \langle T, \check{S} * \varphi\rangle.$$

Since $T \in \mathcal{S}'(\mathbb{R}^d)$, according to proposition 7.8, there exists $C, C' > 0$ and $p, m \in \mathbb{N}$ such that

$$\left|\langle T, \check{S} * \varphi\rangle\right| \leq C\mathbf{N}_p(\check{S} * \varphi) \leq C'\mathbf{N}_{p+m}(\varphi).$$

Consequently, $T * S \in \mathcal{S}'(\mathbb{R}^d)$. □

7.3. Exercises with solutions

EXERCISE 7.1.– Let f be the real function defined by

$$f(x) = e^{-x^2}\cos(e^{2x^2}), \ \forall x \in \mathbb{R}.$$

1) Show that f is a rapidly decreasing function.

2) Does f belong to the Schwartz space $\mathcal{S}(\mathbb{R})$?

SOLUTION 7.1.–

1) From the definition of a rapidly decreasing function, we have to show that for any index $\alpha \in \mathbb{N}$, we have $\lim\limits_{|x| \mapsto +\infty} |x^\alpha f(x)| = 0$. Since, on the one hand, $|\cos(e^{2x^2})| \leq 1$ and, on the other hand, $\lim\limits_{|x| \mapsto +\infty} \left| x^\alpha e^{-x^2} \right| = 0$, then

$$\lim_{|x| \mapsto +\infty} \left| x^\alpha e^{-x^2} \cos(e^{2x^2}) \right| = 0.$$

Therefore, $\lim\limits_{|x| \mapsto +\infty} |x^\alpha f(x)| = 0$. Consequently, f is a rapidly decreasing function.

2) Through a direct and easy calculation, we obtain

$$f'(x) = -2x \left(e^{-x^2} \cos(e^{2x^2}) + 2e^{x^2} \sin(e^{2x^2}) \right).$$

Considering the sequence (x_k) defined by $x_k = \sqrt{\frac{1}{2} \ln \left(2k\pi + \frac{\pi}{2} \right)}$, we have

$$\lim_{k \mapsto +\infty} |f'(x_k)| = +\infty \text{ and } \lim_{k \mapsto +\infty} |x_k| = +\infty.$$

Then, it is impossible to have $\lim\limits_{|x| \mapsto +\infty} |x^\alpha f'(x)| = 0$. Consequently, f' is not a rapidly decreasing function, so f does not belong to the Schwartz space $\mathcal{S}(\mathbb{R})$.

EXERCISE 7.2.–

1) Show that a Dirac mass is a tempered distribution.

2) Show that a polynomial function is a tempered distribution.

3) Let f be a measurable function verifying $x \mapsto (1 + |x|^2)^{-m} f(x) \in L^p(\mathbb{R}^d)$ for a specific integer $m \in \mathbb{N}$ and a specific $p \in [1, +\infty]$. Show that f is a temperate distribution.

4) Show that the function $x \longmapsto \ln |x| \in L^p_{loc}(\mathbb{R})$ is a tempered distribution and compute its derivative. What can be deduced about the principal value of $\frac{1}{x}$? Indication: For the differentiation, $\varphi \in \mathcal{D}(\mathbb{R})$ will be considered and we will integrate the function $x \longmapsto \ln |x| \varphi'(x)$ by parts on a domain such as $\{x \in \mathbb{R} : \varepsilon \leq |x| \leq R\}$ before making R tend to $+\infty$ then ε to 0.

SOLUTION 7.2.–

1) For any rapidly decreasing function $\varphi \in \mathcal{S}(\mathbb{R}^d)$ and any $a \in \mathbb{R}^d$, we have

$$|\langle \delta_a, \varphi \rangle| = |\varphi(a)| \leqslant \|\varphi\|_\infty \cdot$$

Consequently, δ_a is therefore indeed a tempered distribution.

It can also be observed that δ_a is a compact support distribution, so it is tempered.

2) If P is a polynomial function, then there exists $m \in \mathbb{N}$ such that $P(x) \underset{\infty}{=} O\left(1+|x|^2\right)^m$. Let $\varphi \in \mathcal{S}(\mathbb{R}^d)$. Since $x \longmapsto \frac{|P(x)|}{(1+|x|^2)^{m+d+1}} \in L^1(\mathbb{R}^d)$, then

$$\begin{aligned}\left|\int_{\mathbb{R}^d} P(x)\varphi(x)\mathrm{d}x\right| &\leqslant \left\{\int_{\mathbb{R}^d} \frac{|P(x)|}{\left(1+|x|^2\right)^{m+d+1}}\mathrm{d}x\right\} \sup_{x\in\mathbb{R}^d} \left(1+|x|^2\right)^{m+d+1} |\varphi(x)| \\ &\leq C\mathbf{M}_{m+d+1}(\varphi) \\ &\leq C\mathbf{N}_{2(m+d+1)}(\varphi).\end{aligned}$$

Consequently, P is thus indeed a tempered distribution.

3) We denote $Q_m(x) := \left(1+\|x\|^2\right)^m$. We have in this case, for any rapidly decreasing function $\varphi \in \mathcal{S}(\mathbb{R}^d)$, and using the Hölder inequality if $p \in]1,\infty[$, and simply the inequality of the mean if $p \in \{1,\infty\}$,

$$\begin{aligned}\left|\int_{\mathbb{R}^d} f(x)\varphi(x)\mathrm{d}x\right| &\leqslant \int_{\mathbb{R}^d} \left|\frac{f(x)}{Q_m(x)}\right| |\varphi(x)Q_m(x)|\,\mathrm{d}x \\ &\leqslant \left\|\frac{f}{Q_m}\right\|_p \|\varphi Q_m\|_{p'} \left(\frac{1}{p'}+\frac{1}{p}=1\right).\end{aligned}$$

Since $q_d :\longmapsto \left(1+\|x\|^2\right)^{-d-1}$ actually belongs to all spaces $L^p(\mathbb{R}^d)$, we finally have

$$\begin{aligned}\left|\int_{\mathbb{R}^d} f(x)\varphi(x)\mathrm{d}x\right| &\leqslant \left\|\frac{f}{Q_m}\right\|_p \|q_d\|_{p'} \sup_{x\in\mathbb{R}^d} \left(1+|x|^2\right)^m \left(1+|x|^2\right)^{d+1} |\varphi(x)| \\ &\leqslant C \sup_{x\in\mathbb{R}^d} \left(1+|x|^2\right)^{d+m+3} |\varphi(x)| \\ &\leq C\mathbf{M}_{m+d+3}(\varphi) \\ &\leq C\mathbf{N}_{2(m+d+3)}(\varphi).\end{aligned}$$

Hence, f is thus a tempered distribution.

4) First, $x \longmapsto \ln|x|$ is indeed a tempered distribution, since for instance $x \longmapsto \ln|x|\left(1+x^2\right)^{-2}$ is integrable, so that the criterion of the previous question applies. Then, if $\varphi \in \mathcal{D}(\mathbb{R})$, we follow the indication given in the text and it follows that

$$\begin{aligned} I_\epsilon^R &:= \int_{\epsilon \leqslant |x| \leqslant R} \ln|x| \varphi'(x) \mathrm{d}x \\ &= [\ln(-x)\varphi(x)]_{-R}^{-\epsilon} + [\ln(x)\varphi(x)]_R^{\epsilon} - \int_{-R}^{-\epsilon} \frac{\varphi(x)}{x} \mathrm{d}x - \int_\epsilon^R \frac{\varphi(x)}{x} \mathrm{d}x, \end{aligned}$$

where we recall that the function $x \longmapsto \ln|x|$ is differentiable on $\mathbb{R}^*$, and where its derivative is $x \longmapsto \dfrac{1}{x}$. We thus have

$$I_\epsilon^R := \ln(R)\left[\varphi(R) - \varphi(-R)\right] + \ln(\epsilon)\left[\varphi(-\epsilon) - \varphi(\epsilon)\right] - \int_{\epsilon \leqslant |x| \leqslant R} \frac{\varphi(x)}{x} \mathrm{d}x.$$

We first evaluate the limit for $R \longrightarrow +\infty$. The factor term of $\ln(R)$ is zero for $R \geqslant \mathrm{diam}\{\mathrm{supp}(\varphi)\}$ (support which is compact). Since $\varphi \in \mathcal{D}(\mathbb{R})$, then, by dominated convergence, it is obtained that

$$\int_{|x| \geqslant \epsilon} \ln|x| \varphi'(x) \mathrm{d}x = \ln(\epsilon)\left[\varphi(-\epsilon) - \varphi(\epsilon)\right] - \int_{\epsilon \leqslant |x|} \frac{\varphi(x)}{x} \mathrm{d}x. \qquad [7.4]$$

We now evaluate the limit $\epsilon \longrightarrow 0$. By the finite increment theorem, we have

$$|\varphi(-\epsilon) - \varphi(\epsilon)| \leqslant |\varphi(-\epsilon) - \varphi(0)| + |\varphi(0) - \varphi(\epsilon)| \leqslant 2\epsilon||\varphi'||_\infty,$$

so that the term in factor of $\ln(\epsilon)$ tends to 0 (by domination). Finally, by dominated convergence, the left member of the equality [7.4], $\int_{|x| \geqslant \epsilon} \ln|x| \varphi'(x) \mathrm{d}x$ tends to $\langle \ln|x|, \varphi' \rangle$, and by definition of the principal value, the right-hand side member tends to $-\left\langle \mathrm{Vp}(\frac{1}{x}), \varphi \right\rangle$. By definition of the derivative in the sense of distributions, we finally have that the derivative of $x \longmapsto \ln|x|$ is precisely the principal value. It is known that $\mathcal{S}'(\mathbb{R})$ is stable by differentiation, so $\mathrm{Vp}(\frac{1}{x}) \in \mathcal{S}'(\mathbb{R})$.

EXERCISE 7.3.–

1) Show that the locally integrable function $x \mapsto e^x$ is not a tempered distribution.

Hint: Set a non-identically zero positive test function θ with support in $]-1, 1[$ and study the family of square brackets $\langle T_{\exp}, \tau_a \theta \rangle$ for $a \in \mathbb{R}$.

2) Let $f(x) := e^x \cos(e^x)$.

a) Show that f is not bounded by any polynomial.

b) Show, however, that f is a tempered distribution.

Hint: Give a primitive of f.

SOLUTION 7.3.–

1) We follow the hint given above. If $T_{\exp}$ were tempered, then there would exist a constant $C > 0$ and integers $m, p \in \mathbb{N}$ such that for any test function $\varphi \in \mathcal{D}(\mathbb{R})$, we have $|\langle T_{\exp}, \varphi\rangle| \leqslant C \sup_{k \leqslant p} \sup_{x \in \mathbb{R}} \left(1 + x^2\right)^m |\varphi^{(k)}(x)|$. Let $\theta \in \mathcal{D}\left(]-1, 1[\right)$ be a positive function and $a \in \mathbb{R}$. Since $\tau_a \theta$ has support in $I_a =]a - 1, a + 1[$, we have

$$
\begin{aligned}
|\langle T_{\exp}, \tau_a \theta\rangle| &\leqslant C \sup_{k \leqslant p} \sup_{x \in I_a} \left(1 + x^2\right)^m |(\tau_a \theta)^{(k)}(x)| \\
&\leqslant \sup_{k \leqslant p} \sup_{x \in I_a} \left(1 + x^2\right)^m |\tau_a (\theta)^{(k)}(x)| \\
&\leq \sup_{k \leqslant p} \sup_{x \in I_a} \left(1 + x^2\right)^m |\theta^{(k)}(x - a)|.
\end{aligned}
$$

However, $x \longrightarrow \left(1 + x^2\right)^m$ is bounded by $\left(2 + (1 + a)^2\right)^m$ on I_a, so that finally it would follow that $|\langle T_{\exp}, \tau_a \theta\rangle| \leqslant C \left(2 + (1 + a)^2\right)^m \sup_{k \leqslant p} \sup_{x \in \mathbb{R}} |\theta^k|$, but on the other hand, using a change of variable $u = x - a$, it can be noted that $\langle T_{\exp}, \tau_a \theta\rangle = \int_{\mathbb{R}} e^x \theta(x - a) \mathrm{d}x = e^a \langle T_{\exp}, \theta\rangle$ such that the previous inequality is written as $e^a |\langle T_{\exp}, \theta\rangle| \leqslant C \left(2 + (1 + a)^2\right)^m \sup_{k \leqslant p} \sup_{x \in \mathbb{R}} |\theta^{(k)}|$, and since $\theta \neq 0$, we have $\langle T_{\exp}, \theta\rangle \neq 0$, and by dividing it follows that

$$
e^a \leqslant C_\theta \left(2 + (1 + a)^2\right)^m \sup_{k \leqslant p} \sup_{x \in \mathbb{R}} |\theta^{(k)}|
$$

that is, a polynomial bound (at a) for the exponential function, which is obviously contradictory: $\exp \notin \mathcal{S}'(\mathbb{R})$.

We can do better without following the hint!

Simply taking the function $\varphi(x) = \dfrac{1}{\cosh(x)} \in \mathcal{S}(\mathbb{R})$, we then obtain

$$
\int_{-\infty}^{+\infty} \frac{e^x}{\cosh(x)} \mathrm{d}x = +\infty,
$$

which shows that $e^x \notin \mathcal{S}(\mathbb{R})$.

2)a) If f were bounded by a polynomial, by domination, we would have

$$\cos\left(e^{x}\right) \underset{x \longrightarrow +\infty}{\longrightarrow} 0,$$

which is not true, as can be seen by considering the sequence (x_k) defined by $x_k = \ln(2\pi k)$, which tends to $+\infty$ when k tends to $+\infty$, whereas $\cos\left(e^{x_k}\right) = 1$.

b) A primitive of f is the function $F : x \longmapsto \sin(e^{x}) \in L^{\infty}(\mathbb{R})$. Since $L^{\infty}(\mathbb{R}) \subset \mathcal{S}'(\mathbb{R})$ and that the latter space is stable by differentiation, we thus indeed have $f \in \mathcal{S}'(\mathbb{R})$.

EXERCISE 7.4.– Show that if f is locally summable and if it is square-summable, then it determines a tempered distribution.

SOLUTION 7.4.– Let $\varphi \in \mathcal{S}(\mathbb{R})$. From Schwarzt's inequality, we have

$$\begin{aligned}\left(\int_{-\infty}^{+\infty} f(x)\varphi(x)\mathrm{d}x\right)^{2} &\leq \int_{-\infty}^{+\infty} f^{2}(x)\mathrm{d}x \int_{-\infty}^{+\infty} \varphi^{2}(x)\mathrm{d}x \\ &\leq \int_{-\infty}^{+\infty} f^{2}(x)\mathrm{d}x \int_{-\infty}^{+\infty} \frac{\varphi^{2}(x)\left(1+|x|^{2}\right)}{1+|x|^{2}}\mathrm{d}x \\ &\leq \int_{-\infty}^{+\infty} f^{2}(x)\mathrm{d}x \ \pi \sup_{x\in\mathbb{R}} |\varphi(x)|^{2}\left(1+|x|^{2}\right) \\ &\leq \pi \int_{-\infty}^{+\infty} f^{2}(x)\mathrm{d}x \ \sup_{x\in\mathbb{R}} \left(|\varphi(x)|\left(1+|x|^{2}\right)\right)^{2}.\end{aligned}$$

Then:

$$|\langle f, \varphi\rangle| \leq \sqrt{\pi}\|f\|_{2} \sup_{x\in\mathbb{R}} \left|\varphi(x)\left(1+|x|^{2}\right)\right|.$$

Otherwise, we have $|\langle f, \varphi\rangle| \leq \sqrt{\pi}\|f\|_{2}N_{2}(\varphi)$. Consequently, f defines a tempered distribution.

EXERCISE 7.5.– Let $(a_n)_{n\in\mathbb{Z}}$ be at most a polynomially growing sequence (i.e. $\exists p > 0,\ a_n \leq (1+|n|)^{p}, \forall n \in \mathbb{Z}$). We consider T defined by $T = \sum_{n\in\mathbb{Z}} a_n \delta_{na}$, $a > 0$. Show that T is a tempered distribution.

SOLUTION 7.5.– Let $M > 0$ and $\varphi \in \mathcal{D}(\mathbb{R})$ such that $\operatorname{supp}\varphi \Subset [-M, M]$. Let $N \in \mathbb{N}$ such that $N \geq M$. We have

$$\begin{aligned} |\langle T, \varphi\rangle| &\leq \sum_{-N}^{N} |a_k \varphi(ka)| \\ &\leq \|\varphi\|_\infty \sum_{-N}^{N} |a_k| \\ &\leq C \|\varphi\|_\infty . \end{aligned}$$

Therefore, T is indeed a zeroth-order distribution.

Let $\varphi \in \mathcal{S}(\mathbb{R})$. We have

$$\begin{aligned} |\langle a_k \delta_{ka}, \varphi\rangle| &= |a_k \varphi(ka)| \\ &\leq C(1+|k|)^p |\varphi(ka)| \\ &\leq C(1+|k|)^p |\varphi(ka)| \\ &\leq \frac{C(1+|k|)^p}{(1+|ka|)^p}(1+|ka|)^{p+2}\frac{|\varphi(ka)|}{(1+|ka|)^2}. \end{aligned}$$

Since $\varphi \in \mathcal{S}(\mathbb{R})$, then

$$(1+|ka|)^{p+2}|\varphi(ka)| \leq \sup_{x\mathbb{R}}(1+|x|)^{p+2}|\varphi(x)| \leq \mathbf{N}_{P+2}(\varphi).$$

In addition, we have $\dfrac{C(1+|k|)^p}{(1+|ka|)^p} \leq C \sup\limits_{x\mathbb{R}} \dfrac{(1+|x|)^p}{(1+|xa|)^p} = C_1$. Then

$$|\langle a_k \delta_{ka}, \varphi(k)\rangle| \leq C_1 \frac{\mathbf{N}_{P+2}(\varphi)}{(1+|ka|)^2}.$$

Consequently, $|\langle T, \varphi\rangle| \leq C_2 \mathbf{N}_{P+2}(\varphi)$ where $C_2 = C_1 \sum\limits_{n\in\mathbb{Z}} \dfrac{1}{(1+|ka|)^2}$. We then conclude that $T \in \mathcal{S}'(\mathbb{R})$.

Consequence: In the special case when $a_k = 1$, we find that the Dirac comb $\Delta_a = \sum\limits_{n\in\mathbb{Z}} \delta_{na}$ is a tempered distribution.

EXERCISE 7.6.– For all $n\mathbb{N}^*$, consider the function f_n defined by

$$\forall x \in \mathbb{R}^*, f_n(x) = \frac{1-\cos(2n\pi x)}{\pi x}.$$

1) Show that for all $n \in \mathbb{N}^*$, f_n defines a tempered distribution that will be denoted by f_n.

2) Show that the limit of the sequence of tempered distributions (f_n) is converging in $\mathcal{S}'(\mathbb{R})$ to the distribution $\frac{1}{\pi}\mathrm{Vp}(\frac{1}{x})$.

SOLUTION 7.6.–

1) For any $n \geq 1$, the function f_n is continuous in $\mathbb{R}^*$ as the quotient of two continuous functions in $\mathbb{R}^*$, since the function of the denominator does not cancel out in $\mathbb{R}^*$. In the neighborhood of 0, by a limited expansion of the function cos, we have

$$1-\cos(2n\pi x) = \frac{(2n\pi x)^2}{2} - \frac{(2n\pi x)^4}{4!} + o(x^4).$$

We deduce that

$$\lim_{x \longmapsto 0} f_n(x) = \lim_{x \longmapsto 0} \frac{1-\cos(2n\pi x)}{\pi x} = 0.$$

The function f_n is thus extendable by 0 in the neighborhood of 0, and thus defines a continuous function on $\mathbb{R}$. The function f_n is bounded on $[-1,1]$ as a continuous function on a compact. On the other hand, for all $x \in \mathbb{R}$ such that $|x| > 1$, we have $|f_n(x)| \leq \frac{2}{\pi}$. f_n is a function of $L^1_{loc}(\mathbb{R})$ and at most polynomially growing, so $f_n \in \mathcal{S}'(\mathbb{R})$.

2) Let $n \in \mathbb{N}^*$ and $\varphi \in \mathcal{S}(\mathbb{R})$. We start by verifying that

$$\left\langle \mathrm{Vp}(\frac{1}{x}), \varphi \right\rangle = \int_{[-1,1]} \frac{\varphi(x)-\varphi(0)}{x} \mathrm{d}x - \int_{|x|\geq 1} \frac{\varphi(x)}{x} \mathrm{d}x.$$

Then, we show that

$$\begin{aligned}\langle f_n, \varphi \rangle - \left\langle \frac{1}{\pi}\mathrm{Vp}(\frac{1}{x}), \varphi \right\rangle &= \int_{[-1,1]} \frac{\cos(2n\pi x)(\varphi(x)-\varphi(0))}{\pi x} \mathrm{d}x \\ &- \int_{|x|\geq 1} \frac{\cos(2n\pi x)\varphi(x)}{\pi x} \mathrm{d}x.\end{aligned}$$

By using the Taylor formula at the first order at the point 0 applied to the function φ at x, we have $\varphi(x) - \varphi(0) = x\psi(x)$ with $\psi \in \mathcal{C}^\infty$. Therefore,

$$\langle f_n, \varphi\rangle - \left\langle \frac{1}{\pi}\text{Vp}(\frac{1}{x}), \varphi \right\rangle = \frac{1}{\pi}\int_{[-1,1]} \cos(2n\pi x)\psi(x)\text{d}x - \int_{|x|\geq 1} \frac{\cos(2n\pi x)\varphi(x)}{\pi x}\text{d}x.$$

Showing that both integrals tend to 0 when n tends to $+\infty$, we get

$$\lim_{n \longrightarrow +\infty} \langle f_n, \varphi\rangle = \left\langle \frac{1}{\pi}\text{Vp}(\frac{1}{x}), \varphi \right\rangle, \text{ for all } \varphi \in \mathcal{S}(\mathbb{R}).$$

This allows us to conclude.

8

Fourier Transform

The Fourier transform is a particularly important tool in mathematics, either from the fundamental point of view, or from the point of view of its applications (in signal theory for example, or in everything related to wave phenomena). There are therefore many varied motivations underlying its study. Because of the inversion theorem, the Fourier transform will make it possible to write an integrable function and more generally a tempered distribution as a superposition of complex exponential functions. This is a very powerful tool for studying convolution equations, its resolution being reduced to division problems.

As for the most important applications of the Fourier transform, the reader will find them in the study of partial differential equations, which is the subject of Chapter 9 of this book. It is important to be able to use this transformation under the most general conditions possible. We define it within the too restricted framework of integrable functions $L^1(\mathbb{R}^d)$. To extend it to a larger space, it is restricted to the $\mathcal{S}(\mathbb{R}^d)$ space, which is a much smaller space than $L^1(\mathbb{R}^d)$. By duality, this transformation can be finally extended to $\mathcal{S}'(\mathbb{R}^d)$, which presents good invariance properties for the Fourier transform. Unfortunately, it is not possible to define the Fourier transform of any distribution.

8.1. Fourier transform in $L^1(\mathbb{R}^d)$

We begin by recalling the definition and some properties of the Fourier transform in the $L^1(\mathbb{R}^d)$ space.

8.1.1. *Definition and first properties*

DEFINITION 8.1.– *Let $f \in L^1(\mathbb{R}^d)$, its Fourier transform is the function denoted by $\widehat{f}$ (or $\mathcal{F}f$) and is defined by $\forall x \in \mathbb{R}^d$, $\widehat{f}(x) = \int_{\mathbb{R}^d} e^{-i2\pi xy} f(y)dy$, with $xy = \langle x, y \rangle$, which denotes the scalar product in $\mathbb{R}^d$.*

REMARK 8.1.–

1) We should point out that some authors adopt other definitions and the reader should bear them in mind. For example, we cite the following formulas:

$$\widehat{f}(x) = \int_{\mathbb{R}^d} e^{-i\pi xy} f(y)dy \quad or \quad \widehat{f}(x) = \frac{1}{\sqrt{2\pi}} \int_{\mathbb{R}^d} e^{-i\pi xy} f(y)dy.$$

2) Note that the Fourier transform does not always exist.

For example, the function $f(x) = x$ does not have a Fourier transform because the integral $\int_{\mathbb{R}} ye^{-i2\pi xy}dy$ does not exist for any value of x.

3) In order to avoid overburdening the book with further details, the Fourier transform of f is indifferently expressed by $\widehat{f}$ or by $\mathcal{F}f$ independently of the space on which it is defined. For more details concerning these two notations, the interested reader can refer to Dalmasso and Witomski (1996).

EXAMPLE 8.1.– The rectangular function denoted by Π is defined by

$$\begin{cases} \Pi(t) = 1 & \text{if } t \in \left[-\frac{1}{2}, \frac{1}{2}\right] \\ \Pi(t) = 0 & \text{if } t \notin \left[-\frac{1}{2}, \frac{1}{2}\right]. \end{cases}$$

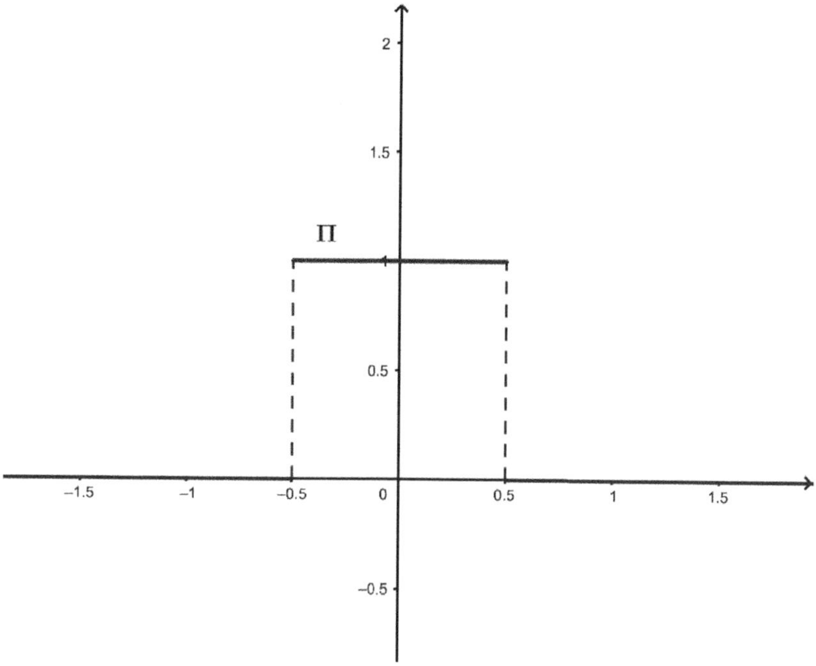

Figure 8.1. *Rectangular function* Π

Given that Π is even, then if $s \neq 0$, we have

$$\mathcal{F}(\Pi)(s) = 2\int_0^{\frac{1}{2}} \Pi(t)\cos(2\pi st)\,\mathrm{d}t = 2\left[\frac{\sin 2\pi st}{2\pi s}\right]_0^{\frac{1}{2}} = \frac{\sin \pi s}{\pi s}.$$

If $s = 0$, then $\mathcal{F}(\Pi)(0) = 1$. The function $\mathcal{F}(\Pi)$ is extendable by continuity at 0.

Therefore, the Fourier transform of the rectangular function, Π, is the function defined on $\mathbb{R}$ by

$$\forall s \in \mathbb{R}, \mathcal{F}(\Pi)(s) = \frac{\sin \pi s}{\pi s}.$$

This function is called cardinal sine function and is denoted by sinc. Its graphical representation can be found in Figure 8.2.

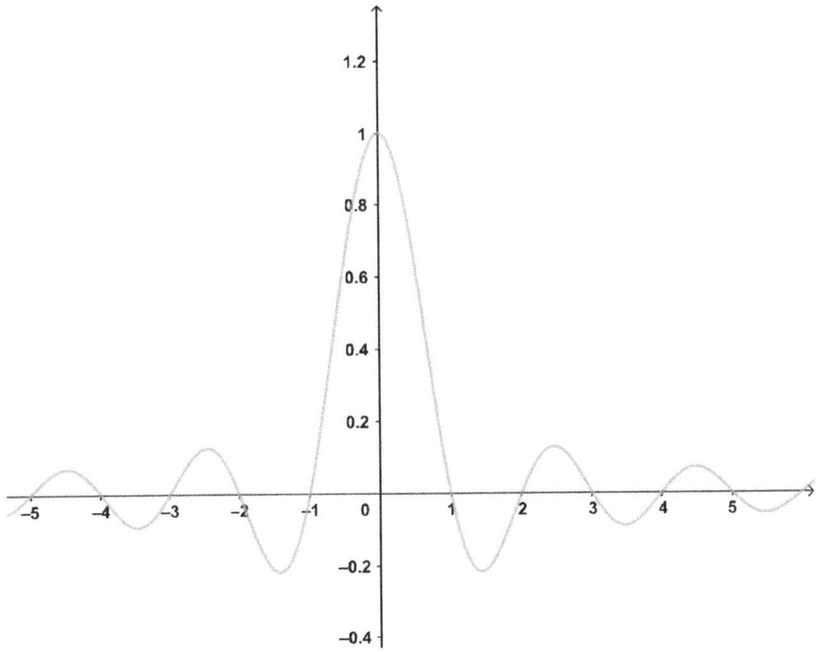

Figure 8.2. *Sinc function: $\mathcal{F}(\Pi)$*

REMARK 8.2.– *The rectangular function Π is an example of an integrable function whose Fourier transform is not integrable! In other words, the Fourier transform does not operate in $L^1(\mathbb{R}^d)$.*

EXAMPLE 8.2 (Poisson density type function).– Consider the function f defined by $f(y) = e^{-a|y|}$ where $a > 0$. We have

$$\mathcal{F}(f)(s) = \frac{2a}{a^2 + 4\pi s^2}.$$

In reality, the function f is even, so $\mathcal{F}(f)(s) = 2\int_0^{+\infty} e^{-ay}\cos(2\pi sy)\mathrm{d}y$. A double integration by part gives $\mathcal{F}(f)(s) = \dfrac{2a}{a^2 + 4\pi s^2}$.

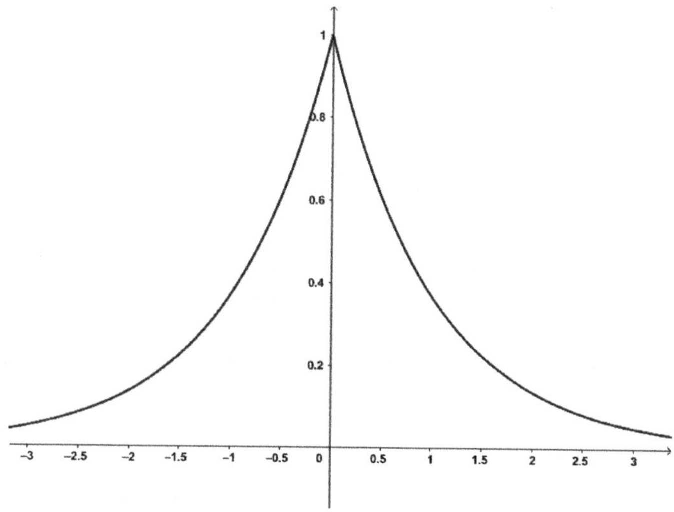

Figure 8.3. *The function* $e^{-|x|}$

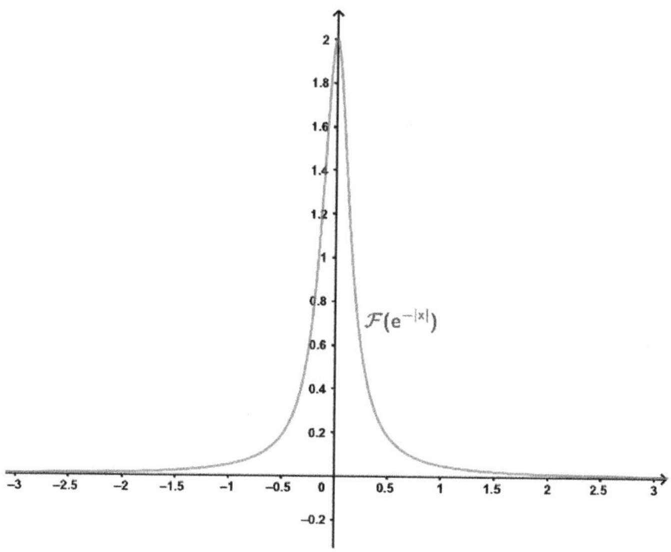

Figure 8.4. *The function* $\mathcal{F}(e^{-|x|})$

The first natural property of the Fourier transform is linearity.

PROPOSITION 8.1 (Linearity).– Let $f, g \in L^1(\mathbb{R}^d)$ and $\alpha, \beta \in \mathbb{C}$. Then,

$$\mathcal{F}(\alpha f + \beta g) = \alpha \mathcal{F}(f) + \beta \mathcal{F}(g).$$

Proof. The proof is derived from the linearity of the integral. □

Among the properties of the Fourier transform, we give a result often used, which is directly derived from the Riemann–Lebesgue lemma.

PROPRIETY 8.1.– *Let $f \in L^1(\mathbb{R}^d)$, then $\widehat{f}$, the Fourier transform of f, is a continuous function, and in addition* $\lim\limits_{\|x\| \to +\infty} \widehat{f}(x) = 0$.

Proof. $\widehat{f}$ is well defined, since $|\widehat{f}(x)| \leqslant \|f\|_1$, then $\widehat{f}$ makes sense.

Continuity: Let $x_0 \in \mathbb{R}^d$ be fixed and let $(s_n)_{n \geq 1}$ be a sequence converging to $x_0 \in \mathbb{R}^d$. We have $\widehat{f}(s_n) = \displaystyle\int_{\mathbb{R}^d} e^{-i2\pi s_n y} f(y) \mathrm{d}y$. Since $s_n \to x_0$, then $e^{-i2\pi s_n y} f(y) \to e^{-i2\pi x_0 y} f(y)$ a.e. $y \in \mathbb{R}^d$. Moreover,

$$\forall y \in \mathbb{R}^d, \left|e^{-i2\pi s_n y} f(y)\right| \leqslant |f(y)|.$$

Now $f \in L^1(\mathbb{R}^d)$, according to Lebesgue's dominated convergence theorem, we have $\widehat{f}(s_n) \underset{n \to +\infty}{\longmapsto} \widehat{f}(x_0)$. Hence, $\widehat{f}$ is continuous at x_0. Consequently, $\widehat{f}$ is continuous on $\mathbb{R}^d$.

To show that $\widehat{f}(x)$ tends to 0 when $\|x\| \to +\infty$, only the scalar case is considered. Furthermore, the space of piecewise continuous functions with bounded support on $\mathbb{R}$ or that of step functions are dense in $L^1(\mathbb{R})$. Any function $f \in L^1(\mathbb{R})$ can be approximated in the sense of the norm of $L^1(\mathbb{R})$ by a sequence of step functions with bounded support. Therefore, a stepped function φ can be found; this means that there exist bounded intervals $[a_{k-1}, a_k]$ and values $c_1, ..., c_n$ such that

$$\varphi = \sum_{k=1}^{n} c_k \mathbb{1}_{[a_{k-1}, a_k]},$$

where $\mathbb{1}_{[a_{k-1},a_k]}$ designates the characteristic function of $[a_{k-1}, a_k]$. As such, φ is written as a finite linear combination of interval characteristic functions and its Fourier transform $\widehat{\varphi}$ can be computed by linearity. More precisely, we have

$$\begin{aligned}\widehat{\varphi}(x) &= \int_{-\infty}^{+\infty} \sum_{k=1}^{n} c_k \mathbb{1}_{[a_{k-1},a_k]}(y) e^{-i2\pi xy} \mathrm{d}y \\ &= \sum_{k=1}^{n} c_k \int_{a_{k-1}}^{a_k} e^{-i2\pi xy} \mathrm{d}y \\ &= \sum_{k=1}^{n} \frac{ic_k}{2\pi x} \left(e^{-i2\pi x a_k} - e^{-i2\pi x a_{k-1}}\right).\end{aligned}$$

Then

$$|\widehat{\varphi}(x)| \leq \frac{1}{\pi|x|} \sum_{k=1}^{n} |c_k|.$$

Let $\varepsilon > 0$. For $|x| \to +\infty$, we have $|\widehat{\varphi}(x)| \leq \frac{\varepsilon}{2}$. Consequently,

$$\left|\widehat{f}(x)\right| \leq \left|\widehat{f}(x) - \widehat{\varphi}(x)\right| + |\widehat{\varphi}(x)| \leq \varepsilon.$$

Then $\lim_{\|x\| \to +\infty} \widehat{f}(x) = 0$. This completes the proof. □

REMARK 8.3.–

1) To show that $\lim_{\|x\| \to +\infty} \widehat{f}(x) = 0$, *we could directly use the Riemann–Lebesgue theorem. The proof given here implicitly reuses the proof of the Riemann–Lebesgue theorem.*

2) Although the Fourier transform of a function $f \in L^1(\mathbb{R}^d)$ *has meaning and is even continuous, in general it is not in* $L^1(\mathbb{R}^d)$.

3) Application of the Fourier transform

$$\mathcal{F} : \left(L^1(\mathbb{R}^d), \|\cdot\|_1\right) \longrightarrow \left(\mathcal{C}_0(\mathbb{R}^d), \|\cdot\|_\infty\right)$$

is a linear and continuous mapping.

It should be noted that $\mathcal{C}_0(\mathbb{R}^d)$ designates the set of continuous functions on $\mathbb{R}^d$ which tends to 0 when $\|x\|\to +\infty$. We have

$$\mathcal{C}_0(\mathbb{R}^d) = \{f \in \mathcal{C}(\mathbb{R}^d), \lim_{\|x\|\to+\infty} \widehat{f}(x) = 0\}.$$

8.1.2. *Fourier transform and operations*

We now present some highly relevant properties of the Fourier transform.

8.1.2.1. *Fourier transform and differentiation*

We start with the scalar case $d = 1$.

8.1.2.1.1. Scalar case

We give a lemma which will be useful in the proof of the following theorem.

LEMMA 8.1.– *Let $f \in L^1(\mathbb{R})$. If $f' \in L^1(\mathbb{R})$, then* $\lim_{x\to\pm\infty} f(x) = 0$.

Proof. For all $x \in \mathbb{R}$, we have $f(x) = f(0) + \displaystyle\int_0^x f'(s)ds$.

Therefore, $\lim_{x\to+\infty} f(x) = f(0) + \displaystyle\int_0^{+\infty} f'(s)ds$. Hence $\lim_{x\to+\infty} f(x)$ does exist.

Let us assume that $\lim_{x\to+\infty} f(x) \neq 0$ and we set $L = \lim_{x\to+\infty} f(x)$. If we take $\epsilon = \dfrac{|L|}{2}$, we get

$$\exists A > 0, \forall x \in \mathbb{R}, \quad x > A \Longrightarrow |f(x)| > \frac{|L|}{2},$$

which is non-sense because $f \in L^1(\mathbb{R})$. Then, $\lim_{x\to+\infty} f(x) = 0$.

Similarly, we also show that $\lim_{x\to-\infty} f(x) = 0$. □

Now we will study the Fourier transform of a derivative.

THEOREM 8.1.– *Let $f \in L^1(\mathbb{R})$. We assume that f is differentiable and that $f' \in L^1(\mathbb{R})$. Then*

$$\mathcal{F}(f')(x) = (2i\pi)x\mathcal{F}f(x) = (2i\pi)x\widehat{f}(x).$$

If, in addition, f admits derivatives up to order n that are in $L^1(\mathbb{R})$, then:

$$\mathcal{F}(f^{(n)})(x) = (2i\pi)^n x^n \mathcal{F}f(x).$$

Proof. We have

$$\begin{aligned}\mathcal{F}(f')(x) &= \int_{\mathbb{R}} e^{-2i\pi xy} f'(y)\mathrm{d}y \\ &= \left[f(x)e^{-2i\pi xy}\right]_{-\infty}^{+\infty} + (2i\pi)x \int_{\mathbb{R}} e^{-2i\pi xy} f(y)\mathrm{d}y.\end{aligned}$$

By hypothesis, $f \in L^1(\mathbb{R})$ and $f' \in L^1(\mathbb{R})$. According to lemma 8.1, we have $\lim\limits_{x\to\pm\infty} f(x) = 0$. It follows that $\lim\limits_{x\to\pm\infty} f(x)e^{-2i\pi xy} = 0$. Then,

$$\mathcal{F}(f')(x) = (2i\pi)x \int_{\mathbb{R}} e^{-2i\pi xy} f(y)\mathrm{d}y = (2i\pi)x\mathcal{F}(f)(x).$$

Since f admits derivatives up to order n then, by repeating the above process, we obtain $\mathcal{F}(f^{(n)})(x) = (2i\pi)^n x^n \mathcal{F}f(x)$. □

The following result shows that the more differentiable the function f is, with derivatives in $L^1(\mathbb{R})$, the more rapidly its Fourier transform $\widehat{f}$ decreases at infinity.

COROLLARY 8.1.– *Let $f \in L^1(\mathbb{R})$. If f admits derivatives up to order n which are in $L^1(\mathbb{R})$, then there exists $C > 0$ such that $\forall x \in \mathbb{R}^*, \left|\widehat{f}(x)\right| \leq \frac{C}{|x|^n}$.*

Proof. From the formula $\mathcal{F}(f^{(n)}(x)) = (2\pi x)^n \widehat{f}(x)$, it is immediately deduced that

$$\forall x \in \mathbb{R}^*, \left|\widehat{f}(x)\right| \leq \frac{1}{(2\pi|x|)^n} \int_{-\infty}^{+\infty} \left|f^{(n)}(y)\right| \mathrm{d}y \leq \frac{C}{|x|^n},$$

with $C = \dfrac{1}{(2\pi)^n} \displaystyle\int_{-\infty}^{+\infty} \left|f^{(n)}(y)\right| \mathrm{d}y$. This completes the proof. □

Observation: *The corollary shows that the more differentiable f is and the more integrable its derivatives are, the more rapidly $\widehat{f}$ decreases at infinity.*

The following result is a result that gives conditions which ensure the regularity of the Fourier transform of a function if this latter multiplied with a polynomial is integrable.

PROPOSITION 8.2.– Let $f \in L^1(\mathbb{R})$. If $yf(y) \in L^1(\mathbb{R})$, then $\widehat{f}$ is differentiable and we have

$$\left(\widehat{f}\right)'(x) = \mathcal{F}\left((-2i\pi x)f\right).$$

Moreover, if $y^n f(y) \in L^1(\mathbb{R})$, then $\widehat{f}$ is n times differentiable and we have

$$\left(\widehat{f}\right)^{(n)} = \mathcal{F}\left((-2i\pi\, x)^n f\right).$$

Proof. We have $\left|(-2i\pi y)f(y)e^{-2i\pi xy}\right| = 2\pi\,|yf(y)|$. Since, by hypothesis, $yf(y) \in L^1(\mathbb{R})$, then according to the differentiation theorem under the integral sign, we have

$$\left(\widehat{f}\right)'(x) = (-2i\pi)\int_{\mathbb{R}} e^{-2i\pi xy} yf(y)\mathrm{d}y.$$

Consequently,

$$\left(\widehat{f}\right)'(x) = \mathcal{F}\left((-2i\pi x)f\right).$$

Since $y^n f(y) \in L^1(\mathbb{R})$, then repeating the above process gives the formula

$$\left(\widehat{f}\right)^{(n)} = \mathcal{F}\left((-2i\pi\, x)^n f\right).$$

□

COROLLARY 8.2.– *Let $f \in L^1(\mathbb{R})$. If $y^n f(y) \in L^1(\mathbb{R})$, then*

$$\forall x \in \mathbb{R}, \left|\left(\widehat{f}\right)^{(n)}(x)\right| \leq (2\pi)^n \int_{\mathbb{R}} |y^n f(y)|\, dy.$$

Observation: *The corollary reflects the fact that the faster f decreases at infinity, the more differentiable $\widehat{f}$ is and its derivatives are bounded.*

8.1.2.1.2. Vector case

The properties given in the two previous propositions generalize to the multidimensional case ($d \geq 1$).

THEOREM 8.2 (Fourier transform and differentiation).– *Let $f \in L^1(\mathbb{R}^d)$ and let $k \geq 1$ be an integer.*

i) If $\partial^\alpha f$ exists for $|\alpha| \leqslant k$ and $\partial^\alpha f \in L^1(\mathbb{R}^d)$, then we have

$$\mathcal{F}(\partial^\alpha f)(x) = (2i\pi)^{|\alpha|} x^\alpha \mathcal{F}f(x).$$

ii) If for any $\alpha : |\alpha| \leqslant k, x^\alpha f \in L^1(\mathbb{R}^d)$, then $\mathcal{F}f$ is of class $\mathcal{C}^k$ and we have

$$\partial^\alpha \mathcal{F}f = \mathcal{F}\left((-2i\pi)x)^\alpha f\right).$$

8.1.2.2. *Fourier transform and translation*

PROPOSITION 8.3 (Fourier transform and translation).– Let $f \in L^1(\mathbb{R}^d)$ and $a \in \mathbb{R}^d$. We have

i) the Fourier transform of the function $\tau_a(f) = f \circ \tau_{-a} : x \to f(x-a)$ is

$$\mathcal{F}(\tau_a(f))(\zeta) = e^{-2i\pi\zeta a}\mathcal{F}(f)(\zeta), \ \forall \zeta \in \mathbb{R}^d.$$

ii) We also have $\tau_a(\mathcal{F}(f)(\zeta)) = \mathcal{F}(f)(\zeta - a) = \mathcal{F}(e^{2ia\pi x} f)(\zeta), \ \ \forall \zeta \in \mathbb{R}^d.$

Proof. Let $\zeta \in \mathbb{R}^d$. We have

$$\begin{aligned}
\mathcal{F}\left(\tau_a(f)\right)(\zeta) &= \int_{\mathbb{R}^d} e^{-2i\pi x\zeta} f(x-a)\mathrm{d}x \\
&= \int_{\mathbb{R}^d} e^{-2i\pi(y+a)\zeta} f(y)\mathrm{d}y \\
&= e^{-2i\pi a\zeta} \int_{\mathbb{R}^d} e^{-2i\pi y\zeta} f(y)\mathrm{d}y \\
&= e^{-2i\pi a\zeta}\mathcal{F}(f)(\zeta).
\end{aligned}$$

Hence, $\forall \zeta \in \mathbb{R}^d, \ \mathcal{F}(\tau_a(f))(\zeta) = e^{-2i\pi\zeta a}\mathcal{F}(f)(\zeta).$

Similarly, we show property (ii). □

8.1.2.3. *Fourier transform and dilation*

PROPOSITION 8.4.– Let $f \in L^1(\mathbb{R}^d)$ and $\lambda \in \mathbb{R}^*$. We have

i) if $g(t) = f(\frac{t}{\lambda})$, then $\mathcal{F}g(x) = |\lambda|^d \mathcal{F}f(\lambda x)$;

ii) $\widehat{\check{f}} = \check{\widehat{f}}$ with $\check{f}(x) = f(-x)$.

Proof.

i) We have

$$\begin{aligned}\widehat{g}(x) &= \int_{\mathbb{R}^d} e^{-i2\pi xy} g(y) \mathrm{d}y \\ &= \int_{\mathbb{R}^d} e^{-i2\pi xy} f(\frac{y}{\lambda}) \mathrm{d}y.\end{aligned}$$

Using the change in variable $y = \lambda u$, we obtain $\mathrm{d}y = |\lambda|^d \mathrm{d}u$. As a result,

$$\begin{aligned}\widehat{g}(x) &= \int_{\mathbb{R}^d} |\lambda|^d e^{-2i\pi\lambda xu} f(u) \mathrm{d}u \\ &= |\lambda|^d \widehat{f}(\lambda x).\end{aligned}$$

Conclusion: $\mathcal{F}(f(\frac{t}{\lambda}))(x) = |\lambda|^d \mathcal{F}f(\lambda x)$.

ii) We have

$$\begin{aligned}\check{\widehat{f}}(x) &= \widehat{f}(-x) \\ &= \int_{\mathbb{R}^d} e^{2i\pi xy} f(y) \mathrm{d}y \\ &= \int_{\mathbb{R}^d} e^{-2i\pi xy} f(-y) \mathrm{d}y \\ &= \widehat{\check{f}}(x).\end{aligned}$$

We can also apply the first point of the proposition by taking $\lambda = -1$. □

Observation: For a small λ, $f(\frac{t}{\lambda})$ is very concentrated while its Fourier transform is very spread out.

COROLLARY 8.3.– *Let $f \in L^1(\mathbb{R}^d)$. We have*

– *if f is even, then $\widehat{f}$ is also even;*

– *if f is odd, then $\widehat{f}$ is also odd.*

Proof.

– If f is even, then $\check{f} = f$. As a result, $\widehat{f} = \widehat{\check{f}} = \check{\widehat{f}}$. Consequently, $\widehat{f}$ is even.

– The proof is similar to that of the first point. □

8.1.2.4. *Exchange formula*

PROPOSITION 8.5 (Transfer formula).– Let $f, g \in L^1(\mathbb{R}^d)$. We have

$$\int_{\mathbb{R}^d} f(x)\widehat{g}(x)\mathrm{d}x = \int_{\mathbb{R}^d} \widehat{f}(x)g(x)\mathrm{d}x.$$

Proof. The proof is mainly based on the Fubini theorem. Furthermore, according to the Fubini–Tonelli theorem, we have

$$\int_{\mathbb{R}^d}\int_{\mathbb{R}^d} \left|f(x)g(y)e^{-2i\pi xy}\right| \mathrm{d}x\mathrm{d}y = \int_{\mathbb{R}^d} |f(x)|\, \mathrm{d}x \int_{\mathbb{R}^d} |g(y)|\, \mathrm{d}y = \|f\|_1\|g\|_1 < \infty.$$

Fubini's theorem then gives

$$\begin{aligned}\int_{\mathbb{R}^d} f(x)\widehat{g}(x)\mathrm{d}x &= \int_{\mathbb{R}^d} f(x)\left(\int_{\mathbb{R}^d} g(y)e^{-2i\pi xy}\mathrm{d}y\right)\mathrm{d}x \\ &= \int_{\mathbb{R}^d} g(y)\left(\int_{\mathbb{R}^d} f(x)e^{-2i\pi xy}\mathrm{d}x\right)\mathrm{d}y \\ &= \int_{\mathbb{R}^d} \widehat{f}(y)g(y)\mathrm{d}y.\end{aligned}$$

Hence the exchange formula, also called the duality formula. □

One of the applications of the exchange (or transfer) formula is the theorem on the injection of the Fourier transform on $L^1(\mathbb{R}^d)$.

THEOREM 8.3.– *(El Amrani 2008) The Fourier transform $\mathcal{F} : L^1(\mathbb{R}^d) \longrightarrow \mathcal{C}_0(\mathbb{R}^d)$ is an injective application.*

8.1.2.5. *Fourier transform and convolution*

PROPOSITION 8.6 (Fourier transform and convolution).– Let $f, g \in L^1(\mathbb{R}^d)$. We have

$$f * g \in L^1(\mathbb{R}^d) \text{ and } \forall x \in \mathbb{R}^d,\ \mathcal{F}(f * g)(x) = \mathcal{F}(f)(x)\mathcal{F}(g)(x).$$

In other words, the Fourier transform of a convolution product is equal to the ordinary product of the Fourier transforms of each factor.

Proof. In order to provide a simple proof, the proposition is presented for the scalar case (Fubini will allow generalizing the proof to the vector case).

We have $\forall x \in \mathbb{R},\ f * g(x) = \int_{-\infty}^{+\infty} f(t)g(x-t)\mathrm{d}t$. Hence, for all $x \in \mathbb{R}$, we have

$$\begin{aligned}\mathcal{F}(f * g)(x) &= \int_{-\infty}^{+\infty} f * g(y)e^{-2i\pi xy}\mathrm{d}y \\ &= \int_{-\infty}^{+\infty}\int_{-\infty}^{+\infty} f(t)g(y-t)e^{-2i\pi xy}\mathrm{d}t\mathrm{d}y \\ &= \int_{-\infty}^{+\infty}\int_{-\infty}^{+\infty} f(t)e^{-2i\pi xt}g(y-t)e^{-2i\pi x(y-t)}\mathrm{d}t\mathrm{d}y.\end{aligned}$$

We set $u = t$ and $v = y - t$. From the change in variables theorem in multiple integrals, we have

$$\mathcal{F}(f * g)(x) = \int_{-\infty}^{+\infty}\int_{-\infty}^{+\infty} f(t)e^{-2i\pi xt}g(v)e^{-2i\pi xv}|J|\mathrm{d}t\mathrm{d}v,$$

with $J = \det\begin{pmatrix}\frac{\partial t}{\partial u} & \frac{\partial t}{\partial v} \\ \frac{\partial y}{\partial u} & \frac{\partial y}{\partial v}\end{pmatrix} = \begin{vmatrix}1 & 0 \\ 1 & 1\end{vmatrix} = 1.$

Then $\forall x \in \mathbb{R},\ \mathcal{F}(f * g)(x) = \int_{-\infty}^{+\infty} f(t)e^{-2i\pi xt}\mathrm{d}t \int_{-\infty}^{+\infty} g(v)e^{-2i\pi xv}\mathrm{d}v.$ Consequently, $\forall x \in \mathbb{R},\ \mathcal{F}(f * g)(x) = \mathcal{F}(f)(x)\mathcal{F}(g)(x)$. □

The operations on the Fourier transform are summarized in Table 8.1.

Function	$f(x)$	$\partial^\alpha f(x)$	$(f * g)(x)$	$f(x+a)$	$f(ax)$
Fourier transform	$\mathcal{F}(f)(\zeta)$	$(2i\pi)^{\lvert\alpha\rvert}x^\alpha\mathcal{F}(f)(\zeta)$	$\mathcal{F}(f)(\zeta)\mathcal{F}(g)(\zeta)$	$e^{i2\pi a\cdot\zeta}\mathcal{F}(f)(\zeta)$	$\frac{1}{\lvert a\rvert^d}\mathcal{F}(f)(\frac{\zeta}{a})$

Table 8.1. *Operations and Fourier transform*

8.1.3. *Fourier transform inversion*

Although the Fourier transform is not bijective onto $L^1(\mathbb{R}^d)$ (it is only injective), when it exists its inverse is defined by the following.

DEFINITION 8.2.– *Let $f \in L^1(\mathbb{R}^d)$. The* **conjugate** *Fourier transform (or inverse) of the function f is the function denoted by $\overline{\mathcal{F}}f$ and it is defined by*

$$\forall x \in \mathbb{R}^d,\ \overline{\mathcal{F}}f(x) = \int_{\mathbb{R}^d} e^{2i\pi xy} f(y)dy.$$

THEOREM 8.4.– *Let $f \in L^1(\mathbb{R}^d)$. If $\mathcal{F}f \in L^1(\mathbb{R}^d)$ and f is a continuous function, then*

$$\overline{\mathcal{F}}\left(\mathcal{F}f\right) = f. \qquad [8.1]$$

Proof. To simplify, the theorem is shown for $d = 1$. Let $x \in \mathbb{R}$. We have

$$\overline{\mathcal{F}}\left(\mathcal{F}(f)\right)(x) = \int_{\mathbb{R}} e^{2i\pi xy}\widehat{f}(y)\mathrm{d}y$$

$$= \lim_{\varepsilon \longrightarrow 0} \int_{\mathbb{R}} e^{2i\pi xy} e^{-\varepsilon|y|}\widehat{f}(y)\mathrm{d}y$$

$$= \lim_{\varepsilon \longrightarrow 0} \int_{\mathbb{R}}\int_{\mathbb{R}} e^{2i\pi y(x-t)} e^{-\varepsilon|y|} f(t)\mathrm{d}t\mathrm{d}y$$

$$= \lim_{\varepsilon \longrightarrow 0} \int_{\mathbb{R}} f(t)\mathrm{d}t \int_{\mathbb{R}} e^{-2i\pi y(t-x)} e^{-\varepsilon|y|}\mathrm{d}y.$$

Now, $\mathcal{F}(e^{-\varepsilon|y|})(s) = \dfrac{2\varepsilon}{\varepsilon^2 + 4\pi^2 s^2}$, thus $\overline{\mathcal{F}}\left(\mathcal{F}(f)\right)(x) = \lim_{\varepsilon \longrightarrow 0} \displaystyle\int_{\mathbb{R}} f(t)\frac{2\varepsilon}{\varepsilon^2 + 4\pi^2(t-x)^2}\mathrm{d}t$. By change in variable $t = x + \varepsilon u$, we obtain

$$\overline{\mathcal{F}}\left(\mathcal{F}(f)\right)(x) = \lim_{\varepsilon \longrightarrow 0} \frac{1}{\pi}\int_{\mathbb{R}} f(x+\varepsilon u)\frac{1}{1+u^2}\mathrm{d}u.$$

We have that f is continuous, so $\lim\limits_{\varepsilon \longrightarrow 0} f(x+\varepsilon u) = f(x)$. By Lebesgue's dominated convergence theorem, we obtain

$$\overline{\mathcal{F}}\left(\mathcal{F}(f)\right)(x) = f(x)\frac{1}{\pi}\int_{\mathbb{R}} \frac{1}{1+u^2}\mathrm{d}u = f(x)\frac{1}{\pi}\left[\arctan(u)\right]_{-\infty}^{+\infty} = f(x).$$

Consequently, $\overline{\mathcal{F}}(\mathcal{F}f) = f$. □

REMARK 8.4.–

1) The identity [8.1] *shows that the Fourier transform can be inverted: there exists therefore an inverse transform which is $\overline{\mathcal{F}}$, which could also be written as $\mathcal{F}^{-1}$.*

2) If $f \in L^1(\mathbb{R})$ is a function that is not continuous in t, then

$$\overline{\mathcal{F}}(\mathcal{F}(f))(t) = \frac{f(t^+) + f(t^-)}{2}.$$

3) These results are particularly significant when the operator $\mathcal{F}$ is used for the solution of partial differential equations.

8.2. Fourier transform in $\mathcal{S}(\mathbb{R}^d)$

As pointed out in theorem 8.3, the Fourier transform is injective on $L^1(\mathbb{R}^d)$ but is not surjective. To overcome this shortcoming, the starting space is reduced to the subspace $\mathcal{S}(\mathbb{R}^d) \subset L^1(\mathbb{R}^d)$ while hoping to ensure the bijection of the Fourier transform on this new space $\mathcal{S}(\mathbb{R}^d)$ and preserving all the properties established in the $L^1(\mathbb{R}^d)$ space.

8.2.1. *Definition and first properties*

We start with a lemma which will be useful to establish a stability result of the Schwartz space with respect to the Fourier transform.

LEMMA 8.2.– *Let (f_n) be a sequence of elements of $\mathcal{S}(\mathbb{R}^d)$.*

If $f_n \underset{n\to+\infty}{\longmapsto} 0$ in $\mathcal{S}(\mathbb{R}^d)$, then $f_n \underset{n\to+\infty}{\longmapsto} 0$ in $L^1(\mathbb{R}^d)$.

Proof. Let (f_n) be a sequence of elements of $\mathcal{S}(\mathbb{R}^d)$ such that $f_n \underset{n\to+\infty}{\longmapsto} 0$ in $\mathcal{S}(\mathbb{R}^d)$. We have: $f_n \underset{n\to+\infty}{\longmapsto} 0$ in $\mathcal{S}(\mathbb{R}^d)$, then $\forall p \in \mathbb{N}, \mathbf{Q}_p(f_n) \underset{n\to+\infty}{\longmapsto} 0$.

Let $p \in \mathbb{N}$ be such that $\displaystyle\int_{\mathbb{R}^d} \frac{dx}{(1+|x|)^p} < +\infty$. We have

$$\begin{aligned}\int_{\mathbb{R}^d} |f_n(x)|dx &= \int_{\mathbb{R}^d} \frac{|f_n(x)|}{(1+|x|)^p}((1+|x|)^p)dx \\ &\leqslant \int_{\mathbb{R}^d} \frac{dx}{(1+|x|)^p} \times \sup_{x\in\mathbb{R}^d} |(1+|x|)^p f_n(x)| \\ &\leqslant \int_{\mathbb{R}^d} \frac{dx}{(1+|x|)^p} \times \mathbf{Q}_p(f_n) \\ &\leq \int_{\mathbb{R}^d} \frac{dx}{(1+|x|)^p} \mathbf{N}_p(f_n) \underset{n\to+\infty}{\longmapsto} 0.\end{aligned}$$

Then, $f_n \underset{n\to+\infty}{\longrightarrow} 0$ in $L^1(\mathbb{R}^d)$. □

We have $\mathcal{S}(\mathbb{R}^d) \subset L^p(\mathbb{R}^d), \forall p \in [1,+\infty]$, in particular, $\mathcal{S}(\mathbb{R}^d) \subset L^1(\mathbb{R}^d)$. Then, for all $f \in \mathcal{S}(\mathbb{R}^d)$, $\widehat{f}$ exists.

THEOREM 8.5.– *The Fourier transform $\mathcal{F}$ is continuous on $\mathcal{S}(\mathbb{R}^d)$ and moreover, $\mathcal{F}(\mathcal{S}(\mathbb{R}^d)) \subset \mathcal{S}(\mathbb{R}^d)$.*

Proof. Let $f \in \mathcal{S}(\mathbb{R}^d)$.

$\mathcal{F}f \in \mathcal{S}(\mathbb{R}^d)$: We have $\widehat{f}(x) = \displaystyle\int_{\mathbb{R}^d} e^{-2i\pi xy} f(y)dy$. According to the derivation property of the Fourier transform of a function, we have

$$\forall \beta \in \mathbb{N}^d,\ \partial^\beta \widehat{f}(x) = (-2i\pi)^{|\beta|}\mathcal{F}(x^\beta f).$$

On the other hand, we know that if $f \in \mathcal{S}(\mathbb{R}^d)$ and $P \in \mathbb{R}[X]$, then $Pf \in \mathcal{S}(\mathbb{R}^d)$.

Hence, $\forall \beta \in \mathbb{N}^d, \partial^\beta \widehat{f}$ exists. As a result, $\widehat{f} \in \mathcal{C}^\infty(\mathbb{R}^d)$.

Let us show that $\displaystyle\sup_{x\in\mathbb{R}^d} \left|x^\alpha \partial^\beta \widehat{f}(x)\right| < +\infty$.

We have $x^\alpha \partial^\beta \widehat{f}(x) = x^\alpha (-2i\pi)^{|\beta|} \mathcal{F}(x^\beta f)$ and $\mathcal{F}(\partial^\alpha (x^\beta f)) = (2i\pi)^{|\alpha|} x^\alpha (\mathcal{F}(x^\beta f))$. Hence, $x^\alpha \partial^\beta \widehat{f} = c.\mathcal{F}\left(\partial^\alpha (x^\beta f)\right)$ where c is a constant. We have

$$\begin{aligned} f \in \mathcal{S}(\mathbb{R}^d) &\Longrightarrow x^\beta f \in \mathcal{S}(\mathbb{R}^d) \\ &\Longrightarrow \partial^\alpha (x^\beta f) \in \mathcal{S}(\mathbb{R}^d) \subset L^1(\mathbb{R}^d) \\ &\Longrightarrow x^\alpha \partial^\beta \widehat{f} \text{ is bounded} \\ &\Longrightarrow \sup_{x \in \mathbb{R}^d} |x^\alpha \partial^\beta \widehat{f}(x)| < +\infty. \end{aligned}$$

Hence $\widehat{f} \in \mathcal{S}(\mathbb{R}^d)$. Consequently, $\forall f \in \mathcal{S}(\mathbb{R}^d)$, $\mathcal{F}(f) \in \mathcal{S}(\mathbb{R}^d)$.

Conclusion: $\mathcal{F}$ is a linear mapping that sends $\mathcal{S}(\mathbb{R}^d)$ in $\mathcal{S}(\mathbb{R}^d)$.

Continuity: Let (f_n) be a sequence of elements of $\mathcal{S}(\mathbb{R}^d)$ such that $f_n \underset{n \to +\infty}{\longmapsto} 0$. Let us show that: $\mathcal{F}(f_n) = \widehat{f_n} \underset{n \to +\infty}{\longmapsto} 0$ in $\mathcal{S}(\mathbb{R}^d)$, that is, $\forall \alpha \in \mathbb{N}^d, \forall \beta \in \mathbb{N}^d, x^\alpha \partial^\beta \widehat{f_n} \to 0$ uniformly on $\mathbb{R}^d$.

We recall that $x^\alpha \partial^\beta \widehat{f_n} = \dfrac{(-2i\pi)^{|\beta|}}{(-2i\pi)^{|\alpha|}} \mathcal{F}(\partial^\alpha (x^\beta f_n))$. We have $f_n \underset{n \to +\infty}{\longmapsto} 0$ in $\mathcal{S}(\mathbb{R}^d)$. Then, $x^\beta f_n \underset{n \to +\infty}{\longmapsto} 0$ in $\mathcal{S}(\mathbb{R}^d)$. As such, $\partial^\alpha (x^\beta f_n) \underset{n \to +\infty}{\longmapsto} 0$ in $\mathcal{S}(\mathbb{R}^d)$. According to lemma 8.2, $\partial^\alpha (x^\beta f_n) \underset{n \to +\infty}{\longmapsto} 0$ in $L^1(\mathbb{R}^d)$. Since

$$\sup_{y \in \mathbb{R}^d} |\mathcal{F}(\partial^\alpha (x^\beta f_n))(y)| \leqslant \|\partial^\alpha (x^\beta f_n)\|_{L^1(\mathbb{R}^d} \underset{n \to +\infty}{\longmapsto} 0,$$

then $\sup_{y \in \mathbb{R}^d} |\mathcal{F}(\partial^\alpha (x^\beta f_n))(y)| \underset{n \to +\infty}{\longmapsto} 0$. As a result, $x^\alpha \partial^\beta \widehat{f_n} \underset{n \to +\infty}{\longmapsto} 0$ uniformly on $\mathbb{R}^d$.

Conclusion: *$\mathcal{F}$ is a continuous linear mapping from $\mathcal{S}(\mathbb{R}^d)$ to $\mathcal{S}(\mathbb{R}^d)$.* □

8.2.2. *Fourier transform and operations*

Here are some elementary but very important properties of the Fourier transform on $\mathcal{S}(\mathbb{R}^d)$.

8.2.2.1. *Fourier transform and differentiation*

PROPOSITION 8.7 (Fourier transform and differentiation).– Let φ be a function of $\mathcal{S}(\mathbb{R}^d)$. We have the following:

i) the function $\mathcal{F}(\varphi)$ is of class $\mathcal{C}^1(\mathbb{R}^d)$, and for all $j = 1, ..., d$, we have

$$\partial_{\xi_j}\mathcal{F}(\varphi)(\xi) = -2i\pi\mathcal{F}(x_j\varphi)(\xi), \forall\xi \in \mathbb{R}^d;$$

ii) for $j = 1, ..., d$, we have $\mathcal{F}(\partial_{x_j}\varphi)(\xi) = 2i\pi\xi_j\mathcal{F}(\varphi)(\xi), \forall\xi \in \mathbb{R}^d$.

Proof. According to theorem 8.5, the conditions of theorem 8.2 are verified by the function $\varphi \in \mathcal{S}(\mathbb{R}^d)$. The function $\mathcal{F}(\varphi)$ is even of class $\mathcal{C}^\infty$ and the differentiation formulas follow therefrom. □

Application: Consider the "Gaussian kernel" function f, defined by $f(x) = e^{-\pi x^2}$. We know that $f \in \mathcal{S}(\mathbb{R}^d)$, and we show that $\widehat{f} = f$.

We set $F(x) := \widehat{f}(x) = \int_{-\infty}^{+\infty} e^{-\pi y^2}e^{-2i\pi xy}dy$ and it can be noted that $F(0) = \int_{-\infty}^{+\infty} e^{-\pi y^2}dy = 1$ (a classical integral that can be found through a little calculation in polar coordinates). A legitimate differentiation with respect to x of the integral in the expression of $F(x)$ gives

$$\begin{aligned} F'(x) &= \int_{-\infty}^{+\infty} \underbrace{(-2i\pi y)e^{-\pi y^2}}_{=if'(y)} e^{-2i\pi xy}dy \\ &= i\int_{-\infty}^{+\infty} f'(y)e^{-2i\pi xy}dy. \end{aligned}$$

A new reading of property (ii) in proposition 8.7 allows us to write

$$\begin{aligned} F'(x) &= i(2i\pi x)\widehat{f}(x) \\ &= -2\pi xF(x). \end{aligned}$$

This ordinary differential equation is easily integrated and we obtain $F(x) = ce^{-\pi x^2}$. Yet, $F(0) = 1$, thus $c = 1$. As a result, $F(x) = e^{-\pi x^2} = f(x)$. Consequently, $\widehat{f} = f$.

8.2.2.2. *Fourier transform and translation*

PROPOSITION 8.8 (Fourier transform and translation).– Let φ be a function of $\mathcal{S}(\mathbb{R}^d)$. We have

i) for all $a \in \mathbb{R}^d$, the function $\tau_a(\varphi) = \varphi \circ \tau_{-a} : x \to \varphi(x-a)$ has the Fourier transform

$$\mathcal{F}(\tau_a(\varphi))(\xi) = e^{-2i\pi\xi a}\mathcal{F}(\varphi)(\xi);$$

ii) for all $a \in \mathbb{R}^d$ and for all $\xi \in \mathbb{R}^d$, we have

$$\mathcal{F}(e^{2ia\pi x}\varphi)(\xi) = \tau_a(\mathcal{F}(\varphi)(\xi)) = \mathcal{F}(\varphi)(\xi - a).$$

8.2.2.3. *Fourier transform and dilation*

PROPOSITION 8.9.– Let $f \in \mathcal{S}(\mathbb{R}^d)$ and $\lambda \in \mathbb{R}^*$. We have

i) $\mathcal{F}(f(\frac{t}{\lambda}))(\zeta) = |\lambda|^d \mathcal{F}f(\lambda\zeta)$;

ii) $\widehat{\check{f}} = \check{\widehat{f}}$ with $\check{f} = f(-x)$.

REMARK 8.5.– *The Fourier transform $\mathcal{F}$ substitutes differentiation and multiplication with x, as we have just seen. Consequently, $\mathcal{F}$ exchanges regularity and decreasing at infinity, that is, the more differentiable a function is (with integrable derivatives on $\mathbb{R}^d$), the more rapidly its Fourier transform decreases at infinity. This fact makes it possible to understand the interest of the Schwartz class with respect to the Fourier transform: since any function of $\mathcal{S}(\mathbb{R}^d)$ is of class $\mathcal{C}^\infty$ with derivatives which are all integrable, it can be deduced that its Fourier transform is rapidly decreasing. And since any function of $\mathcal{S}(\mathbb{R}^d)$ is rapidly decreasing, it is deduced that its Fourier transform is of class $\mathcal{C}^\infty$. Therefore, it can be seen that the Schwartz class is invariant by Fourier transform, which is one of the advantages of this space.*

8.2.3. ***Fourier transform inversion***

This section shows that the Fourier transform is invertible from $\mathcal{S}(\mathbb{R}^d)$ to $\mathcal{S}(\mathbb{R}^d)$.

THEOREM 8.6 (Fourier inversion).– *Let $f \in \mathcal{S}(\mathbb{R}^d)$. We have*

$$f(x) = \int_{\mathbb{R}^d} e^{2i\pi xy}\widehat{f}(y)dy, \quad \forall x \in \mathbb{R}^d, \tag{8.2}$$

that is, $\overline{\mathcal{F}}(\mathcal{F}(f)) = f$.

Proof. Let $\lambda > 0$. We set $\varphi(x) = e^{-|x|^2}$ and $f_\lambda(t) = f(\frac{t}{\lambda})$.

According to the exchange formula, we have

$$\int_{\mathbb{R}^d} \widehat{f_\lambda}(x)\varphi(x)\mathrm{d}x = \int_{\mathbb{R}^d} f_\lambda(x)\widehat{\varphi}(x)\mathrm{d}x = \int_{\mathbb{R}^d} \lambda^d f(u)\widehat{\varphi}(\lambda u)\mathrm{d}u \ (u = \frac{x}{\lambda}).$$

Since $\widehat{f_\lambda}(x) = \lambda^d \widehat{f}(\lambda x)$, then

$$\int_{\mathbb{R}^d} \lambda^d f(u)\widehat{\varphi}(\lambda u)\mathrm{d}x = \int_{\mathbb{R}^d} f(u)\widehat{\varphi_\lambda}(u)\mathrm{d}u = \int_{\mathbb{R}^d} f(x)\widehat{\varphi_\lambda}(x)\mathrm{d}x.$$

On the other hand, we have

$$\int_{\mathbb{R}^d} f_\lambda(x)\widehat{\varphi_\lambda}(x)\mathrm{d}x = \int_{\mathbb{R}^d} \widehat{f}(x)\varphi_\lambda(x)\mathrm{d}x = \int_{\mathbb{R}^d} \widehat{f}(x)\varphi(\frac{x}{\lambda})\mathrm{d}x.$$

We then obtain

$$\int_{\mathbb{R}^d} \widehat{f}(x)\varphi(\frac{x}{\lambda})\mathrm{d}x = \int_{\mathbb{R}^d} f_\lambda(x)\widehat{\varphi}(x)\mathrm{d}x.$$

When $\lambda \to +\infty$, Lebesgue's dominated convergence theorem implies that

$$\int_{\mathbb{R}^d} \widehat{f}(x)\mathrm{d}x = f(0)\int_{\mathbb{R}^d} \widehat{\varphi}(x)\mathrm{d}x.$$

Then, $\int_{\mathbb{R}^d} \widehat{\varphi}(x)\mathrm{d}x = \int_{\mathbb{R}^d} (\pi)^{\frac{d}{2}} e^{-\pi^2(x_1^2+\ldots+x_d^2)}\mathrm{d}x_1\mathrm{d}x_2\ldots\mathrm{d}x_d$. Setting $y = \pi x$, we get

$$\int_{\mathbb{R}} e^{-\pi^2 x^2}\mathrm{d}x = \int_{\mathbb{R}} e^{-y^2}\frac{\mathrm{d}y}{\pi} = \frac{1}{\sqrt{\pi}}.$$

Hence, $\int_{\mathbb{R}^d} \widehat{\varphi}(x)\mathrm{d}x = (\pi)^{\frac{d}{2}}.\frac{1}{\pi^{\frac{d}{2}}} = 1$. As a result, $f(0) = \int_{\mathbb{R}^d} \widehat{f}(x)\mathrm{d}x$. Let $\varphi \in \mathcal{S}(\mathbb{R}^d)$ and we set $f = \tau_{-x}\varphi \in \mathcal{S}(\mathbb{R}^d)$. We have $f(z) = \varphi(z+x)$ and $f(0) = \varphi(x)$. Therefore,

$$f(0) = \int_{\mathbb{R}^d} \mathcal{F}(\tau_{-x}\varphi(y))\mathrm{d}y = \varphi(x).$$

Consequently, $\varphi(x) = \displaystyle\int_{\mathbb{R}^d} e^{2i\pi xy}\widehat{\varphi}(y)\mathrm{d}y, \forall\varphi \in \mathcal{S}(\mathbb{R}^d)$. This completes the demonstration. □

Physical Interpretation: Physically, in one dimension, we often write [8.2] as $f(t) = \int_{-\infty}^{+\infty} e^{2i\pi\nu t}\widehat{f}(\nu)\mathrm{d}\nu$. t denotes time and ν designates frequency.

The term $e^{2i\pi\nu t}$ is called the harmonic component and $\widehat{f}(\nu)\mathrm{d}\nu$ is called the weighting factor. These harmonic components $e^{2i\pi\nu t}$ are trigonometric functions of frequency ν and are the complex equivalent of sine waves. This is the reason why it is sometimes said that any function $f(t)$ (admitting a Fourier transform) can be written as a continuous sum of sinusoids of frequencies ν ranging from $-\infty$ to $+\infty$. The term $\widehat{f}(\nu)$ represents the weight of the harmonic component of frequency ν in the expansion of f.

REMARK 8.6.– *We have*

$$\overline{\mathcal{F}} \circ \mathcal{F} = I_{\mathcal{S}(\mathbb{R}^d)}.$$

THEOREM 8.7.– *$\mathcal{F}$ is a bijection from $\mathcal{S}(\mathbb{R}^d)$ onto $\mathcal{S}(\mathbb{R}^d)$:*

$$(\textit{thatis}\ \forall g \in \mathcal{S}(\mathbb{R}^d), \exists! f \in \mathcal{S}(\mathbb{R}^d) : \mathcal{F}f = g).$$

Proof. Let $f \in \mathcal{S}(\mathbb{R}^d)$. Let us first show that $\widehat{\widehat{f}} = \check{f}$ (we should recall that $\check{f}(x) = f(-x)$).

We have $\widehat{\widehat{f}}(x) = \displaystyle\int_{\mathbb{R}^d} e^{-i2\pi xy}\widehat{f}(y)\mathrm{d}y$. According to the inverse Fourier formula, we have $f(-x) = \displaystyle\int_{\mathbb{R}^d} e^{-i2\pi xy}\widehat{f}(y)\mathrm{d}y$. Since $f(-x) = \check{f}(x)$, thus $\widehat{\widehat{f}}(x) = \check{f}(x)$. It follows that $\widehat{\widehat{f}} = \check{f}$. We have $\widehat{\widehat{f}} = \check{f}$, thus $\widehat{\widehat{\widehat{\widehat{f}}}} = \widehat{\widehat{\check{f}}} = f$. Consequently, $\mathcal{F} \circ \mathcal{F} \circ \mathcal{F} \circ \mathcal{F} = Id_{\mathcal{S}(\mathbb{R}^d)}$. It can be deduced that $\mathcal{F}$ is bijective and $\mathcal{F}^{-1} = \mathcal{F}^3$.

Conclusion: Since $\mathcal{F}$ is also continuous from $\mathcal{S}(\mathbb{R}^d)$ to $\mathcal{S}(\mathbb{R}^d)$, then $\mathcal{F}$ is an isomorphism from $\mathcal{S}(\mathbb{R}^d)$ to $\mathcal{S}(\mathbb{R}^d)$. □

We complete this series of properties of the Fourier transform on the Schwartz class by the Plancherel formula, which can be seen as a trivial consequence of the inversion formula.

THEOREM 8.8 (Plancherel).– *The Fourier transform $\mathcal{F}$ is extended by continuity into an isometry from $L^2(\mathbb{R}^d)$ to $L^2(\mathbb{R}^d)$. That is to say, there exists an extension that is denoted again by $\mathcal{F} : L^2(\mathbb{R}^d) \to L^2(\mathbb{R}^d)$ verifying, in addition, the identity*

$$\langle f, g\rangle_{L^2} = \langle \mathcal{F}f, \mathcal{F}g\rangle_{L^2}, \forall f, g \in L^2(\mathbb{R}^d),$$

and in particular,

$$\| f \|_{L^2} = \| \mathcal{F}f \|_{L^2}, \forall f \in L^2(\mathbb{R}^d).$$

Proof. Let $f, g \in \mathcal{S}(\mathbb{R}^d)$. We have $\langle f, g\rangle = \displaystyle\int_{\mathbb{R}^d} f(x)\overline{g}(x)\mathrm{d}x$. Now

$$g \in \mathcal{S}(\mathbb{R}^d) \Longrightarrow \overline{g}(x) = \overline{\int_{\mathbb{R}^d} e^{2i\pi xy}\widehat{g}(y)\mathrm{d}y} = \int_{\mathbb{R}^d} e^{-2i\pi xy}\overline{\widehat{g}}(y)\mathrm{d}y,$$

thus

$$\langle f, g\rangle = \int_{\mathbb{R}^d} f(x) \int_{\mathbb{R}^d} e^{-2i\pi xy}\overline{\widehat{g}}(y)\mathrm{d}y\mathrm{d}x.$$

According to Fubini's theorem, we have

$$\begin{aligned}
\langle f, g\rangle &= \int_{\mathbb{R}^d} \overline{\widehat{g}}(y) \left(\int_{\mathbb{R}^d} e^{-2i\pi xy} f(x)\mathrm{d}x\right)\mathrm{d}y \\
&= \int_{\mathbb{R}^d} \overline{\widehat{g}}(y)\widehat{f}(y)\mathrm{d}y \\
&= \int_{\mathbb{R}^d} \widehat{f}(y)\overline{\widehat{g}}(y)\mathrm{d}y \\
&= \langle \widehat{f}, \widehat{g}\rangle.
\end{aligned}$$

Conclusion:

$$\forall f, g \in \mathcal{S}(\mathbb{R}^d), \langle f, g\rangle = \langle \mathcal{F}f, \mathcal{F}g\rangle_{L^2}$$

$$\forall f \in \mathcal{S}(\mathbb{R}^d), \| f \|_{L^2} = \| \mathcal{F}f \|_{L^2}.$$

Since $\overline{\mathcal{S}(\mathbb{R}^d)}|_{L^2} = L^2(\mathbb{R}^d)$, then $\mathcal{F}$ is extended by continuity on $L^2(\mathbb{R}^d)$ and again this extension is denoted by $\mathcal{F}$. Moreover, we have

$$\forall f, g \in L^2(\mathbb{R}^d), \langle f, g\rangle_{L^2} = \langle \mathcal{F}f, \mathcal{F}g\rangle_{L^2}, \| f \|_{L^2} = \| \mathcal{F}f \|_{L^2}. \qquad \square$$

REMARK 8.7.– *The extension $\mathcal{F}$ verifying $\|f\|_{L^2}= \|\mathcal{F}f\|_{L^2}$ is and isomorphism from $L^2(\mathbb{R}^d)$ to $L^2(\mathbb{R}^d)$ and $\mathcal{F}^4 = Id_{L^2(\mathbb{R}^d)}$.*

8.2.4. Fourier transform and convolution

Along with the Fourier transform and in connection with it, the convolution product is an essential tool for the study of physical phenomena. Many systems actually operates as filters, in the sense that an input signal is modified after traveling through the system. Under very general assumptions of linearity and time invariance, the output signal is expressed as a convolution product between the input signal and a function that depends on the system under study.

The Fourier transform exchanges convolution product and usual multiplication.

THEOREM 8.9.– *Let $\varphi, \psi \in \mathcal{S}(\mathbb{R}^d)$. We have*

$$\mathcal{F}(\varphi * \psi) = \mathcal{F}\varphi \cdot \mathcal{F}\psi \quad \textit{and} \quad \mathcal{F}(\varphi\psi) = \mathcal{F}\varphi * \mathcal{F}\psi$$

Proof. The use of Fubini's theorem can be easily justified, so we have

1) For all $x \in \mathbb{R}^d$, we have

$$\begin{aligned}
\mathcal{F}(\varphi * \psi)(x) &= \int_{\mathbb{R}^d} e^{-2i\pi xy}(\varphi * \psi)(y)\mathrm{d}y \\
&= \int_{\mathbb{R}^d}\int_{\mathbb{R}^d} e^{-2i\pi xy}\varphi(s)\psi(y-s)ds\mathrm{d}y \\
&= \int_{\mathbb{R}^d} e^{-2i\pi xs}\varphi(s)\left[\int_{\mathbb{R}^d} e^{-2i\pi x(y-s)}\psi(y-s)\mathrm{d}y\right] ds \\
&= \int_{\mathbb{R}^d} e^{-2i\pi xs}\varphi(s)\widehat{\psi}(x)ds \\
&= \widehat{\psi}(x)\int_{\mathbb{R}^d} e^{-2i\pi xs}\varphi(s)ds \\
&= (\widehat{\varphi}\widehat{\psi})(x) \\
&= (\mathcal{F}\varphi \cdot \mathcal{F}\psi)(x).
\end{aligned}$$

Conclusion: $\mathcal{F}(\varphi * \psi) = \mathcal{F}\varphi \cdot \mathcal{F}\psi$.

2) For all $x \in \mathbb{R}^d$, we have

$$\mathcal{F}(\varphi\psi)(x) = \int_{\mathbb{R}^d} e^{-2i\pi xy}(\varphi\psi)(y)\mathrm{d}y = \int_{\mathbb{R}^d} e^{-2i\pi xy}\varphi(y)\psi(y)\mathrm{d}y.$$

Let $\Phi = \widehat{\varphi}$ and $\Psi = \widehat{\psi}$, we have $\varphi(y) = \mathcal{F}^{-1}(\Phi)(y) = \int_{\mathbb{R}^d} e^{2i\pi sy}\Phi(s)\mathrm{d}s$. Therefore,

$$\begin{aligned}
\mathcal{F}(\varphi\psi)(x) &= \int_{\mathbb{R}^d} e^{-2i\pi xy}\Big(\int_{\mathbb{R}^d} e^{2i\pi sy}\Phi(s)\mathrm{d}s\Big)\psi(y)\mathrm{d}y \\
&= \int_{\mathbb{R}^d} \Phi(s)\Big(\int_{\mathbb{R}^d} e^{-2i\pi(x-s)y}\psi(y)\mathrm{d}y\Big)\mathrm{d}s \\
&= \int_{\mathbb{R}^d} \Phi(s)\Psi(x-s)\mathrm{d}s \\
&= \Phi * \Psi(x) \\
&= \mathcal{F}\varphi * \mathcal{F}\psi(x).
\end{aligned}$$

Conclusion: $\mathcal{F}(\varphi\psi) = \mathcal{F}\varphi * \mathcal{F}\psi$. □

PROPOSITION 8.10.– Let $T \in \mathcal{E}'(\mathbb{R}^d)$. For any $\varphi \in \mathcal{S}(\mathbb{R}^d)$, we have

$$\mathcal{F}(\varphi * T) = \mathcal{F}(\varphi)\mathcal{F}(T).$$

Proof. Let $\varphi \in \mathcal{S}(\mathbb{R}^d)$ and $\zeta \in \mathbb{R}^d$. Using integration under the bracket, we get

$$\begin{aligned}
\mathcal{F}(\varphi * T)(\zeta) &= \int_{\mathbb{R}^d} e^{-2i\pi x\zeta}\langle T_y, \varphi(x-y)\rangle \mathrm{d}x \\
&= \Big\langle T_y, \int_{\mathbb{R}^d} e^{-2i\pi x\zeta}\varphi(x-y)\mathrm{d}x\Big\rangle \\
&= \Big\langle T_y, \int_{\mathbb{R}^d} e^{-2i\pi(y+z)\zeta}\varphi(z)\mathrm{d}x\Big\rangle \\
&= \big\langle T_y, e^{-2i\pi y\zeta}\big\rangle\mathcal{F}(\varphi)(\zeta) \\
&= \mathcal{F}(T)(\zeta)\mathcal{F}(\varphi)(\zeta).
\end{aligned}$$

Consequently, $\mathcal{F}(\varphi * T) = \mathcal{F}(\varphi)\mathcal{F}(T)$.

At this point, a justification should be given about how we move from the first line to the second one in the above calculation because the function $(x, y) \mapsto \varphi(x-y)$ does not necessarily have a compact support if we take $\varphi \in \mathcal{S}(\mathbb{R}^d)$. To this end, consider (φ_n) a sequence of $\mathcal{D}(\mathbb{R}^d)$ which converges to φ in $\mathcal{S}(\mathbb{R}^d)$. According to the previous calculation, we have $\mathcal{F}(\varphi_n * T) = \mathcal{F}(\varphi_n)\mathcal{F}(T)$. According to the first point of the proposition, $\varphi_n * T \longrightarrow \varphi * T$ in $\mathcal{S}(\mathbb{R}^d)$. Since the Fourier transform is continuous $\mathcal{S}(\mathbb{R}^d)$, then $\mathcal{F}(\varphi_n * T) \longrightarrow \mathcal{F}(\varphi * T)$ in $\mathcal{S}(\mathbb{R}^d)$.

Finally, we have $\mathcal{F}(\varphi_n)(\zeta)\mathcal{F}(T)(\zeta) \longrightarrow \mathcal{F}(\varphi_n)(\zeta)\mathcal{F}(T)(\zeta)$ in $\mathcal{S}(\mathbb{R}^d)$. It completes the demonstration. □

8.3. Fourier transform in $\mathcal{S}'(\mathbb{R}^d)$

In section 8.2, the Fourier transform has been defined on the Schwartz class $\mathcal{S}(\mathbb{R}^d)$. Unfortunately, this cannot be considered as sufficient, because there are far too few functions in $\mathcal{S}(\mathbb{R}^d)$. We will now extend its definition to the space $\mathcal{S}'(\mathbb{R}^d)$ of tempered distributions by applying the same duality strategy.

8.3.1. *Definition and first properties*

To make sense of the Fourier transform of a tempered distribution, we start with the following result.

PROPOSITION 8.11.– Let $T \in \mathcal{S}'(\mathbb{R}^d)$. The relation $\langle \mathcal{F}T, \varphi\rangle = \langle T, \widehat{\varphi}\rangle, \forall \varphi \in \mathcal{S}(\mathbb{R}^d)$ defines a tempered distribution. The relation $\langle \overline{\mathcal{F}}T, \varphi\rangle = \langle T, \overline{\widehat{\varphi}}\rangle, \forall \varphi \in \mathcal{S}(\mathbb{R}^d)$ also defines a tempered distribution.

Proof. Let $\varphi_n \underset{n\to+\infty}{\longmapsto} 0$ in $\mathcal{S}(\mathbb{R}^d)$. Let us show that $\langle \mathcal{F}T, \varphi_n\rangle \underset{n\to+\infty}{\longmapsto} 0$.

Since $\varphi_n \underset{n\to+\infty}{\longmapsto} 0$ in $\mathcal{S}(\mathbb{R}^d)$, the continuity of the Fourier transform on $\mathcal{S}(\mathbb{R}^d)$ leads to $\widehat{\varphi_n} \underset{n\to+\infty}{\longmapsto} 0$ in $\mathcal{S}(\mathbb{R}^d)$. As a result, $\langle T, \widehat{\varphi_n}\rangle \underset{n\to+\infty}{\longmapsto} 0$. On the other hand, we have $\langle T, \widehat{\varphi_n}\rangle = \langle \mathcal{F}T, \varphi_n\rangle$. Hence, $\langle \mathcal{F}T, \varphi_n\rangle \underset{n\to+\infty}{\longmapsto} 0$. Consequently, $\mathcal{F}T \in \mathcal{S}'(\mathbb{R}^d)$.

Similarly, we show that $\overline{\mathcal{F}}T \in \mathcal{S}'(\mathbb{R}^d)$. □

DEFINITION 8.3.– *Let $T \in \mathcal{S}'(\mathbb{R}^d)$. Then the Fourier transform $\mathcal{F}T$ and the inverse Fourier transform are defined by*

$$\langle \mathcal{F}T, \varphi\rangle = \langle T, \widehat{\varphi}\rangle \text{ and } \langle \overline{\mathcal{F}}T, \varphi\rangle = \langle T, \overline{\widehat{\varphi}}\rangle, \forall \varphi \in \mathcal{S}(\mathbb{R}^d).$$

EXAMPLE 8.3.–

1) Let $f \in L^1(\mathbb{R})$.

We have $T_f \in \mathcal{S}'(\mathbb{R})$, donc $\mathcal{F}(T_f) \in \mathcal{S}'(\mathbb{R})$. Moreover, we have $\mathcal{F}(T_f) = T_{\widehat{f}}$. As a matter of fact, Let $\varphi \in \mathcal{S}(\mathbb{R})$, we have

$$\begin{aligned}\langle \mathcal{F}(T_f), \varphi\rangle &= \langle T_f, \mathcal{F}(\varphi)\rangle \\ &= \int_{\mathbb{R}} f(x)\mathcal{F}(\varphi)(x)\mathrm{d}x \\ &= \int_{\mathbb{R}} f(x)\left(\int_{\mathbb{R}} e^{-2i\pi xt}\varphi(t)\mathrm{d}t\right)\mathrm{d}x.\end{aligned}$$

The function $(x,t) \to f(x)\varphi(t)$ is integrable on $\mathbb{R}^2$. Then, according to Fubini's theorem, we have

$$\begin{aligned}\langle \mathcal{F}(T_f), \varphi\rangle &= \int_{\mathbb{R}} \varphi(t)\left(\int_{\mathbb{R}} e^{-2i\pi xt} f(x)\mathrm{d}x\right)\mathrm{d}t \\ &= \int_{\mathbb{R}} \varphi(t)\widehat{f}(t)\mathrm{d}t \\ &= \langle T_{\widehat{f}}, \varphi\rangle.\end{aligned}$$

Hence, $\widehat{T_f} = T_{\widehat{f}}$.

2) We have $\delta_0 \in \mathcal{S}'(\mathbb{R})$ because $\delta_0 \in \mathcal{E}'(\mathbb{R})$. In addition, $\widehat{\delta_0} = 1$.

In reality, for any $\varphi \in \mathcal{S}(\mathbb{R})$, we have

$$\begin{aligned}\langle \mathcal{F}(\delta), \varphi\rangle &= \langle \delta, \mathcal{F}(\varphi)\rangle \\ &= \mathcal{F}(\varphi)(0) \\ &= \int_{\mathbb{R}} \varphi(t)\mathrm{d}t \\ &= \langle 1, \varphi\rangle \\ &= \langle T_1, \varphi\rangle.\end{aligned}$$

Finally, $\widehat{\delta_0} = 1$.

3) We have $\widehat{\delta_a}(x) = e^{-2i\pi xa}$.

Furthermore, for all $\varphi \in \mathcal{S}(\mathbb{R})$, it follows that

$$\begin{aligned}\langle \mathcal{F}\delta_a, \varphi\rangle &= \langle \delta_a, \mathcal{F}\varphi\rangle \\ &= \mathcal{F}(\varphi)(a) \\ &= \int_{\mathbb{R}} e^{-2i\pi ta}\varphi(t)\mathrm{d}t \\ &= \langle e^{-2i\pi ta}, \varphi\rangle \\ &= \langle T_{e^{-2i\pi ta}}, \varphi\rangle.\end{aligned}$$

Hence, $\widehat{\delta_a} = e^{-2i\pi.a} : t \to e^{-2i\pi ta}$.

REMARK 8.8.– *By duality, the translation property and the dilation property of $\mathcal{F}$ on $\mathcal{S}(\mathbb{R})$ remain valid on $\mathcal{S}'(\mathbb{R})$.*

8.3.2. *Fourier transform and differentiation*

Given its importance, we reconsider the property of the differentiation of the Fourier transform of a tempered distribution.

THEOREM 8.10.– *Let $T \in \mathcal{S}'(\mathbb{R}^d)$ and $\alpha \in \mathbb{N}^d$. Then, we have*

i) $\partial^\alpha(\mathcal{F}T) = \mathcal{F}[(-2i\pi x)^\alpha T] = (-2i\pi)^{|\alpha|}\mathcal{F}[x^\alpha T]$;

ii) $\mathcal{F}(\partial^\alpha T) = (2i\pi x)^\alpha \mathcal{F}T$.

Proof.

i) Let $\varphi \in \mathcal{S}(\mathbb{R}^d)$. We have

$$\begin{aligned}\langle \partial^\alpha(\mathcal{F}T), \varphi\rangle &= (-1)^{|\alpha|}\langle \mathcal{F}T, \partial^\alpha\varphi\rangle \\ &= (-1)^{|\alpha|}\langle T, \mathcal{F}(\partial^\alpha\varphi)\rangle.\end{aligned}$$

Now, $\mathcal{F}(\partial^\alpha\varphi) = (2i\pi)^{|\alpha|}x^\alpha\widehat{\varphi}$, then

$$\begin{aligned}\langle \partial^\alpha(\mathcal{F}T), \varphi\rangle &= (-1)^{|\alpha|}\langle T, (2i\pi x)^\alpha\widehat{\varphi}\rangle \\ &= \langle \mathcal{F}((-2i\pi x)^\alpha T), \varphi\rangle.\end{aligned}$$

Hence, the formula $\partial^\alpha(\mathcal{F}T) = \mathcal{F}[(-2i\pi x)^\alpha T]$.

ii) Let $\varphi \in \mathcal{S}(\mathbb{R}^d)$. We have

$$\begin{aligned}\langle \mathcal{F}(\partial^\alpha T), \varphi\rangle &= \langle \partial^\alpha T, \widehat{\varphi}\rangle \\ &= (-1)^{|\alpha|}\langle T, \partial^\alpha(\mathcal{F}\varphi)\rangle \\ &= (-1)^{|\alpha|}\langle T, (-2i\pi)^{|\alpha|}\mathcal{F}(x^\alpha\varphi)\rangle \\ &= \langle (2i\pi)^{|\alpha|}T, \mathcal{F}(x^\alpha\varphi)\rangle \\ &= \langle \mathcal{F}((2i\pi)^{|\alpha|}T), x^\alpha\varphi\rangle \\ &= \langle x^\alpha\mathcal{F}((2i\pi)^{|\alpha|}T), \varphi\rangle \\ &= \langle (2i\pi x)^\alpha\widehat{T}, \varphi\rangle.\end{aligned}$$

Consequently, $\mathcal{F}(\partial^\alpha T) = (2i\pi x)^\alpha \mathcal{F}T$. □

REMARK 8.9.– *As on the Schwartz class, the Fourier transform $\mathcal{F}$ on the space $\mathcal{S}'(\mathbb{R})$ of the tempered distributions substitute differentiation and multiplication with x up to a factor $\pm 2\pi\, i$. All the interest of the Fourier transform for tempered distributions lies in this simple fact: on the one hand, the framework of the distributions makes it possible to differentiate as many times as necessary; on the other hand, after Fourier transform, any partial derivative of any order corresponds to the multiplication by a monomial.*

8.3.3. *Fourier transform inversion*

THEOREM 8.11.– *$\mathcal{F}$ is a continuous bijection of $\mathcal{S}'(\mathbb{R}^d)$ to $\mathcal{S}'(\mathbb{R}^d)$ and we have:*

$$\mathcal{F}^{-1} = \mathcal{F}^3 = \overline{\mathcal{F}}.$$

Proof.

– *Bijection*: We have $\langle \overline{\mathcal{F}}\mathcal{F}T, \varphi\rangle = \langle T, \overline{\mathcal{F}}\mathcal{F}\varphi\rangle = \langle T, \varphi\rangle$. Thus, $\overline{\mathcal{F}}\mathcal{F}T = T$ *and* $\mathcal{F}\overline{\mathcal{F}}T = T$.

– *Continuity:* Consider a sequence (T_j), $T_j \in \mathcal{S}'(\mathbb{R}^d)$ such that $T_j \underset{j\to+\infty}{\longmapsto} 0$ in $\mathcal{S}'(\mathbb{R}^d)$. We have

$$T_j \overset{\mathcal{S}'(\mathbb{R}^d)}{\underset{j\to+\infty}{\longmapsto}} 0 \iff \forall\varphi \in \mathcal{S}(\mathbb{R}^d), \langle T_j, \varphi\rangle \underset{j\to+\infty}{\longmapsto} 0.$$

Thus,

$$\langle \widehat{T_j}, \varphi \rangle = \langle T_j, \widehat{\varphi} \rangle \underset{j \to +\infty}{\longmapsto} 0 \text{ because } \widehat{\varphi} \in \mathcal{S}(\mathbb{R}^d).$$

Therefore, $\widehat{T_j} \overset{\mathcal{S}'(\mathbb{R}^d)}{\underset{j\to+\infty}{\longmapsto}} 0$. Consequently, $\mathcal{F}$ is continuous (even bicontinuous), that is, $\mathcal{F}^{-1} = \overline{\mathcal{F}}$ is continuous. □

EXAMPLE 8.4.– The constant function 1 belongs neither to $L^1(\mathbb{R})$ nor to $L^2(\mathbb{R})$ and its Fourier transform in the sense of functions does not exist. Its transform in the sense of distributions is given by: $\mathcal{F}1 = \delta$, which can be easily obtained from the fact that $\mathcal{F}(\delta) = 1$ and with Fourier inverse.

Observation: Plancherel's theorem can be obtained by at least three methods:

– *the first* method utilizes the Fourier transform on $L^2(\mathbb{R}^d)$, built by density of $L^2(\mathbb{R}^d) \cap L^1(\mathbb{R}^d)$ in the Hilbert space $L^2(\mathbb{R}^d)$, and the Plancherel formula which can be demonstrated for step functions with compact support;

– *the second* method is the one that appears in theorem 8.8 and which consists of extending the identity provided by the Fourier transform defined on $\mathcal{S}(\mathbb{R}^d)$ to the Hilbert space $L^2(\mathbb{R}^d)$;

– *the third* method consists of showing that $L^2(\mathbb{R}^d)$ is a subspace of $\mathcal{S}'(\mathbb{R}^d)$ that is stable under the action of the Fourier transform $\mathcal{F}$.

THEOREM 8.12 (Plancherel).– *The Fourier transform $\mathcal{F}$ defined on $\mathcal{S}'(\mathbb{R}^d)$ induces a $\mathbb{C}$–vector space isomorphism from $L^2(\mathbb{R}^d)$ in itself, verifying in addition the identity*

$$\langle f, g \rangle_{L^2} = \langle \mathcal{F}f, \mathcal{F}g \rangle_{L^2}, \forall f, g \in L^2(\mathbb{R}^d),$$

and in particular,

$$\|f\|_{L^2} = \|\mathcal{F}f\|_{L^2}, \forall f \in L^2(\mathbb{R}^d).$$

8.3.4. *Fourier transform in $\mathcal{E}'(\mathbb{R}^d)$*

THEOREM 8.13.– *For any distribution $T \in \mathcal{E}'(\mathbb{R}^d)$, the tempered distribution $\mathcal{F}T$ is the function defined by $\mathbb{R}^d \ni \xi \to \langle T, e_{-\xi} \rangle$, with $e_{-\xi} : x \to e^{-2i\pi x\xi}$. This function $g : \xi \to \mathcal{F}(T)(\xi)$ is of class $\mathcal{C}^\infty(\mathbb{R}^d)$ and polynomially increasing as well as all its derivatives.*

We also write $\mathcal{F}(T)(\xi) = g(\xi) = \langle T_x, e^{-2i\pi x\xi} \rangle$.

Proof. Since the distribution T has compact support, then the expression $\langle T_x, e^{-2i\pi x\xi} \rangle$ exists and defines an indefinitely differentiable function that extends into a holomorphic function on the complex plane. We have

$$\begin{aligned}
\langle \mathcal{F}(T), \varphi \rangle &= \langle T, \mathcal{F}\varphi \rangle \\
&= \left\langle T_x, \int_{-\infty}^{+\infty} e^{-2i\pi xy} \varphi(y) \mathrm{d}y \right\rangle \\
&= \langle T_x, \langle \varphi(y), e^{-2i\pi xy} \rangle \rangle \\
&= \langle T_x \varphi(y), e^{-2i\pi xy} \rangle \\
&= \langle \varphi(y), \langle T_x, e^{-2i\pi xy} \rangle \rangle \\
&= \int_{-\infty}^{+\infty} \varphi(y) \langle T_x, e^{-2i\pi xy} \rangle \, \mathrm{d}y \\
&= \langle g, \varphi \rangle .
\end{aligned}$$

This completes the demonstration. □

EXAMPLE 8.5.– We find that the Fourier transform of the Dirac distribution δ is $\widehat{\delta} = 1$. In fact, the support of δ being $\{0\}$, so $\delta \in \mathcal{E}'(\mathbb{R}^d)$ and we have immediately

$$\mathcal{F}(\delta)(\zeta) = g(\zeta) = \langle \delta, e^{-2i\pi x\zeta} \rangle = 1.$$

EXAMPLE 8.6.– The Fourier transform of the *n*th derivative of the Dirac distribution δ is

$$\mathcal{F}\left(\delta^{(n)}\right) = (2i\pi x)^n .$$

REMARK 8.10.– *We recall that, as mentioned above, the transform $\mathcal{F}$ exchanges regularity and decreasing at infinity. A function of $L^1(\mathbb{R}^d)$ being a less irregular object than the Dirac mass, it is not surprising that its Fourier transform tends to 0 at infinity, whereas $\mathcal{F}\delta_0 = 1$ does not tend to 0 at infinity. On the other hand, the Dirac mass tends to 0 at infinity as fast as possible, since its support (reduced to $\{0\}$) is as small as possible: it is therefore not surprising that the Fourier transform of δ_0 is a constant function, constants being the most regular functions possible. In reality, the Fourier transform of a distribution with compact support is much better than of class C^∞: since it is an analytic function. More precisely, if $T \in \mathcal{E}'(\mathbb{R}^d)$, then $\mathcal{F}T$ is analytic.*

THEOREM 8.14 (Paley-Wiener).– *(Dalmasso and Witomski 1996) Let $T \in \mathcal{E}'(\mathbb{R})$ and $M > 0$ be such that supp$T \Subset [-M, M]$. The function $g = \mathcal{F}T \in \mathcal{C}^\infty(\mathbb{R})$ is extended to a function $\tilde{g}$ holomorphic on $\mathbb{C}$. In addition, $\exists m \in \mathbb{N}, \exists C > 0$, such that for all $z \in \mathbb{C}$, we have*

$$|\tilde{g}(z)| \leq C\left(1 + |z|^2\right)^{\frac{m}{2}} e^{2\pi M|\Im m z|}.$$

Proof. We know that $g(\xi) = \mathcal{F}(T)(\xi)$ is a function of class $\mathcal{C}^\infty$ defined by $g(\xi) = \langle T_x, e^{-2i\pi x\xi}\rangle$. For all $z \in \mathbb{C}$, we set $\tilde{g}(z) = \langle T_x, e^{-2i\pi xz}\rangle$. A direct calculation shows that $\tilde{g}$ is holomorphic on $\mathbb{C}$ and the inequality on the modulus of $\tilde{g}$ follows from the continuity of T. □

REMARK 8.11.– *A generalization of theorem 8.14 to vectors is given by the Paley–Wiener–Schwartz theorem, which consists of extending the Fourier transform of any distribution $T \in \mathcal{E}'(\mathbb{R}^d)$ into an analytic function on C^d.*

A very important consequence of theorem 8.14 is the following proposition.

PROPOSITION 8.12.– The Fourier transform of a distribution T with compact support $(T \neq 0)$ cannot have compact support.

Proof. Let $T \in \mathcal{E}'(\mathbb{R})$ and let $g = \mathcal{F}T$. The function g is analytic on $\mathbb{R}$ as a restriction of a holomorphic function on $\mathbb{C}$. If g has compact support, then it will cancel out on a non-empty open set of $\mathbb{R}$. Since it is analytic, it will be identically zero. This contradicts the fact that $T \neq 0$. □

8.3.5. *Fourier transform and Poisson summation formula*

We give a remarkable identity that is often used in Fourier analysis.

THEOREM 8.15 (Poisson summation formula).– *The distribution $T = \sum_{k \in \mathbb{Z}} \delta_k$ belongs to $\mathcal{S}'(\mathbb{R})$ and its Fourier transform is $\mathcal{F}\left(\sum_{k\in\mathbb{Z}} \delta_k\right) = \sum_{k\in\mathbb{Z}} \delta_k$ in $\mathcal{S}'(\mathbb{R})$.*

Proof. Consider the distribution (Dirac comb), $T = \sum_{k\in\mathbb{Z}} \delta_k$.

The distribution T is tempered because the mapping $\varphi \longrightarrow \left\langle \sum_{k\in\mathbb{Z}} \delta_k, \varphi\right\rangle = \sum_{k\in\mathbb{Z}} \varphi(k)$ is linear and continuous from $\mathcal{S}(\mathbb{R})$ to $\mathbb{R}$.

A more detailed proof is exposed in exercise 7.5.

The Fourier transform $\mathcal{F}$ is a continuous operation on $\mathcal{S}'(\mathbb{R})$, therefore

$$\mathcal{F}(\sum_{k\in\mathbb{Z}}\delta_k)(\xi)=\sum_{k\in\mathbb{Z}}\mathcal{F}(\delta_k)(\xi)=\sum_{k\in\mathbb{Z}}e^{-i2\pi k\xi}=:R(\xi).$$

This distribution is tempered and periodic of period 1. Let us determine $R(\xi)$. We have

$$e^{i2\pi\xi}R(\xi)=e^{i2\pi\xi}\sum_{k\in\mathbb{Z}}e^{-i2\pi k\xi}=\sum_{k\in\mathbb{Z}}e^{-i2\pi\xi(k-1)}=\sum_{k\in\mathbb{Z}}e^{-i2\pi k\xi}=R(\xi).$$

Hence, $g(\xi)R(\xi)=0$ avec $g(\xi)=e^{i2\pi\xi}-1$. The function g is indefinitely differentiable and in addition $g(\xi)=0$ if and only if, $\xi\in\mathbb{Z}$ with $g'(\xi)\neq 0$. Therefrom, $R(\xi)=\sum_{k\in\mathbb{Z}}a_k\delta_k(\xi)$. The coefficients a_k are all equal to a same constant c. The fact that the distribution T is 1−periodic means that $\tau_1 T=T$ where the distribution $\tau_1 T$ is the translate of T by the translation 1. Therefore, $R(\xi)=\sum_{k\in\mathbb{Z}}c\delta_k(\xi)$. We take for φ the "rectangular" function defined by $\begin{cases}\varphi(x)=1,|x|<\frac{1}{2},\\ \varphi(x)=0,|x|\geq\frac{1}{2}.\end{cases}$ We have

$$\begin{aligned}\langle R(\xi),\varphi(\xi)\rangle&=\langle c\sum_{k\in\mathbb{Z}}\delta_k(\xi)),\varphi(\xi)\rangle\\&=c\sum_{k\in\mathbb{Z}}\langle\delta_k(\xi)),\varphi(\xi)\rangle\\&=c\varphi(0)=c.\end{aligned}$$

Now

$$\begin{aligned}\langle R(\xi),\varphi(\xi)\rangle&=\int_{-\frac{1}{2}}^{\frac{1}{2}}\sum_{k\in\mathbb{Z}}e^{i2\pi k\xi}d\xi\\&=\sum_{k\in\mathbb{Z}}\int_{-\frac{1}{2}}^{\frac{1}{2}}e^{i2\pi k\xi}d\xi=1,\end{aligned}$$

because

$$\int_{-\frac{1}{2}}^{\frac{1}{2}}e^{i2\pi k\xi}d\xi=\begin{cases}1,k=0\\0,k\neq 0.\end{cases}$$

Then, $c = 1$. Therefore, $R(\xi) = \sum_{k\in\mathbb{Z}} \delta_k(\xi)$. Consequently,

$$\mathcal{F}\Big(\sum_{k\in\mathbb{Z}} \delta_k\Big) = \sum_{k\in\mathbb{Z}} \delta_k \text{ in } \mathcal{S}'(\mathbb{R}). \qquad \square$$

REMARK 8.12.– *For all $a \in \mathbb{R}_+^*$, we have $\mathcal{F}\Big(\sum_{k\in\mathbb{Z}} \delta_{ak}\Big) = \sum_{k\in\mathbb{Z}} \delta_{ak}$.*

PROPOSITION 8.13 (Poisson summation formula).– If $\varphi \in \mathcal{S}(\mathbb{R})$, then

$$\sum_{k\in\mathbb{Z}} \varphi(k) = \sum_{k\in\mathbb{Z}} \widehat{\varphi}(k). \qquad [8.3]$$

Proof. We first observe that these series converge absolutely. As such, we set

$$c_1 = \sup_{x\in\mathbb{R}} x^2 \left|\varphi(x)\right|, c_2 = \sup_{x\in\mathbb{R}} x^2 \left|\varphi'(x)\right|, c_3 = \sup_{\xi\in\mathbb{R}} \xi^2 \left|\widehat{\varphi}(\xi)\right| \text{ and } C = \max(c_1, c_2, c_3).$$

Now, $\varphi \in \mathcal{S}(\mathbb{R})$, thus $\widehat{\varphi} \in \mathcal{S}(\mathbb{R})$ and we have $C < +\infty$. As a result,

$$|\varphi(k)| \leq \frac{C}{k^2} \quad \text{and} \quad |\widehat{\varphi}(k)| \leq \frac{C}{k^2}.$$

Since $\frac{C}{k^2}$ is the general term of an absolutely convergent series, the convergence of series in identity [8.3] derives thereof. According to theorem 8.15, we have

$$\mathcal{F}\Big(\sum_{k\in\mathbb{Z}} \delta_k\Big) = \sum_{k\in\mathbb{Z}} \delta_k \text{ in } \mathcal{S}'(\mathbb{R}).$$

Then, for any $\varphi \in \mathcal{S}(\mathbb{R})$, we have $\big\langle \mathcal{F}\big(\sum_{k\in\mathbb{Z}} \delta_k\big), \varphi\big\rangle = \big\langle \sum_{k\in\mathbb{Z}} \delta_k, \varphi\big\rangle$. Consequently,

$$\forall \varphi \in \mathcal{S}(\mathbb{R}), \quad \sum_{k\in\mathbb{Z}} \varphi(k) = \sum_{k\in\mathbb{Z}} \widehat{\varphi}(k). \qquad \square$$

REMARK 8.13.– *For all $a \in \mathbb{R}_+^*$, and for any function $\varphi \in \mathcal{S}(\mathbb{R})$, the Poisson summation formula is written as $\frac{1}{a}\sum_{k\in\mathbb{Z}} \varphi(\frac{k}{a}) = \sum_{k\in\mathbb{Z}} \widehat{\varphi}(ka)$.*

8.3.6. *Fourier transform and convolution*

The transformation of the convolution product into the classical product, and vice versa, by the Fourier transform on the convolution products of the elements of $\mathcal{S}(\mathbb{R}^d)$ is also valid on the convolution products of distributions, assuming that they are well defined. This relation, reinforced by the Fourier inversion formula in $\mathcal{S}'(\mathbb{R}^d)$,

$$\mathcal{F}(S * T) = \mathcal{F}S \cdot \mathcal{F}T,$$

can be used to provide the definition of the convolution product of certain tempered distributions whose product of Fourier transforms can be defined as a distribution in $\mathcal{S}'(\mathbb{R}^d)$.

PROPOSITION 8.14.– Let S and T denote two distributions with compact supports. Then, the support of the convolution product $S * T$ is compact and we have

$$\mathcal{F}(S * T) = (\mathcal{F}S).(\mathcal{F}T).$$

Proof. It should be recalled that the convolution product of two distributions S and T exists if one of the supports (supp S or supp T) is compact. In this case, supp $(S * T)$ is also compact because it is included in the adhesion of the sum of the support of S and the support of T. By theorem 8.13, $\mathcal{F}(S * T)$ is a function defined by

$$\begin{aligned}\mathcal{F}(S * T)(\xi) &= \left\langle S * T, e^{-2i\pi x\xi} \right\rangle \\ &= \left\langle S_x \otimes T_y, e^{-2i\pi\xi(x+y)} \right\rangle \\ &= \left\langle S_x, e^{-2i\pi x\xi} \right\rangle \left\langle T_y, e^{-2i\pi y\xi} \right\rangle \\ &= (\mathcal{F}S)(\xi).(\mathcal{F}T)(\xi).\end{aligned}$$

The demonstration is complete. □

REMARK 8.14.– *The above result is valid with more general assumptions. On the other hand, the product* $(\mathcal{F}S).(\mathcal{F}T)$ *cannot be defined to most cases. If* S *is a distribution with compact support and* T *any tempered distribution, then the product* $(\mathcal{F}S).(\mathcal{F}T)$ *must be interpreted as the product of a distribution by a function of class* $\mathcal{C}^\infty$. *It should also be noted that if* S *and* T *are tempered distributions and have lower bounded supports, then the product* $S * T$ *is well defined but the product* $(\mathcal{F}S).(\mathcal{F}T)$ *is generally not defined. To better understand this, the case of the Heaviside distribution should be considered.*

PROPOSITION 8.15.– Let S and T denote two distributions with compact supports. Then we have:

$$\mathcal{F}(ST) = (\mathcal{F}S) * (\mathcal{F}T).$$

Proof. By setting $\overline{\mathcal{F}}S = U$ and $\overline{\mathcal{F}}T = V$, we know from the Fourier reciprocity formulas that $\mathcal{F}U = \mathcal{F}\left(\overline{\mathcal{F}}S\right) = S, \quad \mathcal{F}V = \mathcal{F}\left(\overline{\mathcal{F}}T\right) = T$, and we will have

$$\mathcal{F}(ST) = \mathcal{F}S * \mathcal{F}T.$$ □

We conclude this section with Table 8.2, which gives the Fourier transform of some tempered distributions.

T	$\xrightarrow{\mathcal{F}}$	$\mathcal{F}(T)$
$T^{(k)}$	$\longrightarrow$	$(2i\pi\zeta)^k\mathcal{F}(T)$
$(-2i\pi x)^k T$	$\longrightarrow$	$(\mathcal{F}(T))^{(k)}$
$\tau_a T$	$\longrightarrow$	$e^{-2i\pi\zeta a}\mathcal{F}(T)$
$e^{2i\pi xa}T$	$\longrightarrow$	$\tau_a\mathcal{F}(T)$
δ_a	$\longrightarrow$	$e^{-2i\pi\zeta a}$
$e^{2i\pi xa}$	$\longrightarrow$	δ_a
$\delta_a^{(k)}$	$\longrightarrow$	$(2i\pi\zeta)^k$
x^k	$\longrightarrow$	$(-2i\pi x)^k\delta_a^k$
sign_x	$\longrightarrow$	$\frac{1}{i\pi}\text{Vp}(\frac{1}{\zeta})$
$\text{Vp}(\frac{1}{x})$	$\longrightarrow$	$-i\pi\text{sign}$

Table 8.2. *The Fourier transform of a tempered distribution*

8.4. Exercises with solutions

EXERCISE 8.1.– Build a function $f \in \mathcal{S}(\mathbb{R})$ non-identically zero such that, for all $k \in \mathbb{N}$, $\displaystyle\int_{\mathbb{R}} x^k f(x)\mathrm{d}x = 0$.

SOLUTION 8.1.– For all $k \in \mathbb{N}$, we have $\mathcal{F}\left(x^k f\right)(0) = \displaystyle\int_{\mathbb{R}} x^k f(x)\mathrm{d}x = 0$.

Now $(-2i\pi)^k\mathcal{F}\left(x^k f\right)(0) = \dfrac{d^k}{d\zeta^k}\mathcal{F}(f)(0)$, thus $\forall k \in \mathbb{N}$, $\dfrac{d^k}{d\zeta^k}\mathcal{F}(f)(0) = 0$. The condition imposed on f is expressed by the fact that all derivatives of $\mathcal{F}(f)$ cancel out at 0.

We now consider the plateau function χ with support in $]-2, 2[$ and which takes 1 on $[-1, 1]$. Let $g \in \mathcal{S}(\mathbb{R})$. We take $h : \zeta \mapsto h(\zeta) = (1 - \chi(\zeta))\, g(\zeta)$. Since $g \in \mathcal{S}(\mathbb{R})$, then $h \in \mathcal{S}(\mathbb{R})$. Moreover, all derivatives of h are zero at 0.

The function $f = \mathcal{F}^{-1}(h)$ answers the question.

EXERCISE 8.2.– Solve in $\mathcal{D}'(\mathbb{R})$ the differential equation $u' + u = \delta_0$. What are the tempered solutions?

SOLUTION 8.2.– We start by determining the *classical* solutions of the associated homogeneous equation $u' + u = 0$. We find $\forall x \in \mathbb{R}, u(x) = ce^{-x}, c \in \mathbb{R}$. Then, we look for a distribution S, which is a particular solution to $u' + u = \delta_0$ as $S = e^x T$, where T is the solution of $u' + u = \delta_0$.

We then have $S' = (e^x T)' = e^x(T + T') = \delta_0 e^x = \delta_0 e^0 = \delta_0$.

Therefore, $S = H + c, c \in \mathbb{R}$ (H is the Heaviside distribution).

Hence, $T = He^{-x} + ce^{-x}$ in $\mathcal{D}(\mathbb{R})$.

However, the distribution associated with $x \longmapsto e^{-x}$ is not in $\mathcal{S}'(\mathbb{R})$ (see the behavior at $-\infty$), then the only solution of $u' + u = \delta_0$ which is in $\mathcal{S}'(\mathbb{R})$ is the distribution $T = He^{-x}$.

EXERCISE 8.3.– Compute the Fourier transform of the tempered distributions on $\mathbb{R}$ associated with the following functions:

1) e^{iax} ($a \in \mathbb{R}$);

2) $\cos(x)$;

3) $x\sin(x)$;

4) $e^{-a|x|}$ ($a > 0$);

5) $|x|e^{-a|x|}$ ($a > 0$).

SOLUTION 8.3.–

1) Let $a \in \mathbb{R}$. We have $\mathcal{F}(\delta_a) = e^{-i2\pi a\zeta}$. Then, by Fourier inversion, $\mathcal{F}\mathcal{F}(\delta_a) = \mathcal{F}(e^{-i2\pi a\zeta})$ with $\mathcal{F}\mathcal{F} = \sigma$ où $\sigma(f)\zeta = f(-\zeta)$. Hence, $\delta_a = \sigma\mathcal{F}(e^{-i2\pi a\zeta})$. Therefore, $\mathcal{F}(e^{i2\pi ax}) = \delta_a$. By change in variable, we obtain $\mathcal{F}(e^{iax}) = \delta_{\frac{a}{2\pi}}$.

Special case: $\mathcal{F}(\delta) = 1$ and $\mathcal{F}(1) = \delta$.

2) We have $\forall x \in \mathbb{R}, \cos x = \dfrac{e^{ix} + e^{-ix}}{2}$.

Hence, by linearity of $\mathcal{F}$ and from Question 1, we have

$$\mathcal{F}(\cos x) = \frac{1}{2}\left(\mathcal{F}(e^{ix}) + \mathcal{F}(e^{-ix})\right) = \frac{1}{2}\left(\delta_{\frac{1}{2\pi}} + \delta_{\frac{-1}{2\pi}}\right).$$

It should noted that we also have

$$\mathcal{F}(\sin x) = \frac{1}{2i}\left(\mathcal{F}(e^{ix}) - \mathcal{F}(e^{-ix})\right) = \frac{1}{2i}\left(\delta_{\frac{1}{2\pi}} - \delta_{\frac{-1}{2\pi}}\right).$$

3) We have

$$\mathcal{F}(x\sin x) = \frac{i}{2\pi}\partial_\zeta \mathcal{F}(\sin x) = \frac{i}{2\pi}\partial_\zeta \frac{1}{2i}\left(\delta_{\frac{1}{2\pi}} - \delta_{\frac{-1}{2\pi}}\right) = \frac{1}{4\pi}\left(\delta'_{\frac{1}{2\pi}} + \delta'_{\frac{-1}{2\pi}}\right).$$

4) Since $x \longmapsto e^{-a|x|} \in \mathcal{S}$, for $a > 0$, we then have

$$\begin{aligned}
\mathcal{F}\left(e^{-a|x|}\right)(\zeta) &= \int_{-\infty}^{+\infty} e^{-a|x|}e^{-i2\pi x\zeta}\mathrm{d}x \\
&= \int_{-\infty}^{0} e^{ax-i2\pi x\zeta}\mathrm{d}x + \int_{0}^{+\infty} e^{-ax-i2\pi x\zeta}\mathrm{d}x \\
&= \left[\frac{1}{a-i2\pi\zeta}e^{(a-i2\pi\zeta)x}\right]_{-\infty}^{0} + \left[-\frac{1}{a-+2\pi\zeta}e^{-(a+i2\pi\zeta)x}\right]_{0}^{+\infty} \\
&= \frac{1}{a-i2\pi\zeta} + \frac{1}{a+i2\pi\zeta} \\
&= \frac{2a}{a^2+\left(2\pi\zeta\right)^2}.
\end{aligned}$$

Observation: This issue was addressed in example 8.2.

5) Similarly, we have

$$\begin{aligned}
\mathcal{F}\left(|x|e^{-a|x|}\right)(\zeta) &= \int_{-\infty}^{+\infty} |x|e^{-a|x|}e^{-xi2\pi\zeta}\mathrm{d}x \\
&= \int_{-\infty}^{0} -xe^{ax-i2\pi x\zeta}\mathrm{d}x + \int_{0}^{+\infty} xe^{-ax-i2\pi x\zeta}\mathrm{d}x \\
&= \left[-x\frac{1}{a-i\zeta}e^{(a-i2\pi\zeta)x}\right]_{-\infty}^{0} + \int_{-\infty}^{0}\frac{e^{(a-i2\pi\zeta)x}}{a-i2\pi\zeta}\mathrm{d}x
\end{aligned}$$

$$+\left[\frac{-x}{a+i2\pi\zeta}e^{-(a+i2\pi\zeta)x}\right]_0^{+\infty}+\int_0^{+\infty}\frac{e^{-(a+i2\pi\zeta)x}}{a+i2\pi\zeta}\mathrm{d}x$$

$$=0+\left[\frac{e^{(a-i2\pi\zeta)x}}{(a-i2\pi\zeta)^2}\right]_{-\infty}^{0}+0+\left[\frac{e^{-(a+i2\pi\zeta)x}}{(a+i2\pi\zeta)^2}\right]_0^{\infty}$$

$$=\frac{1}{\left(a-i2\pi\zeta\right)^2}+\frac{1}{\left(a+i2\pi\zeta\right)^2}$$

$$=\frac{\left(a+i2\pi\zeta\right)^2+\left(a-i2\pi\zeta\right)^2}{\left(a^2+\left(2\pi\zeta\right)^2\right)^2}$$

$$=\frac{2\left(a^2-\left(2\pi\zeta\right)^2\right)}{\left(a^2+\left(2\pi\zeta\right)^2\right)^2}.$$

Therefore: $\mathcal{F}(|x|e^{-a|x|})=\dfrac{2\left(a^2-\left(2\pi\zeta\right)^2\right)}{\left(a^2+\left(2\pi\zeta\right)^2\right)^2}.$

EXERCISE 8.4.– Compute the Fourier transform of the tempered distributions on $\mathbb{R}$ associated with the following functions:

1) $\dfrac{\sin(x)}{x}$;

2) $\sin(|x|)$.

SOLUTION 8.4.–

1) **Direct method**: If $u=\mathcal{F}(\dfrac{\sin x}{x})$, since $\sin x=\dfrac{\sin x}{x}\times x$, then we have $\frac{i}{2\pi}u'=\mathcal{F}(\sin x)$.

Hence, $u'=\mathcal{F}\big(-i2\pi x\dfrac{\sin x}{x}\big)$ and according to Question 2 of exercise 8.3, we obtain

$$u'=-\pi\big(\delta_{\frac{1}{2\pi}}-\delta_{\frac{-1}{2\pi}}\big). \qquad [8.4]$$

So

$$u'|_{]-\infty,\frac{-1}{2\pi}[}=0\Rightarrow u|_{]-\infty,\frac{-1}{2\pi}[}=a$$

$$u'|_{]\frac{-1}{2\pi},\frac{1}{2\pi}[}=0\Rightarrow u|_{]\frac{-1}{2\pi},\frac{1}{2\pi}[}=b$$

and $u'|_{]\frac{1}{2\pi},+\infty[}=0\Rightarrow u|_{]\frac{1}{2\pi},+\infty[}=c.$

If we set $\tilde{u} = \begin{cases} a \text{ if } x < \frac{-1}{2\pi}, \\ b \text{ if } \frac{-1}{2\pi} < x < \frac{1}{2\pi}, \\ c \text{ if } x > \frac{1}{2\pi}, \end{cases}$ then by the jump formula, we have

$$\tilde{u}' = (b-a)\delta_{\frac{-1}{2\pi}} + (c-b)\delta_{\frac{1}{2\pi}},$$

and we want a, b, c such that $b - a = \pi$ and $c - b = -\pi$; this leads us into taking $c = a$ and $b = \pi + a$.

Then $\tilde{u} = a + \pi\, \mathbb{1}_{]\frac{-1}{2\pi}, \frac{1}{2\pi}[}$ is a solution to equation [8.4].

Hence, the general solution to equation [8.4] is

$$u = \tilde{u} + \text{ constant } = A + \pi \mathbb{1}_{]\frac{-1}{2\pi}, \frac{-1}{2\pi}[} \text{ with } A \in \mathbb{R}.$$

Yet, $\dfrac{\sin x}{x} \in L^2(\mathbb{R})$, then by Plancherel's theorem, $\mathcal{F}(\frac{\sin x}{x}) \in L^2(\mathbb{R})$.

So, since the constant A is not in $L^2(\mathbb{R})$, unless it is zero, we must have $A = 0$. Finally, $\mathcal{F}(\frac{\sin x}{x}) = \pi\, \mathbb{1}_{]\frac{-1}{2\pi}, \frac{-1}{2\pi}[}$.

Inverse Fourier method: We have

$$\mathcal{F}\left(\mathbb{1}_{]\frac{-1}{2\pi}, \frac{1}{2\pi}[}\right)(\zeta) = \int_{\frac{-1}{2\pi}}^{\frac{1}{2\pi}} e^{-i2\pi x\zeta} \mathrm{d}x = \frac{2\sin(\zeta)}{2\pi\zeta}.$$

Then, $\mathcal{F}\mathcal{F}\left(\mathbb{1}_{]\frac{-1}{2\pi}, \frac{1}{2\pi}[}(x)\right) = \frac{1}{\pi}\mathcal{F}(\frac{\sin x}{x}) = \mathbb{1}_{]\frac{-1}{2\pi}, \frac{1}{2\pi}[}(x)$.

Hence, $\mathcal{F}\left(\frac{\sin x}{x}\right) = \pi\, \mathbb{1}_{]\frac{-1}{2\pi}, \frac{1}{2\pi}[}$. Using the same calculation, we get the following result:

$$\mathcal{F}\left(\frac{\sin(2\pi R x)}{x}\right) = \pi\, \mathbb{1}_{]-R, R[}.$$

2) Since $\sin 0 = 0$, we have $\sin|x| = \sin x\, \mathbb{1}_{x>0} - \sin x\, \mathbb{1}_{x<0}$.

Now, $\sin x\, \mathbb{1}_{x>0} = \lim\limits_{\varepsilon \longrightarrow 0} \sin x\, e^{-\varepsilon x}\, \mathbb{1}_{x>0}$ in $\mathcal{S}'(\mathbb{R})$ $\qquad (*)$.

Moreover, $\forall\varepsilon > 0$, we have

$$\begin{aligned}\mathcal{F}\big(\sin x\, e^{-\varepsilon x}\, \mathbb{1}_{x>0}\big)(\zeta) &= \int_0^{+\infty} \sin x e^{-\varepsilon x - i2\pi x\zeta} \mathrm{d}x \\ &= \frac{1}{2i}\int_0^{+\infty} e^{x(i-\varepsilon-2\pi i\zeta)}\mathrm{d}x - \frac{1}{2i}\int_0^{+\infty} e^{-x(i+\varepsilon+i2\pi\zeta)}\mathrm{d}x \\ &= \frac{1}{2i}\left[\frac{1}{i-\varepsilon-2\pi i\zeta} - \frac{1}{-(i+\varepsilon+i2\pi\zeta)}\right] \\ &= \frac{-1}{2}\left[\frac{1}{2\pi\zeta-1-i\varepsilon} - \frac{1}{2\pi\zeta+1-i\varepsilon}\right].\end{aligned}$$

A classical result on $\mathrm{Vp}(\frac{1}{x})$, in $\mathcal{D}'(\mathbb{R})$, allows us to deduce that

$$\frac{1}{y+a-i\varepsilon} \longrightarrow \mathrm{Vp}(\frac{1}{y+a}) + i\pi\delta_a,$$

when ε tends to 0^+, $\forall a \in \mathbb{R}$. Therefore,

$$\mathcal{F}\left(\sin x\, e^{-\varepsilon x}\, \mathbb{1}_{x>0}\right) \longrightarrow \frac{-1}{2}\left[\mathrm{Vp}\left(\frac{1}{2\pi\zeta-1}\right) - \mathrm{Vp}\left(\frac{1}{2\pi\zeta+1}\right)\right] + \frac{i\pi}{2}(\delta_1-\delta_{-1}),$$

when ε tends to 0^+. Similarly,

$$\mathcal{F}\left(\sin x\, e^{\varepsilon x}\, \mathbb{1}_{x<0}\right) \longrightarrow \frac{1}{2}\left[\mathrm{Vp}\left(\frac{1}{2\pi\zeta+1}\right) - \mathrm{Vp}\left(\frac{1}{2\pi\zeta-1}\right)\right] + \frac{i\pi}{2}(\delta_1-\delta_{-1}),$$

when ε tends to 0^+. Hence, since the convergence of (*) takes place in $\mathcal{S}'(\mathbb{R})$, we have

$$\begin{aligned}\mathcal{F}(\sin|x|)(\zeta) &= \lim_{\varepsilon\longrightarrow 0^+}\left(\mathcal{F}(\sin x\, e^{-\varepsilon x}\, \mathbb{1}_{x>0}) - \mathcal{F}(\sin x\, e^{\varepsilon x}\, \mathbb{1}_{x<0})\right)(\zeta) \\ &= \mathrm{Vp}\left(\frac{1}{2\pi\zeta+1}\right) - \mathrm{Vp}\left(\frac{1}{2\pi\zeta-1}\right).\end{aligned}$$

EXERCISE 8.5.–

1) Determine $\mathcal{F}(\mathrm{Vp}\left(\frac{1}{x}\right))$ the Fourier transform of the distribution $\mathrm{Vp}\left(\frac{1}{x}\right)$.

2) Determine $\mathcal{F}(H(x))$ the Fourier transform of the Heaviside distribution H.

SOLUTION 8.5.–

1) Recall that $\text{Vp}\left(\frac{1}{x}\right)$ is a tempered distribution (see exercise 7.2); we can then mention its Fourier transform.

We have $x\text{Vp}\left(\frac{1}{x}\right) = 1$. Hence, $\mathcal{F}\left(x\text{Vp}\left(\frac{1}{x}\right)\right) = \mathcal{F}(1)$.

Now, $\mathcal{F}(1) = \delta = \frac{\mathrm{d}}{\mathrm{d}x}H(x)$ where H is the Heaviside distribution and moreover,

$$\mathcal{F}\left(\text{Vp}\left(\frac{1}{x}\right)\right) = -\frac{1}{2i\pi}\frac{\mathrm{d}}{\mathrm{d}x}\mathcal{F}\left(\text{Vp}\left(\frac{1}{x}\right)\right),$$

therefore

$$-\frac{1}{2i\pi}\frac{\mathrm{d}}{\mathrm{d}x}\mathcal{F}\left(\text{Vp}\left(\frac{1}{x}\right)\right) = \frac{\mathrm{d}}{\mathrm{d}x}H(x).$$

Then, $\mathcal{F}\left(\text{Vp}\left(\frac{1}{x}\right)\right) = -2i\pi H(x) + C$, where C is a constant to be determined. The distribution $\text{Vp}\left(\frac{1}{x}\right)$ being odd since $\int_{|x|\geq\varepsilon}\varphi(-x)\mathrm{d}x = -\int_{|x|\geq\varepsilon}\varphi(x)\mathrm{d}x$. Then, its Fourier transform is also odd. Hence, $-2i\pi + C = -C$ that is, $C = i\pi$. Consequently,

$$\mathcal{F}\left(\text{Vp}\left(\frac{1}{x}\right)\right) = -2i\pi H(x) + i\pi.$$

It can also be written that

$$\mathcal{F}\left(\text{Vp}\left(\frac{1}{x}\right)\right) = i\pi\mathbb{1}_{]-\infty,0[} - i\pi\mathbb{1}_{]0,+\infty[}.$$

We can also show that

$$\overline{\mathcal{F}}\left(\text{Vp}\left(\frac{1}{x}\right)\right) = 2i\pi H(x) - i\pi.$$

2) The function $H(x)$ is not summable and does not have a Fourier transform in the sense of functions. Yet, this function is slowly increasing at infinity (because it is bounded), therefore it determines a tempered distribution and as we know, the latter has a Fourier transform.

From the previous question, we have $\mathcal{F}\left(\text{Vp}\left(\frac{1}{x}\right)\right) = -2i\pi H(x) + i\pi$. Applying $\mathcal{F}$ to this equality, we obtain $\mathcal{F}\mathcal{F}\left(\text{Vp}\left(\frac{1}{x}\right)\right) = -2i\pi\mathcal{F}H(x) + i\pi\delta$ because $\mathcal{F}1 = \delta$. Hence,

$$\mathcal{F}H(x) = \frac{\delta}{2} - \frac{1}{2i\pi}\mathcal{F}\mathcal{F}\left(\text{Vp}\left(\frac{1}{x}\right)\right).$$

However, $\overline{\mathcal{F}}\left(\text{Vp}\left(\frac{1}{x}\right)\right) = 2i\pi H(x) - i\pi = -\mathcal{F}\left(\text{Vp}\left(\frac{1}{x}\right)\right)$, thus

$$\mathcal{F}\mathcal{F}\left(\text{Vp}\left(\frac{1}{x}\right)\right) = -\mathcal{F}\overline{\mathcal{F}}\left(\text{Vp}\left(\frac{1}{x}\right)\right) = -\text{Vp}\left(\frac{1}{x}\right).$$

Consequently,

$$\mathcal{F}H(x) = \frac{\delta}{2} + \frac{1}{2i\pi}\text{Vp}\left(\frac{1}{x}\right).$$

Similarly, we show that

$$\overline{\mathcal{F}}H(x) = \frac{\delta}{2} - \frac{1}{2i\pi}\text{Vp}\left(\frac{1}{x}\right).$$

EXERCISE 8.6.– Compute the Fourier transform of the function defined by

$$f(x) = e^{-\pi x^2}.$$

SOLUTION 8.6.– We have $f \in \mathcal{S}$ and $\mathcal{F}(f)(\zeta) = \int_{\mathbb{R}} e^{-\pi x^2} e^{-2i\pi x\zeta}\mathrm{d}x$. Differentiating with respect to ζ, we get $(\mathcal{F}(f)(\zeta))' = -2i\pi\mathcal{F}(h)(\zeta)$ where $h(x) = xf(x)$. Then

$$(\mathcal{F}(f)(\zeta))' = -2i\pi\int_{\mathbb{R}} xe^{-\pi x^2} e^{-2i\pi x\zeta}\mathrm{d}x.$$

An integration by parts gives

$$(\mathcal{F}(f)(\zeta))' = -2\pi\zeta\mathcal{F}(f)(\zeta).$$

The solution to this differential equation is immediate and gives $\mathcal{F}(f)(\zeta) = Ce^{-\pi\zeta^2}$, where C is a constant. To determine the latter, we proceed as follows:

$$\mathcal{F}(f)(0) = C = \int_{\mathbb{R}} e^{-\pi x^2}\mathrm{d}x = \frac{1}{\sqrt{\pi}}\int_{\mathbb{R}} e^{-t^2}\mathrm{d}t = 1.$$

Consequently, $\mathcal{F}(f)(\zeta) = e^{-\pi\zeta^2}$.

Generally speaking, denoting by $f_\alpha : x \mapsto e^{-\alpha x^2}$ $(\alpha > 0)$ the Gaussian of a variable, we have $\widehat{f_\alpha} = \sqrt{\frac{\pi}{\alpha}} f_{\frac{\pi^2}{\alpha}}$.

On $\mathbb{R}^d$, we have

$$\mathcal{F}\left(e^{-\alpha\|x\|^2}\right)(\zeta) = \left(\frac{\pi}{\alpha}\right)^{\frac{d}{2}} e^{-\frac{\pi^2}{\alpha}\|\zeta\|^2}.$$

EXERCISE 8.7.– Consider the two functions Π and $\triangle$ defined by

$$\Pi(x) = \begin{cases} 1 \text{ if } |x| \leq \frac{1}{2} \\ 0 \text{ if } |x| > \frac{1}{2} \end{cases} \quad \text{and } \triangle(x) = \begin{cases} 1 - |x| \text{ if } |x| \leq 1 \\ 0 \text{ if } |x| > 1. \end{cases}$$

1) Compute the Fourier transform of the function Π.

2) Compute the Fourier transform of the function $\triangle$.

3) Compute and give $\triangle'(x)$ as a function of $\Pi(x)$.

4) Using Question 3, find the result of Question 2.

5) Determine $\Pi * \Pi$, and then deduce the result of Question 2.

SOLUTION 8.7.–

1) In example 8.1, it has been shown that

$$\forall \zeta \in \mathbb{R}, \mathcal{F}(\Pi)(\zeta) = \frac{\sin(\pi\zeta)}{\pi\zeta}.$$

2) For any $\zeta \in \mathbb{R}$, we have

$$\begin{aligned} \mathcal{F}(\triangle)(\zeta) &= \int_{-\infty}^{+\infty} \triangle(x) e^{-2i\pi x\zeta} \mathrm{d}x \\ &= 2\int_0^{+\infty} \triangle(x)\cos(2i\pi x\zeta)\mathrm{d}x \\ &= 2\int_0^1 (1-x)\cos(2i\pi x\zeta)\mathrm{d}x. \end{aligned}$$

By integration by parts, it is obtained that

$$\forall \zeta \in \mathbb{R}, \mathcal{F}(\triangle)(\zeta) = \frac{1-\cos(2\pi\zeta)}{\pi^2\zeta^2} = \frac{\sin^2(\pi\zeta)}{\pi^2\zeta^2}.$$

3) It is easily shown that

$$\triangle'(x) = \begin{cases} -1 \text{ if } 0 < x \leq 1 \\ 1 \text{ if } -1 \leq x < 0 \\ 0 \text{ if } |x| > 1. \end{cases}$$

Otherwise, we have $\triangle'(x) = \Pi(x + \frac{1}{2}) - \Pi(x - \frac{1}{2})$.

4) We have $\triangle'(x) = \Pi(x + \frac{1}{2}) - \Pi(x - \frac{1}{2})$. Therefore,

$$\begin{aligned} \mathcal{F}(\triangle')(\zeta) &= \mathcal{F}(\Pi(x + \frac{1}{2}))(\zeta) - \mathcal{F}(\Pi(x - \frac{1}{2}))(\zeta) \\ &= e^{\pi i \zeta}\mathcal{F}(\Pi(x))(\zeta) - e^{-\pi i \zeta}\mathcal{F}(\Pi(x))(\zeta) \\ &= 2i \sin(\pi\zeta)\mathcal{F}(\Pi(x))(\zeta) \\ &= 2i\frac{\sin^2(\pi\zeta)}{\pi\zeta}. \end{aligned}$$

On the other hand, from the Fourier transform property and the derivation, it is known that $\mathcal{F}(\triangle')(\zeta) = 2i\pi\zeta\mathcal{F}(\triangle(x))(\zeta)$. Consequently,

$$\mathcal{F}(\triangle)(\zeta) = \frac{\sin^2(\pi\zeta)}{\pi^2\zeta^2}.$$

5) For any $\zeta \in \mathbb{R}$, we have

$$\begin{aligned} \Pi * \Pi(x) &= \int_{-\infty}^{+\infty} \Pi(t)\Pi(x-t)dt \\ &= \int_{-\frac{1}{2}}^{\frac{1}{2}} \Pi(x-t)dt \\ &= \int_{x+\frac{1}{2}}^{x-\frac{1}{2}} \Pi(s)ds \\ &= \begin{cases} 1 - x \text{ if } 0 \leq x \leq 1 \\ 1 + x \text{ if } -1 \leq x \leq 0 \\ 0 \text{ if } |x| > 1 \end{cases} \\ &= \triangle(x). \end{aligned}$$

We know that $\mathcal{F}(\Pi * \Pi) = \mathcal{F}(\Pi)\mathcal{F}(\Pi)$, thus $\mathcal{F}(\triangle)(\zeta) = \dfrac{\sin^2(\pi\zeta)}{\pi^2\zeta^2}$.

EXERCISE 8.8.–

1) Determine the Fourier transform of the function f defined by $f(x) = \dfrac{1}{1+x^2}$.

2) Deduce the Fourier transform of the function $g(x) = \dfrac{x}{(1+x^2)^2}$.

SOLUTION 8.8.–

1) Using the residual method, we obtain

$$\mathcal{F}(f)(\zeta) = \int_{\mathbb{R}} \frac{e^{-2i\pi x\zeta}}{1+x^2} dx = \pi e^{-2\pi|\zeta|}.$$

2) We have

$$\begin{aligned}
\mathcal{F}\left(\frac{x}{(1+x^2)^2}\right)(\zeta) &= \mathcal{F}\left(-\frac{1}{2}\left(\frac{1}{1+x^2}\right)'\right)(\zeta) \\
&= -\frac{1}{2}\mathcal{F}\left(\left(\frac{1}{1+x^2}\right)'\right)(\zeta) \\
&= -i\pi\zeta\, \mathcal{F}\left(\frac{1}{1+x^2}\right)(\zeta) \\
&= -i\pi\zeta\mathcal{F}\left(\frac{1}{1+x^2}\right)(\zeta) \\
&= -\pi^2 i\zeta e^{-2\pi|\zeta|}.
\end{aligned}$$

Then, $\mathcal{F}\left(\dfrac{x}{(1+x^2)^2}\right)(\zeta) = -\pi^2 i\zeta e^{-2\pi|\zeta|}$.

EXERCISE 8.9.– Let $k > 0$ and $u \in \mathcal{S}'(\mathbb{R})$ be such that $\dfrac{d^4 u}{dx^4} + ku \in L^2(\mathbb{R})$.

Show that $\dfrac{d^j u}{dx^j} \in L^2(\mathbb{R})$ for $0 \leq j \leq 4$.

SOLUTION 8.9.– By the Plancherel theorem, since $\dfrac{d^4 u}{dx^4} + ku \in L^2(\mathbb{R})$, we have $((i2\pi\zeta)^4 + k)\,\widehat{u}(\zeta) \in L^2(\mathbb{R})$. Therefore, $\widehat{u}$ is measurable. Since $k > 0$, then we have $(i2\pi\zeta)^4 + k = k + (2\pi\zeta)^4 > 0$. We set $\xi = 2\pi\zeta$.

– If $|\xi| \leqslant 1$, then $\forall j \in \{0, ..., 4\}, |\xi|^j \leqslant 1 \leqslant C_1 k$ if $C_1 \geqslant \frac{1}{k}$.

Therefore, $\exists C_1 > 0, |\xi|^j \leqslant C_1 k \leqslant C_1(k + \xi^4)$.

– If $|\xi \geqslant 1$, then for $C_2 \geqslant 1, |\xi|^j \leqslant C_2|\xi|^4 = C_2\xi^4$.

Hence, $\exists C_2 \geqslant 1, |\xi|^j \leqslant C_2(\xi^4 + k)$.

– If $C = \max(C_1, C_2)$, we have just shown that in all cases,

$$\exists C > 0, \forall \xi \in \mathbb{R}, |\xi|^j \leqslant C(k + \xi^4), \text{ thus } |\xi|^j|\widehat{u}| \leqslant C(k + \xi^4)|\widehat{u}|,$$

that is, $|2\pi\zeta|^j|\widehat{u}| \leqslant k + (2\pi\zeta)^4\widehat{u}$, for all $j \in \{0, ..., 4\}$.

Placing an upper bound, since $\left(k + (2\pi\zeta)^4\right)\widehat{u} \in L^2(\mathbb{R})$, then $(2\pi\zeta)^j\widehat{u} \in L^2(\mathbb{R})$ for all $j \in \{0, ..., 4\}$.

By Plancherel, $\dfrac{\mathrm{d}^j u}{\mathrm{d}x^j} \in L^2(\mathbb{R}), \forall j \in \{0, ..., 4\}$.

EXERCISE 8.10.– Let $P(\xi)$ be a polynomial in $\mathbb{R}^d$ not identically zero. Let $P(D)$ designate the differential operator with constant coefficients associated with $P(\xi)$.

Show that if $u \in \mathcal{E}'(\mathbb{R}^d)$ is such that $P(D)u = 0$, then $u = 0$.

SOLUTION 8.10.– First, we should observe that since u has compact support, $\widehat{u}$ is a function $\mathcal{C}^\infty$ on $\mathbb{R}^n$.

Let us apply $\mathcal{F}$ to the equation $P(\partial)u = 0$, we obtain $P(2\pi i\zeta)\widehat{u}(\zeta) = 0, \forall \zeta \in \mathbb{R}^n$.

We set $Z = \{\zeta \in \mathbb{R}^n, P(2\pi i\zeta) = 0\}$. Then $\forall \zeta \in \mathbb{R}^n \backslash Z, \widehat{u}(\zeta) = 0$.

However, since P is a polynomial, Z is a closed set with empty interior because P is non-identically null. Therefore, $\mathbb{R}^n \backslash Z$ is dense in $\mathbb{R}^n$. Since $\widehat{u}$ is continuous and null on the dense set $\mathbb{R}^n \backslash Z$, $\widehat{u}$ is null everywhere.

By inverse Fourier transform, we obtain that $u = 0$.

EXERCISE 8.11.– Let $P(D) = \sum\limits_{|\alpha| \leq m} a_\alpha \partial^\alpha$ be a differential operator on $\mathbb{R}^d$ with constant coefficients such that: $\Sigma := \{\xi \in \mathbb{R}^d;\ P(\xi) = \sum\limits_{|\alpha| \leq m} a_\alpha \xi^\alpha = 0\} = \{0\}$. Show that the kernel of $P(D)$ in $\mathcal{S}'(\mathbb{R}^d)$ consists of polynomials.

SOLUTION 8.11.– Let $u \in \mathcal{S}'(\mathbb{R}^n), u \in \ker P(D)$. Then $P(D)u = 0$.

Since $u \in \mathcal{S}'(\mathbb{R}^n)$ and does not necessarily have a compact support, the argument of exercise 8.10 should be adapted.

Since $u \in \mathcal{S}'(\mathbb{R}^n)$ and $P(D)u = 0$, by Fourier transform, $P(2\pi i\zeta)\widehat{u} = 0$ in $\mathcal{S}'(\mathbb{R}^n)$. Therefore, $\widehat{u}$ is null in $\mathbb{R}^n \backslash \Sigma$.

Since $\Sigma = \{0\}$, we deduce that $\operatorname{supp} \widehat{u} \subset \Sigma = \{0\}$. Hence $\widehat{u} = \sum_{|\alpha| \leqslant m} a_\alpha \delta_0^{(\alpha)}$. By inverse Fourier transform, $u = \sum_{|\alpha| \leqslant m} a_\alpha (2\pi i x)^\alpha$. Consequently, u is indeed a polynomial.

EXERCISE 8.12 (Dirac comb Fourier transform).– We consider the Dirac comb $\triangle_a = \sum_{k=-\infty}^{+\infty} \delta_{ka}$, where $a > 0$.

1) Show that

$$\triangle_a = \sum_{k=-\infty}^{+\infty} \delta_{ka} = \frac{1}{a} \sum_{k=-\infty}^{+\infty} e^{2i\pi k \frac{x}{a}}. \qquad [8.5]$$

2) Compute the Fourier transform of $\triangle_a$.

3) Deduce that $\mathcal{F}(\triangle_a) = \frac{1}{a} \triangle_{\frac{1}{a}}$.

SOLUTION 8.12.–

1) Considering the periodic function f of period $a > 0$ defined on $[0, a[$ by $f(x) = \frac{x}{a}$, we recall that identity [4.2] gives

$$f'(x) = \frac{1}{a} - \sum_{k=-\infty}^{+\infty} \delta_{ka}. \qquad [8.6]$$

A Fourier series expansion of the function f gives

$$f(x) = \frac{1}{2} + \frac{i}{2\pi} \sum_{k \neq 0} \frac{1}{k} e^{2i\pi k \frac{x}{a}}.$$

Obviously, the convergence of this series is in $L^2(0,a)$, but it can be shown that it also takes place in $\mathcal{D}'(\mathbb{R})$. By continuity of the derivation operator, we can differentiate term by term:

$$f'(x) = -\frac{1}{a}\sum_{k\neq 0} e^{2i\pi k\frac{x}{a}}. \qquad [8.7]$$

The two formulas [8.6]–[8.7] ensure that

$$\triangle_a = \sum_{k=-\infty}^{+\infty} \delta_{ka} = \frac{1}{a}\sum_{k=-\infty}^{+\infty} e^{2i\pi k\frac{x}{a}}. \qquad [8.8]$$

2) From exercise 7.5, $\triangle_a$ is a tempered distribution. The continuity of the Fourier transform on the space $\mathcal{S}'(\mathbb{R})$ leads to $\mathcal{F}(\triangle_a) = \sum_{k=-\infty}^{+\infty} \mathcal{F}(\delta_{ka})$. We know that $\forall \zeta \in \mathbb{R}, \mathcal{F}(\delta_{ka})(\zeta) = e^{2i\pi ka\zeta}$, then $\mathcal{F}(\triangle_a)(\zeta) = \sum_{k=-\infty}^{+\infty} e^{2i\pi ka\zeta}$.

3) According to identity [8.8], we have $\sum_{k=-\infty}^{+\infty} e^{2i\pi ka\zeta} = \frac{1}{a}\triangle_{\frac{1}{a}}$. Consequently,

$$\mathcal{F}(\triangle_a) = \frac{1}{a}\triangle_{\frac{1}{a}}.$$

Special case: If $a = 1$, again we find $\mathcal{F}(\sum_{k=-\infty}^{+\infty} \delta_{ka}) = \sum_{k=-\infty}^{+\infty} \delta_{ka}$.

EXERCISE 8.13.– In the following, we will denote by δ_a the Dirac distribution at point a. We define by induction the sequence of distributions $(T_k)_{k>0}$ as follows:

$$T_1 = \frac{1}{2}(\delta_1 + \delta_{-1}) \text{ and } \forall k \geqslant 2,\ T_k = T_{k-1} * T_1.$$

1) Express T_k as a finite linear combination of distributions with point support.

2) Compute the Fourier transform $\widehat{T_k}$ of the distribution with compact support T_k.

3) For $k \geqslant 1$, we set $f_k(\xi) = \widehat{T_k}(\frac{\xi}{\sqrt{k}})$. Show that $f_k \in \mathcal{S}'(\mathbb{R})$ and the sequence $(f_k)_{k\geqslant 1}$ converges in $\mathcal{S}'(\mathbb{R})$ to a distribution that we will determine.

4) Let g_k denote the distribution of which f_k is the Fourier transform. Show that the sequence $(g_k)_{k\geqslant 1}$ converges in $\mathcal{S}'(\mathbb{R})$ to a distribution that we will determine.

Hint: The fact that for $a > 0$, $\mathcal{F}(e^{-ax^2}) = \frac{\sqrt{\pi}}{\sqrt{a}} e^{-\frac{\pi^2 x^2}{a}}$ can be used.

SOLUTION 8.13.–

1) We first point out that since $\text{supp}(T * S) \subset \text{supp}(T) + \text{supp}(S)$, given that T_1 has compact support, we show by induction that all T_k also have compact support.

Recall also that for a and b in $\mathbb{R}$, for all $\varphi \in C_c^\infty(\mathbb{R})$, we have

$$\begin{aligned}\langle \delta_a * \delta_b, \varphi \rangle &= \langle \delta_a, \langle \delta_b, \varphi^\Delta \rangle \rangle \\ &= \langle \delta_a, \varphi(x+b) \rangle \\ &= \varphi(a+b) \\ &= \langle \delta_{a+b}, \varphi \rangle .\end{aligned}$$

Therefore, $\delta_a * \delta_b = \delta_{a+b}$.

We can then calculate T_k.

Let $k \geqslant 2$: $T_k = T_1 * \ldots.. * T_1 = T_1^{*k} = \frac{1}{2^k}(\delta_1 + \delta_{-1})^{*k}$.

Now, the law $*$ is commutative and associative, so we can apply Newton's binomial formula to obtain

$$\begin{aligned}T_k &= \frac{1}{2^k} \sum_{j=0}^{k} \binom{k}{j} \delta_1^{*j} * \delta_{-1}^{*(k-j)} \\ &= \frac{1}{2^k} \sum_{j=0}^{k} \binom{k}{j} \delta_j * \delta_{j-k} \\ &= \frac{1}{2^k} \sum_{j=0}^{k} \binom{k}{j} \delta_{2j-k}.\end{aligned}$$

Therefore, $\forall k \geqslant 2, T_k = \frac{1}{2^k} \sum_{j=0}^{k} \binom{k}{j} \delta_{2j-k}$.

2) Since $T_k = T_1^{*k}$, we have $\widehat{T_k}(\zeta) = (\widehat{T_1}(\zeta))^k$.

Now, $\widehat{T_1}(\zeta) = \frac{1}{2}(\widehat{\delta_1} + \widehat{\delta_{-1}}) = \frac{1}{2}(e^{-i2\pi\zeta} + e^{i2\pi\zeta}) = \cos(2\pi\zeta)$, thus

$\widehat{T_k}(\zeta) = (\cos(2\pi\zeta))^k$ and $\widehat{T_k}$ is indeed of class C^∞ (which we already knew because $T_k \in \mathcal{E}'(\mathbb{R})$).

3) According to Question 2, we have $\forall k \geqslant 1, \forall \zeta \in \mathbb{R}, f_k(\zeta) = \left(\cos\left(\frac{\zeta}{\sqrt{k}}\right)\right)^k$.

Then, $f_k \in L^\infty(\mathbb{R}) \subset \mathcal{S}'(\mathbb{R})$, thus $f_k \in \mathcal{S}'(\mathbb{R})$.

To show the convergence of the sequence (f_k) in $\mathcal{S}'(\mathbb{R})$, we set $\varphi \in \mathcal{S}(\mathbb{R})$ and look for the possible limit of the sequence $\langle f_k, \varphi\rangle$.

Let $\varphi \in \mathcal{S}(\mathbb{R})$, we have $\forall k \geqslant 1, \langle f_k, \varphi\rangle = \displaystyle\int_{\mathbb{R}} f_k(\zeta)\varphi(\zeta)\mathrm{d}\zeta$.

We set $\zeta \in \mathbb{R}$. For $k \geqslant 16|\zeta|^2$, we have $2\pi\frac{|\zeta|}{\sqrt{k}} \leqslant \frac{\pi}{2}$ and $\cos\left(2\pi\frac{\zeta}{\sqrt{k}}\right) \geqslant 0$.

On the other hand, we have $\cos\left(2\pi\frac{\zeta}{\sqrt{k}}\right) = 1 - 2\pi^2\frac{\zeta^2}{k} + o\left(\frac{\zeta^2}{k}\right)$ and we have

$$
\begin{aligned}
\ln f_k(\zeta) &= k\ln\left(\cos\left(2\pi\frac{\zeta}{\sqrt{k}}\right)\right)\\
&= k\ln\left(1 - \frac{2\pi^2\zeta^2}{k} + o(\frac{\zeta^2}{k})\right)\\
&= -2\pi^2\zeta^2 + o(\zeta^2).
\end{aligned}
$$

Now, $\forall \zeta \in \mathbb{R}, |f_k(\zeta)\varphi(\zeta)| \leqslant |\varphi(\zeta)| \in L^1(\mathbb{R})$, so by the dominated convergence theorem, we get

$$
\begin{aligned}
\lim_{k\longrightarrow+\infty} \langle f_k, \varphi\rangle &= \lim_{k\longrightarrow+\infty} \int_{\mathbb{R}} f_k(\zeta)\varphi(\zeta)\mathrm{d}\zeta\\
&= \int_{\mathbb{R}} \lim_{k\longrightarrow+\infty} f_k(\zeta)\varphi(\zeta)\mathrm{d}\zeta\\
&= \int_{\mathbb{R}} e^{-2\pi^2\zeta^2}\varphi(\zeta)\mathrm{d}\zeta\\
&= \left\langle e^{-2\pi^2\zeta^2}, \varphi\right\rangle.
\end{aligned}
$$

Hence, $\left(f_k\right)_{k\geqslant 1}$ converges in $\mathcal{S}'(\mathbb{R})$ to $e^{-2\pi^2\zeta^2}$.

4) Let $g_k \in \mathcal{S}'(\mathbb{R})$ such that $f_k = \mathcal{F}(g_k)$. Then

$$
\mathcal{F}(f_k) = \sigma(g_k) \text{ and } g_k = (\mathcal{F}\circ\sigma)(f_k), \forall k \geqslant 1.
$$

By continuity of $\mathcal{F}$ in $\mathcal{S}'(\mathbb{R})$, it is deduced from the convergence in $\mathcal{S}'(\mathbb{R})$ of $(f_k)_{k\geqslant 1}$, the convergence of $(g_k)_{k\geqslant 1}$ in $\mathcal{S}'(\mathbb{R})$. Moreover, $(g_k)_{k\geqslant 1}$ converges in $\mathcal{S}'(\mathbb{R})$ to $(\mathcal{F}\circ\sigma)(e^{-2\pi^2\zeta^2})$, and we have

$$\forall x \in \mathbb{R}, (\mathcal{F}\circ\sigma)(e^{-2\pi^2\zeta^2})(x) = \sqrt{\frac{1}{2\pi}}e^{-\frac{x^2}{2}}.$$

Therefore, $g_k : x \longrightarrow \frac{1}{\sqrt{2\pi}}e^{-\frac{x^2}{2}}$ in $\mathcal{S}'(\mathbb{R})$.

EXERCISE 8.14.– For ψ differentiable on $\mathbb{R}$, we write $D_x\psi = \dfrac{1}{i}\dfrac{d\psi}{dx}$.

For $\varphi \in \mathcal{S}(\mathbb{R})$ and $s \in \mathbb{R}$, we set $e^{-is\Delta}\varphi = \mathcal{F}^{-1}(e^{is\xi^2}\mathcal{F}\varphi)$.

1) Demonstrate that for any $\psi \in \mathcal{S}(\mathbb{R})$ and all $t \in \mathbb{R}$, $e^{-it\Delta}e^{it\Delta}\psi = \psi$.

2) Let $\psi \in \mathcal{S}(\mathbb{R})$. Show that for all $t \in \mathbb{R}$, we have

$$e^{it\Delta}xe^{-it\Delta}\psi = 4\pi^2 x\psi + 2itD_x\psi.$$

3) Thereof deduce that if $P(x) = \sum_{k=0}^{m} a_k x^k$ with $a_k \in \mathbb{C}$, then for all $\psi \in \mathcal{S}(\mathbb{R})$, we have $\forall t \in \mathbb{R},\ e^{it\Delta}P(x)e^{-it\Delta}\psi = \sum_{k=0}^{m} a_k(4\pi^2 x + 2itD_x)^k\psi$.

SOLUTION 8.14.–

1) Let $t \in \mathbb{R}$ and $\varphi \in \mathcal{S}(\mathbb{R})$. We have

$$\begin{aligned}
\mathcal{F}\big(e^{-it\Delta}e^{it\Delta}\psi\big) &= \mathcal{F}\mathcal{F}^{-1}\big(e^{it\zeta^2}\mathcal{F}\big(e^{it\Delta}\psi\big)\big)\\
&= e^{it\zeta^2}\mathcal{F}\big(\mathcal{F}^{-1}\big(e^{-it\zeta^2}\mathcal{F}(\psi)\big)\big)\\
&= e^{it\zeta^2}e^{-it\zeta^2}\mathcal{F}(\psi)\\
&= \mathcal{F}(\psi).
\end{aligned}$$

Then, by the inverse Fourier formula, we obtain: $e^{-it\Delta}e^{it\Delta}\psi = \psi$.

2) By definition of $e^{is\Delta}$, for all $\psi \in \mathcal{S}(\mathbb{R})$ and any real t we have

$$\begin{aligned}
\mathcal{F}(e^{it\Delta}xe^{-it\Delta}\psi) &= \mathcal{F}\mathcal{F}^{-1}\big(e^{it\zeta^2}\mathcal{F}(xe^{-it\Delta}\psi)\big)\\
&= e^{-it\zeta^2}\mathcal{F}(xe^{-it\Delta}\psi).
\end{aligned}$$

Then

$$\begin{aligned}\mathcal{F}\big(xe^{-it\Delta}\psi\big) &= \big(\mathcal{F}(x)\big) * \mathcal{F}(e^{-it\Delta}\psi)\\ &= \big(i2\pi D_\zeta \delta_0\big) * \mathcal{F}\mathcal{F}^{-1}\big(e^{it\zeta^2}\mathcal{F}\psi\big)\\ &= i2\pi\delta_0' * e^{it\zeta^2}\mathcal{F}\psi(\zeta).\end{aligned}$$

Hence

$$\begin{aligned}\mathcal{F}\big(e^{it\Delta}xe^{-it\Delta}\psi\big) &= i2\pi e^{-it\zeta^2}\big(\delta_0' * e^{it\zeta^2}\widehat{\psi}(\zeta)\big)\\ &= i2\pi e^{-it\zeta^2}\big(\delta_0 * \big(e^{it\zeta^2}\widehat{\psi}(\zeta)\big)'\big)\\ &= i2\pi e^{-it\zeta^2}\big(e^{it\zeta^2}\widehat{\psi}(\zeta)\big)'\\ &= i2\pi e^{-it\zeta^2}\big(2it\zeta e^{it\zeta^2}\widehat{\psi}(\zeta) + e^{it\zeta^2}\frac{d}{d\zeta}\widehat{\psi}(\zeta)\big)\\ &= i2\pi\big(2it\zeta\widehat{\psi}(\zeta) + \frac{d}{d\zeta}\widehat{\psi}(\zeta)\big).\end{aligned}$$

Now, $\mathcal{F}(D_x\psi) = -i2\pi\zeta\mathcal{F}(\psi)$ and $\mathcal{F}(x\psi) = \frac{i}{2\pi}D_\zeta\mathcal{F}(\psi)$, thus

$$\begin{aligned}\mathcal{F}\big(e^{it\Delta}xe^{-it\Delta}\psi\big) &= 2it\mathcal{F}(D_x\psi) + 4\pi^2\mathcal{F}(x\psi)\\ &= \mathcal{F}\big(2itD_x\psi + 4\pi^2 x\psi\big) \ \text{ (from the linearity of } \mathcal{F}).\end{aligned}$$

Hence, by inverse Fourier transform, we obtain

$$\forall\psi \in \mathcal{S}(\mathcal{R}), \forall t \in \mathbb{R}, e^{it\Delta}xe^{-it\Delta}\psi = 2itD_x\psi + 4\pi^2 x\psi.$$

3) By linearity, it suffices to show that for all $k \geqslant 0$, we have

$$\forall t \in \mathbb{R}, \forall\psi \in \mathcal{S}(\mathbb{R}), e^{it\Delta}x^k e^{-it\Delta}\psi = \big(2itD_x + 4\pi^2 x\big)^k\psi. \qquad [8.9]$$

For $k = 0$, this is Question 1.

For $k = 1$, this is Question 2.

We can proceed by induction on k, the initialization at $k = 0$ or $k = 1$ being done. We assume that the result holds at rank $k \geqslant 1$.

Then, by Question 1, as $e^{it\Delta}e^{-it\Delta} = \mathrm{Id}$, for all $t \in \mathbb{R}$ and for all $\psi \in \mathcal{S}(\mathbb{R})$, we have

$$\begin{aligned} e^{it\Delta}x^{k+1}e^{-it\Delta}\psi &= e^{it\Delta}xe^{-it\Delta}e^{it\Delta}x^k e^{-it\Delta}\psi \\ &= e^{it\Delta}x\,e^{-it\Delta}\big((2itD_x + 4\pi^2 x)^k\psi\big) \text{ (by assumption of induction)} \\ &= \big(2itD_x + 4\pi^2 x\big)\big(x - 2tD_x\big)^k\psi \text{ (by Question 2)} \\ &= \big(2itD_x + 4\pi^2 x\big)^{k+1}\psi. \end{aligned}$$

Therefore, identity [8.9] holds at rank $k+1$. By induction, we obtain

$$\forall k \geqslant 0, \forall t \in \mathbb{R}, \forall \psi \in \mathcal{S}(\mathbb{R}),\ e^{it\Delta}x^k e^{-it\Delta}\psi = \big(2itD_x + 4\pi^2 x\big)^k\psi.$$

Then, if $P(x) = \sum_{k=0}^{m} a_k x^k$, by linearity, for all $\psi \in \mathcal{S}(\mathbb{R})$ and for all $t \in \mathbb{R}$, we have

$$\begin{aligned} e^{it\Delta}P(x)e^{-it\Delta}\psi &= \sum_{k=0}^{m} a_k e^{it\Delta}x^k e^{-it\Delta}\psi \\ &= \sum_{k=0}^{m} a_k\big(2itD_x + 4\pi^2 x\big)^k\psi \\ &= P\big(2itD_x + 4\pi^2 x\big)\psi. \end{aligned}$$

EXERCISE 8.15.– For $\varphi \in \mathcal{D}(\mathbb{R}^d)$ and $t > 0$, we denote by φ_t the function defined by $\varphi_t(x) = \varphi(\frac{x}{t})$. Let $T \in \mathcal{S}'(\mathbb{R}^d)$. It is assumed that $\widehat{T} \in L^\infty$. Using the fact that $T = \mathcal{F}^{-1}\mathcal{F}T$, show that, $\forall t > 0, |\langle T, \varphi_t\rangle| \leq \|\widehat{T}\|_{L^\infty}\|\widehat{\varphi}\|_{L^1}$.

SOLUTION 8.15.– Let $\psi \in \mathcal{D}(\mathbb{R}^d)$ and $t > 0$. For $T \in \mathcal{S}'(\mathbb{R}^d)$ such that $\widehat{T} \in L^\infty(\mathbb{R}^d)$, we have

$$\begin{aligned} |\langle T, \varphi_t\rangle| &= \left|\langle \mathcal{F}^{-1}\mathcal{F}T, \varphi_t\rangle\right| \\ &= |\langle \mathcal{F}T, (\mathcal{F}\circ\sigma)\varphi_t\rangle| \\ &= \left|\int_{\mathbb{R}^d} \widehat{T}(\zeta)(\mathcal{F}\circ\sigma)\varphi_t)(\zeta)\mathrm{d}\zeta\right| \\ &\leqslant \|\widehat{T}\|_{L^\infty(\mathbb{R}^d)} \times \int_{\mathbb{R}^d} |((\mathcal{F}\circ\sigma)\varphi_t)(\zeta)|\,\mathrm{d}\zeta. \end{aligned}$$

Now

$$((\mathcal{F}\circ\sigma)\varphi_t)(\zeta)=\int_{\mathbb{R}^d}\varphi(\frac{x}{t})e^{2\pi i x\zeta}\mathrm{d}x=\int_{\mathbb{R}^d}\varphi(y)e^{2\pi i t y\zeta}t^d\mathrm{d}y=t^d\widehat{\varphi}(-t\zeta).$$

Hence

$$\int_{\mathbb{R}^d}|((\mathcal{F}\circ\sigma)\varphi_t)(\zeta)|\,\mathrm{d}\zeta=\int_{\mathbb{R}^d}\left|t^d\widehat{\varphi}(-t\zeta)\right|\mathrm{d}\zeta=\int_{\mathbb{R}^d}|\widehat{\varphi}(-\zeta)|\,\mathrm{d}\zeta=\|\,\widehat{\varphi}\,\|_{L^1(\mathbb{R}^d)}\,.$$

Consequently, $|\langle T,\varphi_t\rangle|\leqslant\|\,\widehat{T}\,\|_{L^\infty(\mathbb{R}^d)}\|\,\widehat{\varphi}\,\|_{L^1(\mathbb{R}^d)}$.

EXERCISE 8.16 (Diagonal distribution).– Consider the distribution T of $\mathcal{D}'(\mathbb{R}^2)$ defined by

$$\begin{aligned}T:\ \mathcal{D}(\mathbb{R}^2)\ &\longrightarrow\ \mathbb{R}\\ \varphi\ &\longmapsto\ \int_{\mathbb{R}}\varphi(t,t)\mathrm{d}t.\end{aligned}$$

1) Show that T is tempered.

2) Show that $\partial_{x_1}T+\partial_{x_2}T=0$.

3) We wish to calculate the Fourier transform of T.

a) We set, for $\varphi\in\mathcal{S}(\mathbb{R}^2)$, $I_\epsilon(\varphi):=\displaystyle\int_{\mathbb{R}}e^{-\epsilon t^2}\varphi(t,t)\mathrm{d}t.$

Show that $\left\langle\widehat{T},\varphi\right\rangle=\lim\limits_{\epsilon\to0}I_\epsilon(\widehat{\varphi})$.

b) Applying the Fubini theorem and an appropriate change in variable, deduce that $\left\langle\widehat{T},\varphi\right\rangle=\lim\limits_{\epsilon\to0}\dfrac{1}{\sqrt{\pi}}\displaystyle\int_{\mathbb{R}^2}\varphi(x,\frac{\sqrt{\varepsilon}}{\pi}z-x)e^{-z^2}\mathrm{d}x\mathrm{d}z$ and conclude.

Hint: By denoting $f_\alpha:x\mapsto e^{-\alpha x^2}$ $(\alpha>0)$ the Gaussian of a variable, we have $\widehat{f_\alpha}=\sqrt{\frac{\pi}{\alpha}}f_{\frac{\pi^2}{\alpha}}$.

SOLUTION 8.16.–

1) It should be noted that if $\varphi\in\mathcal{D}(\mathbb{R}^2)$, $\langle T,\varphi\rangle$ is well defined, since the integrand is regular and has compact support in t. Then, we have for example,

$$\begin{aligned}\left|\int_{\mathbb{R}}\varphi(t,t)\mathrm{d}t\right|&\leq\Big(\int_{\mathbb{R}}\frac{1}{1+t^2}\mathrm{d}t\Big)\sup_{t\in\mathbb{R}}\left(1+t^2\right)|\varphi(t,t)|\\&\leq\Big(\int_{\mathbb{R}}\frac{1}{1+t^2}\mathrm{d}t\Big)\sup_{(x_1,x_2)\in\mathbb{R}^2}\left(1+x_1^2+x_2^2\right)|\varphi(x_1,x_2)|\,.\end{aligned}$$

So T is indeed a tempered distribution.

2) For $\varphi \in \mathcal{D}(\mathbb{R}^2)$, we set $\psi(t) = \varphi(t,t)$. We have

$$\psi'(t) = \partial_{x_1}\varphi(t,t) + \partial_{x_2}\varphi(t,t).$$

Since ψ has compact support, we thus have

$$\begin{aligned} 0 &= \int_{\mathbb{R}} \psi'(t)\mathrm{d}t \\ &= \int_{\mathbb{R}} \partial_{x_1}\varphi(t,t)\mathrm{d}t + \int_{\mathbb{R}} \partial_{x_2}\varphi(t,t)\mathrm{d}t = \langle T, \partial_{x_1}\varphi\rangle + \langle T, \partial_{x_2}\varphi\rangle . \\ &= -\langle \partial_{x_1}T + \partial_{x_2}T, \varphi\rangle \end{aligned}$$

Finally, $\partial_{x_1}T + \partial_{x_2}T = 0$.

3)a) Let $\varphi \in \mathcal{S}(\mathbb{R}^2)$. We have $t \longmapsto \varphi(t,t) \in L^1(\mathbb{R})$, because $|\varphi(t,t)| \leq C\left(1+t^2\right)^{-1} \in L^1(\mathbb{R})$. Therefore, the expression of T defined on the test functions extends naturally (and in a way that is unique) to rapidly decreasing functions. Then, since $\left|e^{-\varepsilon t^2}\varphi(t,t)\right| \leq |\varphi(t,t)|$, and $e^{-\varepsilon t^2}$ converges simply to 1 on $\mathbb{R}$, the expected convergence results from Lebesgue's theorem of dominated convergence (since the Fourier transform sends $\mathcal{S}(\mathbb{R}^2)$ onto itself).

b) Using the notation of the statement $f_\alpha : x \longmapsto e^{-\alpha x^2}$, we denote

$$L_\varepsilon(\varphi) := I_\varepsilon(\widehat{\varphi}) = \int_{\mathbb{R}} f_\varepsilon(t) \left[\int_{\mathbb{R}^2} \varphi(x,y)e^{-i2\pi t(x+y)}\mathrm{d}x\,\mathrm{d}y\right] \mathrm{d}t.$$

The function $(x,y,t) \longmapsto f_\varepsilon(t)\,|\varphi(x,y)| \in L^1(\mathbb{R}^3)$; by application of Fubini's theorem, we get

$$\begin{aligned} L_\varepsilon(\varphi) &= \int_{\mathbb{R}^2} \varphi(x,y) \left[\int_{\mathbb{R}} f_\varepsilon(t)e^{-i2\pi t(x+y)}\mathrm{d}t\right] \mathrm{d}x\,\mathrm{d}y \\ &= \int_{\mathbb{R}^2} \varphi(x,y)\widehat{f_\varepsilon}(x+y)\mathrm{d}x\,\mathrm{d}y. \end{aligned}$$

and according to the hint in the text, we have $\widehat{f_\varepsilon} = \sqrt{\frac{\pi}{\varepsilon}} f_{\frac{\pi^2}{\varepsilon}}$, then

$$L_\varepsilon(\varphi) = \sqrt{\frac{\pi}{\varepsilon}} \int_{\mathbb{R}^2} \varphi(x,y)e^{-\frac{\pi^2(x+y)^2}{\varepsilon}}\mathrm{d}x\,\mathrm{d}y.$$

However, for a fixed x, by change in variable $z = \frac{\pi(x+y)}{\sqrt{\varepsilon}}$, we have

$$\sqrt{\frac{\pi}{\varepsilon}} \int_{\mathbb{R}} \varphi(x,y) e^{-\frac{\pi^2(x+y)^2}{\varepsilon}} \mathrm{d}y = \sqrt{\frac{\pi}{\varepsilon}} \int_{\mathbb{R}} \varphi\left(x, \frac{\sqrt{\varepsilon}}{\pi} z - x\right) e^{-z^2} \frac{\sqrt{\varepsilon}}{\pi} \mathrm{d}z$$
$$= \frac{1}{\sqrt{\pi}} \int_{\mathbb{R}} \varphi\left(x, \frac{\sqrt{\varepsilon}}{\pi} z - x\right) e^{-z^2} \mathrm{d}z.$$

Finally,

$$L_\varepsilon(\varphi) = \frac{1}{\sqrt{\pi}} \int_{\mathbb{R}^2} \varphi\left(x, \frac{\sqrt{\varepsilon}}{\pi} z - x\right) e^{-z^2} \mathrm{d}z\, \mathrm{d}x.$$

Moreover, we have seen in the previous question that $\left\langle \widehat{T}, \varphi \right\rangle = \lim_{\varepsilon \longrightarrow 0} L_\varepsilon(\varphi)$. Since, on the other hand, $\varphi\left(x, \frac{\sqrt{\varepsilon}}{\pi} z - x\right) \underset{\varepsilon \longrightarrow 0}{\longrightarrow} \varphi(x,-x)$ and, on the other hand,

$$\left| \varphi\left(x, \frac{\sqrt{\varepsilon}}{\pi} z - x\right) e^{-z^2} \right| \le C \frac{1}{1+x^2} e^{-z^2} \in L^1(\mathbb{R}^2).$$

then using the fact that $\widehat{f_1}(0) = \sqrt{\pi} f_\pi(0) = \sqrt{\pi}$, Lebesgue's dominated convergence theorem yields

$$L_\varepsilon(\varphi) \underset{\varepsilon \longrightarrow 0}{\longrightarrow} \int_{\mathbb{R}} \varphi(x,-x) \mathrm{d}x.$$

Consequently, $\widehat{T}$ is the tempered distribution with support on the second bisector defined as follows:

$$\begin{aligned} \widehat{T} : \mathcal{S}(\mathbb{R}^2) &\longrightarrow \mathbb{R} \\ \varphi &\longmapsto \int_{\mathbb{R}} \varphi(x,-x) \mathrm{d}x. \end{aligned}$$

In particular, the inversion formula is verified: $\widehat{\widehat{T}} = T$, since T is an even distribution.

EXERCISE 8.17 (Convolution equation).–

1) Let $f, g \in \mathcal{S}(\mathbb{R}^d)$ denote two functions such that $f * g = 0$. Can we say that $f = 0$ or $g = 0$?

2) Same question if $f \in \mathcal{D}(\mathbb{R})$ and $g \in \mathcal{D}(\mathbb{R})$?

SOLUTION 8.17.–

1) By applying the Fourier transform to the equality $f * g = 0$, this equality becomes $\widehat{f}\widehat{g} = 0$; this thus just means that the supports of $\widehat{f}$ and $\widehat{g}$ are disjoint, and it cannot be deduced that $f = 0$ or $g = 0$: the ring $\left(\mathcal{S}(\mathbb{R}^d), +, *\right)$ is not integral. If we are looking for an explicit counterexample, we simply have to take $\varphi \in \mathcal{D}(\mathbb{R}^d)$ non-null with support in $B(0, 1)$ and $\psi \in \mathcal{D}(\mathbb{R}^d)$ non-null with support in $\{x : |x| > 1\}$, for instance, and to consider $f := \mathcal{F}^{-1}(\varphi)$ and $g := \mathcal{F}^{-1}(\psi)$, we have in particular $f, g \in \mathcal{S}(\mathbb{R}^d)$ (since this space is stable by inverse Fourier transform and contains $\mathcal{D}(\mathbb{R}^d)$. We then have $\widehat{f * g} = \widehat{f}\widehat{g} = \varphi\psi$, therefore $f * g = 0$ by injectivity of the Fourier transform, without having $f = 0$ or $g = 0$ (since their Fourier transforms are non-zero).

2) In this case, we know that $\widehat{f}$ and $\widehat{g}$ both admit two extensions holomorphic to $\mathbb{C}$, and their zeros are therefore isolated (unless they are zero). If these two functions are non-zero, this implies in particular that the set of their zeros is countable, which contradicts the fact that the product of these two functions is identically null on $\mathbb{R}$.

9

Applications to ODEs and PDEs

In a number of situations, and especially for the study of evolutionary partial differential equations, it proves to be helpful to perform a partial Fourier transform, which do not involve all the variables, but only some of them. This process will be briefly described, using the framework of evolutionary problems, which is its main application.

ODEs are Ordinary Differential Equations.

PDEs are Partial Differential Equations.

9.1. Partial Fourier transform

DEFINITION 9.1 (Partial Fourier transform on $\mathcal{S}(\mathbb{R}_t \times \mathbb{R}_x^d)$).– *Let $\varphi : (t, x) \longmapsto \varphi(t, x) \in \mathcal{S}(\mathbb{R}_t \times \mathbb{R}_x^d)$. The partial Fourier transform at x of the function φ is defined by the formula $\mathcal{F}_x\varphi(t, \xi) = \displaystyle\int_{\mathbb{R}^d} e^{-i2\pi\xi x}\varphi(t, x)dx$.*

This partial Fourier transform is endowed of the same continuity and invertibility properties as the usual Fourier transform.

Consider now the case of distributions.

DEFINITION 9.2 (Partial Fourier transform on $\mathcal{S}'(\mathbb{R}_t \times \mathbb{R}_x^d)$).– *Let $T \in \mathcal{S}'(\mathbb{R}_t \times \mathbb{R}_x^d)$ be a distribution. The partial Fourier transform at x of the distribution T is defined by the formula $\langle \mathcal{F}_x T, \varphi\rangle = \langle T, \mathcal{F}_x\varphi\rangle, \forall\varphi \in \mathcal{S}(\mathbb{R}_t \times \mathbb{R}_x^d)$.*

The task of verifying that the right-hand side of the above equality defines a tempered distribution on $\mathbb{R}_t \times \mathbb{R}_x^d$ will be left to the reader.

9.2. Tempered solutions of differential equations

Let $\omega > 0$ and let g be a distribution with compact support.

The following differential equation, where we are looking for $y \in \mathcal{S}'(\mathbb{R})$:

$$y'' - \omega^2 y = g \qquad [9.1]$$

can be written in terms of convolution as follows:

$$(\delta'' - \omega^2 \delta) * y = g. \qquad [9.2]$$

To solve it, it is therefore sufficient to have a distribution E such that

$$E'' - \omega^2 E = \delta. \qquad [9.3]$$

As a matter of fact, by convolving the two members of [9.2] by E on the left, and assuming that associativity is permissible, we obtain

$$E*g = E*\big((\delta''-\omega^2\delta)*y\big) = \big(E*(\delta''-\omega^2\delta)\big)*y = \big(E''-\omega^2 E\big)*y = \delta*y = y.$$

We verify that $y_g = E * g$ is indeed a solution of equation [9.1].

DEFINITION 9.3.– *The distribution $E \in \mathcal{S}'(\mathbb{R})$ as a solution of equation* [9.3] *is called elementary solution of the differential equation $y'' - \omega^2 y = g$.*

The equation $E'' - \omega^2 E = \delta$ of unknown $E \in \mathcal{S}'(\mathbb{R})$ is equivalent to

$$-4\pi^2 x^2 \widehat{E} - \omega^2 \widehat{E} = 1.$$

Therefore, $\widehat{E} = -\big(\omega^2 + 4\pi^2 x^2\big)^{-1}$.

As a result, by inverse transform, we have $E(t) = -\frac{e^{-\omega|t|}}{2\omega}$, and it is the unique solution in $\mathcal{S}'(\mathbb{R})$ of the equation $E'' - \omega^2 E = \delta$. The solution of equation [9.1] is thus, if g is a function,

$$y_g(t) = -\int_{\mathbb{R}} \frac{e^{-\omega|s-t|}}{2\omega} g(s)ds.$$

9.3. Fundamental solutions of certain PDEs

Let $P(x, \partial) = \sum_{|\alpha| \leq m} a_\alpha(x) \partial^\alpha$ be a differential operator on $\mathbb{R}^d$, with coefficient $a_\alpha \in \mathcal{C}^\infty(\mathbb{R}^d)$. If the coefficients a_α are constants with respect to x, it is said that the differential operator has constant coefficients and we denote $P(\partial) = \sum_{|\alpha| \leq m} a_\alpha \partial^\alpha$. A distribution $E \in \mathcal{D}'(\Omega)$ is said to be an elementary solution of $P(\partial)$ if it verifies

$$P(\partial)E = \delta, \text{ in } \mathcal{D}'(\Omega).$$

It should be noted that an elementary solution, when it exists, is not unique. To understand it better, we merely have to add a solution of the homogeneous equation

$$P(\partial)T = 0, \text{ in } \mathcal{D}'(\Omega).$$

The distribution $E+T$ is still an elementary solution. It is necessary that conditions be imposed to characterize one of the solutions and to prove the uniqueness of the solution. The existence of the solution is ensured by the following famous result.

THEOREM 9.1 (Malgrange-Ehrenpreis).– *Any differential operator with constant coefficients $P(\partial)$ on $\mathbb{R}^d$ admits an elementary solution E in $\mathcal{D}'(\mathbb{R}^d)$.*

The notion of elementary solution is employed to find the solutions to some partial differential equations in the following way.

THEOREM 9.2.– *Let $P(\partial)$ be a differential operator with constant coefficients on $\mathbb{R}^d$, and let $E \in \mathcal{D}'(\mathbb{R}^d)$ be an elementary solution to $P(\partial)$. Then, for any $f \in \mathcal{E}'(\mathbb{R}^d)$, the equation $P(\partial)u = f$ has at least one solution $u \in \mathcal{D}'(\mathbb{R}^d)$ of the form $u = E * f$.*

Proof. Since f has compact support, the expression for u makes sense. We then have

$$P(\partial)u = P(\partial)(E * f) = \big(P(\partial)E\big) * f = \delta * f = f.$$

□

9.3.1. *Heat equation*

The heat equation is the partial differential equation

$$\frac{\partial u}{\partial t} = \frac{\partial^2 u}{\partial x^2}, \; x \in \mathbb{R}, t \in \mathbb{R}^+.$$

The associated Cauchy problem consists in finding the solution $u(x,t)$ to this equation which verifies the initial condition $u(x,0) = f(x)$.

The solution, in the sense of distributions, is $u \in \mathcal{D}'_{x,t}(\mathbb{R}^2)$. We set

$$D = \frac{\partial}{\partial t} - \frac{\partial^2}{\partial x^2}.$$

Let us verify that the regular distribution associated with the locally integrable function $E : \mathbb{R}^2 \longrightarrow \mathbb{R}$ defined by $E(x,t) = \frac{1}{2\sqrt{\pi t}} \exp\big(-\frac{x^2}{4t}\big)$ is an elementary solution. We must then verify that $DE = \delta_{(0,0)}$. Actually,

$$\begin{aligned}
\left\langle \Big(\frac{\partial}{\partial t} - \frac{\partial^2}{\partial x^2}\Big) E, \varphi \right\rangle &= -\left\langle E, \frac{\partial \varphi}{\partial t} \right\rangle - \left\langle E, \frac{\partial^2 \varphi}{\partial x^2} \right\rangle \\
&= -\lim_{\varepsilon \longmapsto 0} \int_{-\infty}^{+\infty} \left(\int_{t>\varepsilon} \frac{1}{2\sqrt{\pi t}} \exp\left(-\frac{x^2}{4t}\right) \frac{\partial \varphi}{\partial t} \mathrm{d}t \right) \mathrm{d}x \\
&\quad - \lim_{\varepsilon \longmapsto 0} \int_{-\infty}^{+\infty} \left(\int_{t>\varepsilon} \frac{1}{2\sqrt{\pi t}} \exp\left(-\frac{x^2}{4t}\right) \frac{\partial^2 \varphi}{\partial x^2} \mathrm{d}t \right) \mathrm{d}x \\
&= \varphi(0,0) \\
&= \left\langle \delta_{(0,0)}, \varphi \right\rangle.
\end{aligned}$$

In the general case, let $x = (x_1, \cdots, x_d) \in \mathbb{R}^d$ and $t \in \mathbb{R}$ be the time variable. The Laplace operator is $D = \dfrac{\partial}{\partial t} - \Delta$. It is also called the heat operator. The distribution

$$E(x,t) = \Big(\frac{1}{2\sqrt{\pi t}}\Big)^d H(t) \exp\Big(-\frac{|x|^2}{4t}\Big)$$

is a fundamental solution to D.

9.3.2. *Wave equation*

The wave equation is the partial differential equation

$$\frac{\partial^2 u}{\partial t^2} = \frac{\partial^2 u}{\partial x^2}, \; x \in \mathbb{R}, t \in \mathbb{R}^+.$$

The associated Cauchy problem consists of finding the solution $u(x,t)$ to this equation that verifies the initial conditions:

$$u(x,0) = f_0(x), \quad \frac{\partial u}{\partial t}(x,0) = f_1(x).$$

where $f_0 \in \mathcal{C}^2(\mathbb{R})$ and $f_1 \in \mathcal{C}^1(\mathbb{R})$. The elementary solution to the Cauchy problem is the regular distribution g_t associated with the locally integrable function

$$g_t(x) = \frac{1}{2}\big(H(x+t) - H(x-t)\big),$$

and we have $\lim\limits_{t\longmapsto 0} g_t = 0$ and $\lim\limits_{t\longmapsto 0} \dfrac{\partial g_t}{\partial t} = \delta$. The general solution to the wave equation is written as

$$u_t = f_0 * \frac{\partial g_t}{\partial t} + f_1 * g_t.$$

In fact, the solution to the wave equation is analogous to

$$u(x,t) = f(x+t) + g(x-t),$$

where f and g are continuous functions. Furthermore, for any $\varphi \in \mathcal{D}(\mathbb{R})$,

$$\begin{aligned}
\left\langle \frac{\partial^2 u}{\partial t^2}, \varphi \right\rangle &= \frac{\partial^2}{\partial t^2} \int_{\mathbb{R}} \big(f(x+t) + g(x-t)\big)\varphi(x)\mathrm{d}x \\
&= \int_{\mathbb{R}} f(u)\varphi\text{"}(u-t)\mathrm{d}u + \int_{\mathbb{R}} g(u)\varphi\text{"}(u+t)\mathrm{d}u \\
&= \int_{\mathbb{R}} f(x+t)\varphi\text{"}(x)\mathrm{d}x + \int_{\mathbb{R}} g(x-t)\varphi\text{"}(x)\mathrm{d}x \\
&= \left\langle u, \frac{\partial^2 \varphi}{\partial x^2} \right\rangle \\
&= \left\langle \frac{\partial^2 u}{\partial x^2}, \varphi \right\rangle.
\end{aligned}$$

Appendix

Evaluations

This appendix provides exercises for student self-assessment.

EXERCISE A.1.–

1) Let $(f_n)_{n\geq 1}$ be the sequence of functions from $\mathbb{R}$ into $\mathbb{R}$ defined by

$$\forall n \geq 0, \forall x \in \mathbb{R}, f_n(x) = n\cos(nx).$$

a) Justify that for all $n \geq 0$, $f_n \in L^1_{loc}(\mathbb{R})$. We denote by T_{f_n} the distribution associated with f_n.

b) Show that $(T_{f_n})_{n\geq 1}$ converges in $\mathcal{D}'(\mathbb{R})$ and determine its limit.

2) Consider the linear form T defined on $\mathcal{D}(\mathbb{R})$ by

$$T : \varphi \longmapsto \sum_{k=1}^{+\infty} \left(\varphi\left(\frac{1}{k}\right) - \varphi\left(-\frac{1}{k+1}\right)\right).$$

a) Show that T is a distribution.

b) Determine the order of T.

3) Let f be the function defined on $\mathbb{R}$ by: $f(x) = x\sin(x)$.

a) Justify that the distribution associated with f is tempered.

b) Determine the Fourier transform in $\mathcal{S}'(\mathbb{R})$ of the distribution associated with the function f.

4)a) Solve in $\mathcal{D}'(\mathbb{R})$ the differential equation $T' + T = 0$.

b) Solve in $\mathcal{D}'(\mathbb{R})$ the differential equation $T' + T = \delta_0$.

c) Deduce the set of tempered solutions of the differential equation $T' + T = \delta_0$.

EXERCISE A.2.– For all $n \geq 1$, we consider u_n the function defined on $\mathbb{R}$ by

$$\forall x \in \mathbb{R}, u_n(x) = \begin{cases} x^2 \text{ if } x \leq n \\ n^2 \text{ if } x > n. \end{cases}$$

1) Show that u_n defines a distribution T_{u_n}.

Let $T_n = T_{u_n}$ be the distribution associated with u_n.

2) Calculate $T_n^{'}, T_n^{''}$ and $T_n^{(3)}$.

3) What is the order of $T_n^{(3)}$?

4) Determine $\lim\limits_{n \longrightarrow +\infty} T_n$ in $\mathcal{D}'(\mathbb{R})$.

EXERCISE A.3.–

1) Let $\varphi \in \mathcal{D}(\mathbb{R})$.

a) Show that there exists a function $\psi \in \mathcal{C}^\infty(\mathbb{R})$ such that

$$\forall x \in \mathbb{R}, \varphi(x) = \varphi(0) + x\varphi'(0) + x^2\psi(x).$$

b) Therefrom, deduce the existence of the limit

$$\lim_{\varepsilon \longrightarrow 0^+} \int_{|x|>\varepsilon} \left(\frac{\varphi(x)}{x^2} - 2\frac{\varphi(0)}{\varepsilon}\right) \mathrm{d}x.$$

We denote this limit as

$$\left\langle \mathrm{Pf}\left(\frac{1}{x^2}\right), \varphi \right\rangle := \lim_{\varepsilon \longrightarrow 0^+} \int_{|x|>\varepsilon} \left(\frac{\varphi(x)}{x^2} - 2\frac{\varphi(0)}{\varepsilon}\right) \mathrm{d}x.$$

2) Justify that the linear form on $\mathcal{D}(\mathbb{R})$, $\mathrm{Pf}\left(\frac{1}{x^2}\right) : \varphi \longmapsto \left\langle \mathrm{Pf}\left(\frac{1}{x^2}\right), \varphi \right\rangle$ is a distribution of order 2 at most.

3) Let $\mathrm{Vp}\left(\frac{1}{x}\right)$ be the principal value distribution of $\dfrac{1}{x}$ defined by

$$\forall \varphi \in \mathcal{D}(\mathbb{R}), \left\langle \mathrm{Vp}\left(\frac{1}{x}\right), \varphi \right\rangle := \lim_{\varepsilon \longrightarrow 0^+} \int_{|x|>\varepsilon} \frac{\varphi(x)}{x} \mathrm{d}x.$$

a) Show that in the sense of the distributions, $x\text{Vp}\left(\dfrac{1}{x}\right) = 1$.

b) Justify briefly that $\text{Vp}\left(\dfrac{1}{x}\right) \in \mathcal{S}'(\mathbb{R})$.

c) Show that $\text{Vp}\left(\frac{1}{x}\right)$ is an odd distribution.

d) Deduce that the Fourier transform $\mathcal{F}\left(\text{Vp}\left(\frac{1}{x}\right)\right)$ is an odd distribution.

e) Show that $\left[\mathcal{F}\left(\text{Vp}\left(\frac{1}{x}\right)\right)\right]' = -2i\pi\delta_0$.

f) Then deduce that $\mathcal{F}\left(\text{Vp}\left(\frac{1}{x}\right)\right) = -2i\pi H + C$, where H is the distribution associated with the Heaviside function and C is a constant (which by abuse of notation is confused with the distribution associated with it).

4) Show that, in $\mathcal{D}'(\mathbb{R})$, $\text{Vp}\left(\frac{1}{x}\right)' = \text{Pf}\left(\frac{1}{x^2}\right)$.

5) Can we say that $\text{Pf}\left(\frac{1}{x^2}\right) \in \mathcal{S}'(\mathbb{R})$?

6) What can it be said about the parity of $\text{Pf}\left(\frac{1}{x^2}\right)$?

EXERCISE A.4.–

1) Let $f : \mathbb{R} \longrightarrow \mathbb{R}$ be the function defined by

$$\forall x \in \mathbb{R}, f(x) = \begin{cases} -2 \text{ if } x \leq 0 \\ 2 \text{ if } x > 0. \end{cases}$$

a) Justify that the function f defines a distribution on $\mathbb{R}$ that will be denoted by T_f.

b) Compute the derivative of T_f in $\mathcal{D}'(\mathbb{R})$.

2) Let $\left(f_n\right)_{n\geq 1}$ be the sequence of functions from $\mathbb{R}$ into $\mathbb{R}$ defined by

$$\forall n \geq 0,\ \forall x \in \mathbb{R}, f_n(x) = \cos^2(nx).$$

a) Justify that, for all $n \geq 0$, $f_n \in L^1_{loc}(\mathbb{R})$. We denote by T_{f_n} the distribution associated with f_n.

b) Show that $\left(T_{f_n}\right)_{n\geq 1}$ converges in $\mathcal{D}'(\mathbb{R})$ and determine its limit.

3) Consider the linear form T defined on $\mathcal{D}(\mathbb{R})$ by

$$T : \varphi \longmapsto \sum_{k=1}^{+\infty} \frac{1}{k}\left(\varphi\left(\frac{1}{k}\right) - \varphi(0)\right).$$

a) Show that T is a distribution of order 1 at most.

b) Show that T is of order 1.

4) Show that a Dirac mass δ_a is a tempered distribution.

5) a) Show that the function $x \longmapsto \ln|x|$ defines a tempered distribution and compute its derivative.

b) Is $\mathrm{Vp}\left(\frac{1}{x}\right)$ (the principal value of $\frac{1}{x}$) a tempered distribution? Justify.

6) a) Let f be the function defined on $\mathbb{R}$ by $f(x) = \cos(2x)$.

Justify that the distribution associated with f is tempered.

b) Determine the Fourier transform in $\mathcal{S}'(\mathbb{R})$ of the distribution associated with the function f.

EXERCISE A.5.– For all $n \geq 0$, we set $T_n = \sum_{k=0}^{n} \delta_k$, where δ_k is the Dirac distribution at the point $x = k$.

1)a) Show that T_n is a distribution with compact support in $\mathbb{R}$.

b) What is its support?

c) Explain why T_n is tempered.

2)a) Verify that $T = \sum_{k=0}^{+\infty} \delta_k$ is a tempered distribution.

b) Show that the sequence of distributions $\left(T_n\right)_n$ converges in $\mathcal{S}'(\mathbb{R})$ to T.

3) Show that the sequence of convoluted $\left(T_n * T_n\right)_n$ converges in $\mathcal{S}'(\mathbb{R})$ to $T * T$.

(Hint: We can use the properties of the Fourier transform and its inverse and obtain the result without calculating $T_n * T_n$).

4) We assume that $T * T = \sum_{k=0}^{+\infty} (k+1)\delta_k$.

a) Compute $\mathcal{F}(T_n)(\lambda)$ the Fourier transform of T_n in $\lambda \in \mathbb{R}$.

b) Compute $\mathcal{F}(T * T)(\lambda)$, $\lambda \in \mathbb{R}$.

c) Deduce the multiplication of series formula

$$\left(\sum_{k=0}^{+\infty} e^{-2i\pi k\lambda}\right)\left(\sum_{k=0}^{+\infty} e^{-2i\pi k\lambda}\right) = \sum_{k=0}^{+\infty} (k+1)e^{-2i\pi k\lambda}.$$

EXERCISE A.6.–

1) Let $\psi \in \mathcal{C}^\infty(\mathbb{R})$ and let $Y = \mathbb{1}_{\mathbb{R}^+}$ be the indicator of $\mathbb{R}^+$. We set $E = T_{Y\psi}$.

a) Verify that $E \in \mathcal{D}'(\mathbb{R})$.

b) Show that for all $p \geq 1$,

$$E^{(p)} = T_{Y\psi^{(p)}} + \psi^{(p-1)}(0)\delta_0 + \psi^{(p-2)}(0)\delta_0' + ... + \psi(0)\delta_0^{(p-1)}.$$

2) Let ψ be a solution to $\sum_{j=0}^{k} a_j\psi^{(j)}(x) = 0$ for $x \in \mathbb{R}$, where $a_j \in \mathbb{R}$ and $a_k \neq 0$, satisfying

$$\psi(0) = \psi'(0) = ... = \psi^{k-2}(0) = 0 \quad \text{and} \quad \psi^{k-1}(0) = \frac{1}{a_k}.$$

a) Give the expression of $\sum_{j=0}^{k} a_j E^{(j)}$ where $E = T_{Y\psi}$.

b) Compute ψ and then deduce a solution in $\mathcal{D}'(\mathbb{R})$ of the equation

$$E'' + E' + E = \delta_0.$$

References

Bony, J-M. (2001). *Cours d'analyse. Théorie des distributions et analyse de Fourier*. Ellipses, Paris.

Boumaza, H. (2018). Théorie des distributions. Duplicate copy, Université Sorbonne Paris Nord.

Dalmasso, R. and Witomski, P. (1996). *Analyse de Fourier et applications*. Masson, Paris.

El Amrani, M. (2008). *Analyse de Fourier dans les espaces fonctionnels*. Ellipses, Paris.

Gasquet, C. and Witomski, P. (2001). *Analyse de Fourier et applications*. Dunod, Paris.

Guillopé, L. (2008). Analyse fonctionnelle. Duplicate copy, École polytechnique de l'Université de Nantes.

Hirsch, F. and Lacombe, G. (1999). *Éléments d'analyse fonctionnelle*. Dunod, Paris.

Khoan, V. (1972). *Distributions analyse de Fourier opérateurs aux derivées partielles*. Durand, Paris.

Lesfari, A. (2012). *Distributions, analyse de Fourier et transformation de Laplace*. Ellipses, Paris.

Pathak, R.S. (2001). *A Course in Distribution Theory and Applications*. Alpha Science, London.

Schwartz, L. (1966). *Théorie des distributions*. Hermann, Paris.

Schwartz, L. (1983). *Méthodes mathématiques pour les sciences physiques*. Hermann, Paris.

Strichartz, R. (2003). *A Guide to Distribution Theory and Fourier Transforms*. World Scientific Publishing Co. Inc., Singapore.

Zuily, C. (1988). *Éléments de distributions et d'équations aux dérivées partielles*. Hermann, Paris.

Zuily, C. (2002). *Problems in Distributions and Partial Differential Equations*. Dunod, Paris.

Index

S

T, U

Other titles from

iSTE

in

Mathematics and Statistics

2023

PARROCHIA Daniel
Graphs, Orders, Infinites and Philosophy

2022

CHAKRAVARTHY Srinivas R.
Introduction to Matrix-Analytic Methods in Queues 1: Analytical and Simulation Approach – Basics
Introduction to Matrix-Analytic Methods in Queues 2: Analytical and Simulation Approach – Queues and Simulation

DE SAPORTA Benoîte, ZILI Mounir
Martingales and Financial Mathematics in Discrete Time

LESFARI Ahmed
Integrable Systems

RADCHENKI Vadym M.
General Stochastic Measures: Integration, Path Properties and Equations

SIMON Jacques
Distributions
(Analysis for PDEs Set – Volume 3)

2021

KOROLIOUK Dmitri, SAMOILENKO Igor
Random Evolutionary Systems: Asymptotic Properties and Large Deviations

MOKLYACHUK Mikhail
Convex Optimization: Introductory Course

POGORUI Anatoliy, SWISHCHUK Anatoliy, RODRÍGUEZ-DAGNINO Ramón M.
Random Motions in Markov and Semi-Markov Random Environments 1: Homogeneous Random Motions and their Applications
Random Motions in Markov and Semi-Markov Random Environments 2: High-dimensional Random Motions and Financial Applications

PROVENZI Edoardo
From Euclidean to Hilbert Spaces: Introduction to Functional Analysis and its Applications

2020

BARBU Vlad Stefan, VERGNE Nicolas
Statistical Topics and Stochastic Models for Dependent Data with Applications

CHABANYUK Yaroslav, NIKITIN Anatolii, KHIMKA Uliana
Asymptotic Analyses for Complex Evolutionary Systems with Markov and Semi-Markov Switching Using Approximation Schemes

KOROLIOUK Dmitri
Dynamics of Statistical Experiments

MANOU-ABI Solym Mawaki, DABO-NIANG Sophie, SALONE Jean-Jacques
Mathematical Modeling of Random and Deterministic Phenomena

2019

BANNA Oksana, MISHURA Yuliya, RALCHENKO Kostiantyn, SHKLYAR Sergiy
Fractional Brownian Motion: Approximations and Projections

GANA Kamel, BROC Guillaume
Structural Equation Modeling with lavaan

KUKUSH Alexander
Gaussian Measures in Hilbert Space: Construction and Properties

LUZ Maksym, MOKLYACHUK Mikhail
Estimation of Stochastic Processes with Stationary Increments and Cointegrated Sequences

MICHELITSCH Thomas, PÉREZ RIASCOS Alejandro, COLLET Bernard, NOWAKOWSKI Andrzej, NICOLLEAU Franck
Fractional Dynamics on Networks and Lattices

VOTSI Irene, LIMNIOS Nikolaos, PAPADIMITRIOU Eleftheria, TSAKLIDIS George
Earthquake Statistical Analysis through Multi-state Modeling (Statistical Methods for Earthquakes Set – Volume 2)

2018

AZAÏS Romain, BOUGUET Florian
Statistical Inference for Piecewise-deterministic Markov Processes

IBRAHIMI Mohammed
Mergers & Acquisitions: Theory, Strategy, Finance

PARROCHIA Daniel
Mathematics and Philosophy

2017

CARONI Chysseis
First Hitting Time Regression Models: Lifetime Data Analysis Based on Underlying Stochastic Processes
(Mathematical Models and Methods in Reliability Set – Volume 4)

CELANT Giorgio, BRONIATOWSKI Michel
Interpolation and Extrapolation Optimal Designs 2: Finite Dimensional General Models

CONSOLE Rodolfo, MURRU Maura, FALCONE Giuseppe
Earthquake Occurrence: Short- and Long-term Models and their Validation
(Statistical Methods for Earthquakes Set – Volume 1)

D'AMICO Guglielmo, DI BIASE Giuseppe, JANSSEN Jacques, MANCA Raimondo
Semi-Markov Migration Models for Credit Risk
(Stochastic Models for Insurance Set – Volume 1)

GONZÁLEZ VELASCO Miguel, del PUERTO GARCÍA Inés, YANEV George P.
Controlled Branching Processes
(Branching Processes, Branching Random Walks and Branching Particle Fields Set – Volume 2)

HARLAMOV Boris
Stochastic Analysis of Risk and Management
(Stochastic Models in Survival Analysis and Reliability Set – Volume 2)

KERSTING Götz, VATUTIN Vladimir
Discrete Time Branching Processes in Random Environment
(Branching Processes, Branching Random Walks and Branching Particle Fields Set – Volume 1)

MISHURA YULIYA, SHEVCHENKO Georgiy
Theory and Statistical Applications of Stochastic Processes

NIKULIN Mikhail, CHIMITOVA Ekaterina
Chi-squared Goodness-of-fit Tests for Censored Data
(Stochastic Models in Survival Analysis and Reliability Set – Volume 3)

SIMON Jacques
Banach, Fréchet, Hilbert and Neumann Spaces
(Analysis for PDEs Set – Volume 1)

2016

CELANT Giorgio, BRONIATOWSKI Michel
Interpolation and Extrapolation Optimal Designs 1: Polynomial Regression and Approximation Theory

CHIASSERINI Carla Fabiana, GRIBAUDO Marco, MANINI Daniele
Analytical Modeling of Wireless Communication Systems
(Stochastic Models in Computer Science and Telecommunication Networks Set – Volume 1)

GOUDON Thierry
Mathematics for Modeling and Scientific Computing

KAHLE Waltraud, MERCIER Sophie, PAROISSIN Christian
Degradation Processes in Reliability
(Mathematial Models and Methods in Reliability Set – Volume 3)

KERN Michel
Numerical Methods for Inverse Problems

RYKOV Vladimir
Reliability of Engineering Systems and Technological Risks
(Stochastic Models in Survival Analysis and Reliability Set – Volume 1)

2015

DE SAPORTA Benoîte, DUFOUR François, ZHANG Huilong
Numerical Methods for Simulation and Optimization of Piecewise Deterministic Markov Processes

DEVOLDER Pierre, JANSSEN Jacques, MANCA Raimondo
Basic Stochastic Processes

LE GAT Yves
Recurrent Event Modeling Based on the Yule Process
(Mathematical Models and Methods in Reliability Set – Volume 2)

2014

COOKE Roger M., NIEBOER Daan, MISIEWICZ Jolanta
Fat-tailed Distributions: Data, Diagnostics and Dependence (Mathematical Models and Methods in Reliability Set – Volume 1)

MACKEVIČIUS Vigirdas
Integral and Measure: From Rather Simple to Rather Complex

PASCHOS Vangelis Th
Combinatorial Optimization – 3-volume series – 2nd edition
Concepts of Combinatorial Optimization / Concepts and Fundamentals – volume 1
Paradigms of Combinatorial Optimization – volume 2
Applications of Combinatorial Optimization – volume 3

2013

COUALLIER Vincent, GERVILLE-RÉACHE Léo, HUBER Catherine, LIMNIOS Nikolaos, MESBAH Mounir
Statistical Models and Methods for Reliability and Survival Analysis

JANSSEN Jacques, MANCA Oronzio, MANCA Raimondo
Applied Diffusion Processes from Engineering to Finance

SERICOLA Bruno
Markov Chains: Theory, Algorithms and Applications

2012

BOSQ Denis
Mathematical Statistics and Stochastic Processes

CHRISTENSEN Karl Bang, KREINER Svend, MESBAH Mounir
Rasch Models in Health

DEVOLDER Pierre, JANSSEN Jacques, MANCA Raimondo
Stochastic Methods for Pension Funds

2011

MACKEVIČIUS Vigirdas
Introduction to Stochastic Analysis: Integrals and Differential Equations

MAHJOUB Ridha

Recent Progress in Combinatorial Optimization – ISCO2010

RAYNAUD Hervé, ARROW Kenneth
Managerial Logic

2010

BAGDONAVIČIUS Vilijandas, KRUOPIS Julius, NIKULIN Mikhail
Nonparametric Tests for Censored Data

BAGDONAVIČIUS Vilijandas, KRUOPIS Julius, NIKULIN Mikhail
Nonparametric Tests for Complete Data

IOSIFESCU Marius *et al.*
Introduction to Stochastic Models

VASSILIOU PCG
Discrete-time Asset Pricing Models in Applied Stochastic Finance

2008

ANISIMOV Vladimir
Switching Processes in Queuing Models

FICHE Georges, HÉBUTERNE Gérard
Mathematics for Engineers

HUBER Catherine, LIMNIOS Nikolaos *et al.*
Mathematical Methods in Survival Analysis, Reliability and Quality of Life

JANSSEN Jacques, MANCA Raimondo, VOLPE Ernesto
Mathematical Finance

2007

HARLAMOV Boris
Continuous Semi-Markov Processes

2006

CLERC Maurice
Particle Swarm Optimization

Printed and bound by CPI Group (UK) Ltd, Croydon, CR0 4YY
16/09/2026
14963082-0007